Functionalized Gels for Environmental Applications

Editors

Luca Burratti
Paolo Prosposito
Iole Venditti

Basel • Beijing • Wuhan • Barcelona • Belgrade • Novi Sad • Cluj • Manchester

Editors
Luca Burratti
Department of Sciences
University of Roma Tre
Rome, Italy

Paolo Prosposito
Industrial Engineering
University of Rome "Tor
Vergata"
Rome, Italy

Iole Venditti
Department of Sciences
University of Roma Tre
Rome, Italy

Editorial Office
MDPI
St. Alban-Anlage 66
4052 Basel, Switzerland

This is a reprint of articles from the Special Issue published online in the open access journal *Gels* (ISSN 2310-2861) (available at: www.mdpi.com/journal/gels/special_issues/gels_environmental).

For citation purposes, cite each article independently as indicated on the article page online and as indicated below:

Lastname, A.A.; Lastname, B.B. Article Title. *Journal Name* **Year**, *Volume Number*, Page Range.

ISBN 978-3-0365-9501-6 (Hbk)
ISBN 978-3-0365-9500-9 (PDF)
doi.org/10.3390/books978-3-0365-9500-9

© 2023 by the authors. Articles in this book are Open Access and distributed under the Creative Commons Attribution (CC BY) license. The book as a whole is distributed by MDPI under the terms and conditions of the Creative Commons Attribution-NonCommercial-NoDerivs (CC BY-NC-ND) license.

Contents

Luca Burratti, Paolo Prosposito and Iole Venditti
Functionalized Gels for Environmental Applications
Reprinted from: *Gels* **2023**, *9*, 818, doi:10.3390/gels9100818 1

Mariana Chelu, Monica Popa, Jose Calderon Moreno, Anca Ruxandra Leonties, Emma Adriana Ozon, Jeanina Pandele Cusu et al.
Green Synthesis of Hydrogel-Based Adsorbent Material for the Effective Removal of Diclofenac Sodium from Wastewater
Reprinted from: *Gels* **2023**, *9*, 454, doi:10.3390/gels9060454 5

Martina Klučáková
Effect of Chitosan as Active Bio-colloidal Constituent on the Diffusion of Dyes in Agarose Hydrogel
Reprinted from: *Gels* **2023**, *9*, 395, doi:10.3390/gels9050395 27

Luca Burratti, Marco Zannotti, Valentin Maranges, Rita Giovannetti, Leonardo Duranti, Fabio De Matteis et al.
Poly(ethylene glycol) Diacrylate Hydrogel with Silver Nanoclusters for Water Pb(II) Ions Filtering
Reprinted from: *Gels* **2023**, *9*, 133, doi:10.3390/gels9020133 41

Martina Klučáková
How the Addition of Chitosan Affects the Transport and Rheological Properties of Agarose Hydrogels
Reprinted from: *Gels* **2023**, *9*, 99, doi:10.3390/gels9020099 54

Paul Chesler, Cristian Hornoiu, Mihai Anastasescu, Jose Maria Calderon-Moreno, Marin Gheorghe and Mariuca Gartner
Cobalt- and Copper-Based *Chemiresistors* for Low Concentration Methane Detection, a Comparison Study
Reprinted from: *Gels* **2022**, *8*, 721, doi:10.3390/gels8110721 70

Anna Fortunato and Miriam Mba
A Peptide-Based Hydrogel for Adsorption of Dyes and Pharmaceuticals in Water Remediation
Reprinted from: *Gels* **2022**, *8*, 672, doi:10.3390/gels8100672 80

E. Molly Frazar, Anicah Smith, Thomas Dziubla and J. Zach Hilt
Thermoresponsive Cationic Polymers: PFAS Binding Performance under Variable pH, Temperature and Comonomer Composition
Reprinted from: *Gels* **2022**, *8*, 668, doi:10.3390/gels8100668 94

Silvia Sfameni, Giulia Rando, Maurilio Galletta, Ileana Ielo, Marco Brucale, Filomena De Leo et al.
Design and Development of Fluorinated and Biocide-Free Sol–Gel Based Hybrid Functional Coatings for Anti-Biofouling/Foul-Release Activity
Reprinted from: *Gels* **2022**, *8*, 538, doi:10.3390/gels8090538 105

Hojung Choi, Sanghwa Lee, SeongUk Jeong, Yeon Ki Hong and Sang Youl Kim
Synthesis and CO_2 Capture of Porous Hydrogel Particles Consisting of Hyperbranched Poly(amidoamine)s
Reprinted from: *Gels* **2022**, *8*, 500, doi:10.3390/gels8080500 131

Zenab Darban, Syed Shahabuddin, Rama Gaur, Irfan Ahmad and Nanthini Sridewi
Hydrogel-Based Adsorbent Material for the Effective Removal of Heavy Metals from Wastewater: A Comprehensive Review
Reprinted from: *Gels* **2022**, *8*, 263, doi:10.3390/gels8050263 . . . . **140**

Editorial

Functionalized Gels for Environmental Applications

Luca Burratti [1], Paolo Prosposito [2,*] and Iole Venditti [1]

[1] Sciences Department, Roma Tre University, Via della Vasca Navale 79, 00146 Rome, Italy; luca.burratti@uniroma3.it (L.B.); iole.venditti@uniroma3.it (I.V.)
[2] Department of Industrial Engineering, University of Rome Tor Vergata, Via del Politecnico 1, 00133 Rome, Italy
* Correspondence: paolo.prosposito@uniroma2.it

A gel is a type of material that exhibits a semi-solid, jelly-like state, characterized by a three-dimensional network of interconnected particles or molecules dispersed within a liquid or solid medium. This network structure provides the gel with unique physical properties, including the ability to maintain its shape while possessing a high degree of flexibility and deformability. They are distinguished by their capacity to hold a significant amount of liquid within their structure, making them valuable in applications ranging from personal care products (such as hair gel and skincare formulations) to biomedical materials (like hydrogels for tissue engineering) and technological applications (such as aerogels for insulation and catalyst supports). Gel-based systems have demonstrated their utility in diverse environmental contexts, such as the treatment of contaminated water bodies, the decontamination of soil, and the purification of air. The wide-ranging versatility of gel in environmental applications is exemplified by their efficacy in removing a diverse array of contaminants, such as heavy metals, organic pollutants, dyes, and even emerging contaminants like pharmaceuticals and microplastics. The selective adsorption of these pollutants arises from a combination of chemical interactions, such as ion exchange, coordination, and electrostatic attraction, as well as physical mechanisms like diffusion and adsorption.

This Special Issue, entitled "Functionalized Gels for Environmental Applications", in the *Gels* journal includes nine original articles and one review. These manuscripts address different aspects of environmental monitoring/remediation and they provide an interesting overview of the fields covered by research on gels, offering valuable insights into the challenges and prospects that lie ahead for future advancements in the field.

The adsorption capacity of hydrogels toward contaminants can be enhanced by modifying them via various techniques; for example, in the work conducted by M. Klučáková, an agarose hydrogel is enriched by chitosan as an active compound for the adsorption of dyes [1]. The author reports that the adsorption capacity of modified agarose is several times higher in comparison with pure agarose hydrogel and that this behavior is also found at different pH values (3, 7 and 11), certifying the success of the modification. The author asserts that the increase in the dyes' adsorption is due to the electrostatic interactions between the amino group of chitosan and the sulfonic group of dyes, which leads to the formation of distinct dye layers on the surface of the modified hydrogel. From the perspective of rheology, the addition of chitosan results in changes in storage and loss moduli, which can be exhibited on a "more liquid" character of enriched hydrogels. This can contribute to an increase in the effective diffusion coefficients for hydrogels with a higher content of chitosan [2].

Another example of how a properly designed hydrogel is able to remove methylene blue and diclofenac from water is represented by the research of A. Fortunato and M. Mba [3]. In this work, a tetrapeptide–pyrene conjugate is designed to form hydrogels under controlled acidic conditions. The main results indicate that the methylene blue adsorption is guided by the availability of adsorption sites, while in the case of diclofenac, the

concentration is the driving force of the process. In the case of methylene blue, the nature of the dye–hydrogel interactions are explained: first, the dye is adsorbed as a monomer (the authors hypothesize an electrostatic interaction) and successively, at increasing concentrations as the electrostatic adsorption sites are depleted, dimerization on the hydrogel surface occurs. The removal efficiencies (that depend on the initial concentration of the pollutants) for methylene blue and diclofenac are in the range of 90–100% and 53–89%, respectively.

The presence of diclofenac as a water contaminant poses a risk to human health; therefore, it is vital that it can be easily removed from polluted water. The research carried out by M. Chelu et al. [4] constitutes a valid in-depth study on the removal of this pharmaceutical product. In this case, the hydrogel is based on a mixture of chitosan, polyethylene glycol and xanthan gum (CPX), and it is prepared via a green method. The authors characterize the materials well and find that the swelling properties remain unchanged in a wide range of pH (3–9). Moreover, the adsorption capacity study reveals that the adsorbent hydrogel reaches the adsorption capacity (172.41 mg/g) at the highest adsorbent amount (200 mg) after 350 min. Finally, the data obtained in the kinetic study reveal that the used adsorbent may possess great applicative potential in environmental applications as a water cleaning agent.

Heavy metal ions are another class of water pollutants, and their presence in drinkable water poses serious problems to the health of living beings (plants and animals); therefore, it is essential that water can be purified in a simple and effective way. Burratti et al. [5] successfully develop a hydrogel-based filter for removing Pb(II) ions from water. Poly(ethylene glycol) diacrylate (PEGDA) hydrogels, modified with luminescent silver nanoclusters (AgNCs), are synthesized through a photo-crosslinking procedure. This type of filter is able to remove between 80% to 90% of the lead impurity. Their experimental findings suggest that the adsorption of Pb(II) onto the modified filter is predominantly driven by favorable chemisorption. Considering its exceptional contaminant uptake capacity and cost-effectiveness, this hybrid system exhibits significant potential as an adsorbent material for the efficient removal of Pb(II) ions from aqueous environments.

The versatility and unique qualities of thermo-responsive polymeric systems have led to the application of these materials in a multitude of fields. Environmental remediation can significantly benefit from the utilization of innovative and smart materials. Notably, the multifaceted nature of poly(N-isopropylacrylamide) (PNIPAAm) systems, incorporating PNIPAAm copolymerized with diverse cationic comonomers, holds the potential to selectively target and attract negatively charged contaminants, such as perfluorooctanoic acid (PFOA). In the study presented by E. M. Frazar et al. [6], the synthesis of a variety of thermo-responsive cationic hydrogels is carried out. In this work, the effect of pH on the hydrogel swelling behavior is studied and found to be insignificant for PNIPAAm and hydrogels containing loading percentages of 1 and 5 mol% cationic comonomer. The inclusion of cationic comonomers, however, alters the hydrogel swelling capacity, mostly due to losses in the thermo-responsive behavior as the comonomer amount is increased. The adsorption of PFOA is inversely related to buffered aqueous pH, while the cationic monomer type has little noticeable consequence in the buffered solutions. These insights gained from hydrogel performance under variable pH, buffer, temperature, and comonomer compositions provide us with a deeper understanding of which polymer functionalities are most beneficial when designing materials for the remediation of perfluoroalkyl substances in aqueous environments.

Air contaminants, also known as air pollutants, are substances or particles present in the air that can have harmful effects on human health, the environment, or both. Efforts to reduce air contaminants often involve regulatory measures, technological advancements, and changes in behavior to minimize emissions and exposure. A very interesting study is presented by P. Chesler et al. [7], who develop two sensors, one based on cobalt and one based on copper, to detect low concentrations of methane. The authors synthesize sensitive films on an alumina substrate, with gold or platinum interdigital electrodes (IDE) printed onto the alumina surface, using the sol–gel technique. The fabricated sensors exhibit stability, partial selectivity, and the ability to detect low concentrations of methane (5 ppm) with a rapid response time of 250 s and complete recovery within the same timeframe. Some response to interfering species (CO_2 and humidity) is observed, but it is relatively modest, counting for approximately 50% of the sensor's response to methane. The cobalt-based sensor demonstrates superior selectivity, particularly at elevated methane concentrations.

Nowadays, the detection and removal of the CO_2 that naturally occurs in the Earth's atmosphere and is produced abundantly by many industrial processes in high demand, since excessive concentrations can pose hazards to both human health and the environment. Within this framework, H. Choi et al. [8] successfully fabricate novel macroporous hydrogel particles comprising hyperbranched poly(amidoamine)s (HPAMAM) utilizing the oil-in-water-in-oil (O/W/O) suspension polymerization technique. This method, known for conferring a porous architecture to microparticles, results in hydrogel particles with a rich abundance of amine groups embedded within the polymer matrix. Therefore, these synthesized hydrogel particles demonstrate remarkable CO_2 absorption capabilities, with an absorption capacity of 104 mg/g, and exhibit rapid absorption rates in rigorous packed-column tests.

Until now, the topics addressed have concerned the removal of pollutants of various kinds, but sometimes the material employed for purification can be fouled by organic/inorganic substances, consequently reducing the performance of the material enormously. Developing solutions to avoid this type of problem is desirable. An interesting study carried out by S. Sfameni et al. [9] concerns the development of hybrid functional coatings for anti-biofouling and foul-release activity. Here, silica-based materials are prepared using two alkoxysilane cross-linkers containing epoxy and amine groups in combination with two functional fluoro-silanes, featuring well-known hydro repellent and anti-corrosion properties. The efficacy of fouling the release properties is assessed by subjecting the material to various microbial suspensions in seawater-based solutions and within natural seawater microcosms. The newly formulated fluorinated coatings exhibit antimicrobial capabilities. Notably, no biocidal effects are observed on the microorganisms, specifically bacteria.

Finally, in their review, Z. Darban et al. [10] describe the methods employed in order to recycle wastewater by exploiting hydrogel-based adsorbent materials. The synthesis techniques and adsorption mechanisms are also explored, focusing on the regeneration, recovery, and reuse of modified hydrogels.

In conclusion, because the evolution of technology and materials in this area is rapid and extensive, as Guest Editors we realize that it is limiting to present them in a single volume. However, the multidisciplinary nature and high quality of the articles allow us to provide readers with an updated and broad panorama regarding functionalized gels for environmental applications, which we are certain will arouse great interest.

Conflicts of Interest: The authors declare no conflict of interest.

References

1. Klučáková, M. Effect of Chitosan as Active Bio-colloidal Constituent on the Diffusion of Dyes in Agarose Hydrogel. *Gels* **2023**, *9*, 395. [CrossRef] [PubMed]
2. Klučáková, M. How the Addition of Chitosan Affects the Transport and Rheological Properties of Agarose Hydrogels. *Gels* **2023**, *9*, 99. [CrossRef] [PubMed]
3. Fortunato, A.; Mba, M. A Peptide-Based Hydrogel for Adsorption of Dyes and Pharmaceuticals in Water Remediation. *Gels* **2022**, *8*, 672. [CrossRef]
4. Chelu, M.; Popa, M.; Calderon Moreno, J.; Leonties, A.R.; Ozon, E.A.; Pandele Cusu, J.; Surdu, V.A.; Aricov, L.; Musuc, A.M. Green Synthesis of Hydrogel-Based Adsorbent Material for the Effective Removal of Diclofenac Sodium from Wastewater. *Gels* **2023**, *9*, 454. [CrossRef] [PubMed]
5. Burratti, L.; Zannotti, M.; Maranges, V.; Giovannetti, R.; Duranti, L.; De Matteis, F.; Francini, R.; Prosposito, P. Poly(ethylene glycol) Diacrylate Hydrogel with Silver Nanoclusters for Water Pb(II) Ions Filtering. *Gels* **2023**, *9*, 133. [CrossRef] [PubMed]
6. Frazar, E.M.; Smith, A.; Dziubla, T.; Hilt, J.Z. Thermoresponsive Cationic Polymers: PFAS Binding Performance under Variable pH, Temperature and Comonomer Composition. *Gels* **2022**, *8*, 668. [CrossRef] [PubMed]

7. Chesler, P.; Hornoiu, C.; Anastasescu, M.; Calderon-Moreno, J.M.; Gheorghe, M.; Gartner, M. Cobalt- and Copper-Based Chemiresistors for Low Concentration Methane Detection, a Comparison Study. *Gels* **2022**, *8*, 721. [CrossRef] [PubMed]
8. Choi, H.; Lee, S.; Jeong, S.; Hong, Y.K.; Kim, S.Y. Synthesis and CO_2 Capture of Porous Hydrogel Particles Consisting of Hyperbranched Poly(amidoamine)s. *Gels* **2022**, *8*, 500. [CrossRef] [PubMed]
9. Sfameni, S.; Rando, G.; Galletta, M.; Ielo, I.; Brucale, M.; De Leo, F.; Cardiano, P.; Cappello, S.; Visco, A.; Trovato, V.; et al. Design and Development of Fluorinated and Biocide-Free Sol–Gel Based Hybrid Functional Coatings for Anti-Biofouling/Foul-Release Activity. *Gels* **2022**, *8*, 538. [CrossRef] [PubMed]
10. Darban, Z.; Shahabuddin, S.; Gaur, R.; Ahmad, I.; Sridewi, N. Hydrogel-Based Adsorbent Material for the Effective Removal of Heavy Metals from Wastewater: A Comprehensive Review. *Gels* **2022**, *8*, 263. [CrossRef] [PubMed]

Disclaimer/Publisher's Note: The statements, opinions and data contained in all publications are solely those of the individual author(s) and contributor(s) and not of MDPI and/or the editor(s). MDPI and/or the editor(s) disclaim responsibility for any injury to people or property resulting from any ideas, methods, instructions or products referred to in the content.

Article

Green Synthesis of Hydrogel-Based Adsorbent Material for the Effective Removal of Diclofenac Sodium from Wastewater

Mariana Chelu [1,*,†], Monica Popa [1,†], Jose Calderon Moreno [1,*], Anca Ruxandra Leonties [1], Emma Adriana Ozon [2], Jeanina Pandele Cusu [1], Vasile Adrian Surdu [3], Ludmila Aricov [1] and Adina Magdalena Musuc [1,*]

1 "Ilie Murgulescu" Institute of Physical Chemistry, 202 Spl. Independentei, 060021 Bucharest, Romania; pmonica@icf.ro (M.P.); aleonties@icf.ro (A.R.L.); jeanina@icf.ro (J.P.C.); laricov@icf.ro (L.A.)
2 Department of Pharmaceutical Technology and Biopharmacy, Faculty of Pharmacy, Carol Davila University of Medicine and Pharmacy, 6 Traian Vuia Street, 020945 Bucharest, Romania; emma.budura@umfcd.ro
3 Department of Science and Engineering of Oxide Materials and Nanomaterials, Faculty of Applied Chemistry and Materials Science, University Politehnica of Bucharest, 060042 Bucharest, Romania; adrian.surdu@upb.ro
* Correspondence: mchelu@icf.ro (M.C.); calderon@icf.ro (J.C.M.); amusuc@icf.ro (A.M.M.)
† These authors contributed equally to this work.

Abstract: The removal of pharmaceutical contaminants from wastewater has gained considerable attention in recent years, particularly in the advancements of hydrogel-based adsorbents as a green solution for their ease of use, ease of modification, biodegradability, non-toxicity, environmental friendliness, and cost-effectiveness. This study focuses on the design of an efficient adsorbent hydrogel based on 1% chitosan, 40% polyethylene glycol 4000 (PEG4000), and 4% xanthan gum (referred to as CPX) for the removal of diclofenac sodium (DCF) from water. The interaction between positively charged chitosan and negatively charged xanthan gum and PEG4000 leads to strengthening of the hydrogel structure. The obtained CPX hydrogel, prepared by a green, simple, easy, low-cost, and ecological method, has a higher viscosity due to the three-dimensional polymer network and mechanical stability. The physical, chemical, rheological, and pharmacotechnical parameters of the synthesized hydrogel were determined. Swelling analysis demonstrated that the new synthesized hydrogel is not pH-dependent. The obtained adsorbent hydrogel reached the adsorption capacity (172.41 mg/g) at the highest adsorbent amount (200 mg) after 350 min. In addition, the adsorption kinetics were calculated using a pseudo first-order model and Langmuir and Freundlich isotherm parameters. The results demonstrate that CPX hydrogel can be used as an efficient option to remove DCF as a pharmaceutical contaminant from wastewater.

Keywords: hydrogel-based materials; adsorbents; removal of pollutants; isotherm; kinetic; diclofenac sodium; water sustainability; adsorption

1. Introduction

Fresh water is a unique natural resource, essential for life and particularly precious for the daily existence of humanity and for the surrounding flora and fauna. The global development of the human activities determined by the continuous growth of the population and the increase in pollution are great challenges that require imperative measures regarding the decontamination of water [1]. Pharmaceuticals represent a category of emerging pollutants that have revealed a potential risk to human health and the environment [2,3]. A variety of active ingredients (bisphenol-A, carbamazepine, clofibric acid, diclofenac, ibuprofen, iopamidol, phthalates, polycyclic siloxanes, triclosan) are found in some of the most widely used pharmaceuticals and personal care products (around 3000 registered) [4]. These compounds can contribute to various hormonal abnormalities in humans [5].

Wastewater treatment can be performed in a conventional way, mainly for the removal of the total suspended solids and organic matter [6]. However, following the classical methods, only 94% of the total suspended solids are removed [7]. Therefore, pharmaceuticals can be eliminated by the tertiary steps of adsorption, membrane separation, ozonation, and advanced oxidation processes, or by the use of chemical disinfectants [8,9]. These procedures also have disadvantages because they involve environmentally hazardous chemicals (chemical disinfectants) or high costs. Modern membrane techniques have experienced exponential development in recent years, facilitating the removal of micro- to nano-sized contaminants [10,11] through methods such as electrodialysis, distillation, and direct osmosis [12–15] using various filtration characteristics (hydrophobicity, surface charge, pore size). Adsorption-based pharmaceutical pollutants removal methods are becoming very attractive for wastewater treatment, presenting advantages such as high efficiency, ease-of-operation mode with fast response on a wide range of adsorbents (natural and artificial), low costs, and non-toxic byproducts. As a result, various adsorption systems based on materials such as clays, alumina, biomass, activated carbon, agricultural waste, silica gel, polysaccharides, and zeolites have been studied and developed [16].

Recent developments in the biosorption processes use many sustainable biomaterials as effective materials in the form of hydrogels, especially natural biopolymers, which are bioavailable, renewable, and biodegradable [17]. Their main advantage is a high adsorption capacity for contaminants from water and friendliness to the environment [18,19]. The 3D polymer networks of hydrogels make them flexible, multifunctional, reusable, and possess good physical and chemical stability [20,21]. They can be adapted to be more efficient in several uses and to increase the adsorption speed, swelling, durability, porosity, and stability [22–24].

Among the different biosorbents, chitosan (C) is an abundant and versatile natural biopolymer with an essential contribution for wastewater treatment. It has in its composition two types of monomeric units, one with an amino group and other with an acetamido group, as well as a considerable number of primary amines ($-NH_2$) and hydroxyl groups ($-OH$). These groups provide active sites for the efficient adsorption of anionic and cationic contaminants. The use of chitosan in its native form as an adsorbent has some disadvantages, such as low porosity and strength and a reduced surface area [25,26]. Therefore, the properties of chitosan-based adsorbent materials have been improved through different strategies to overcome these shortcomings [27]. Xanthan gum (X), an extracellular anionic polysaccharide produced by the fermentation of glucose, sucrose, or lactose by the bacterium *Xanthomonas campestris* (a Gram-negative bacterium), is currently used as a thickening agent and emulsion stabilizer due to its thermal stability and pseudoplastic behavior. In addition, it is a biodegradable and biocompatible biopolymer, and it has been widely used to remove contaminants alone and in combination with other natural or synthetic polymers [28].

Although in the literature there are many materials used for the adsorption of pharmaceuticals from wastewater [29–32], the scientific novelty of the present research is the design of a green hydrogel as an economic adsorbent, easy to be prepared and easy to be modified, and cost-effective, but with high efficiency for removing the pharmaceutical contaminants from water. For this purpose, diclofenac sodium was chosen as a "model-drug system". In this regard, to achieve a hydrogel-based adsorbent with suitable properties and a good performance, chitosan, polyethylene glycol 4000 (PEG4000), and xanthan gum were chosen to achieve a crosslinked hydrogel network.

Worldwide, the global consumption of diclofenac (DCF), especially used in the treatment of inflammation and pain [33], has been estimated at approximately 940 tons per year, from which approximately 65% of the oral dose of this drug is released through urine and feces, along with its active metabolites, and passes through conventional wastewater treatment plants [34]. DCF is also found in waters due to improper disposal as solid waste or due to ineffective conventional treatments through effluents from industrial and urban wastewater treatment plants [35]. The European Commission defined limits of chronic toxicity with respect to the annual average and acute toxicity for DCF, establishing the maximum allowable concentrations between 0.1 and 0.01 µg/L for surface inland waters and between 75 and 7.5 µg/L for coastal waters [8,36]. Recently, eliminating DCF from the aqueous environment has become a challenge for the scientific community, especially in the context of the establishment by the United Nations organization of the positive impact on many of the Sustainable Development Goals, due to its long-term importance for people and the environment [6].

Adsorbent materials based on various hydrogels have been developed and studied to remove sodium diclofenac from water. A hydrogel composed from bio-based egg albumin (ALB) functionalized with a high density of amine adsorption sites via polyethyleneimine (PEI) demonstrated excellent DCF removal capacity, i.e., 232.5 mg/g under optimal experimental conditions (pH~6; contact time~180 min; adsorbent dose~0.5 g/L) [37]. A poly(methacrylic acid)/montmorillonite (PMA/nMMT)-based nanocomposite hydrogel showed good adsorption capacity for the removal of amoxicillin (152.65 mg/g) and diclofenac (DCF) (152.86 mg/g) and an efficient regeneration [38]. Macroporous chitosan hydrogels were synthesized by crosslinking with genipin and incorporated n-GO as effective adsorbents for DCF. The addition of n-GO has promoted the DCF adsorption process and led to 100% removal of DCF after only 5 h [39]. Self-assembled reduced graphene oxide (rGO) three-dimensional hydrogels demonstrated a removal efficiency of naproxen (NPX), Ibuprofen (IBP), and diclofenac (DCF) between 70 and 80% at acidic pH and showed fast adsorption kinetics [40]. GO nanoparticles were shown to act as both adsorbents and crosslinking agents [41]. Reduced graphene oxide magnetite (r-GOM) has also demonstrated efficacy in removing diclofenac sodium (5.249 mg/g adsorption capacity) and aspirin (23.59 mg/g) from wastewater [42].

Here, the green synthesis of a hydrogel based on crosslinked chitosan and PEG4000 and using xanthan gum as a thickening agent was pursued as a new environmentally friendly, efficient adsorbent material used for the removal of DCF from aqueous media. Furthermore, the physical, chemical, rheological, and pharmacotechnical parameters were determined, and the adsorption kinetics of DCF were analyzed to evaluate the drug adsorption efficiency from the aqueous media.

2. Results and Discussion
2.1. Visual Examination

The hydrogel formation capability of xanthan gum is well known. Complex gel polymers of chitosan and xanthan gum with enhanced properties have been reported previously [43–49]. Figure 1 shows the CPX hydrogel obtained after the reaction between chitosan, xanthan gum, and PEG4000, as prepared (wet) and dry. The dry hydrogel is yellowish–white and has a gelatinous aspect. The complexation reaction between chitosan, xanthan gum, and polyethylene glycol occurs due to interactions among the opposite charges presented in the biopolymers: NH_3^+ groups of chitosan, COO^- groups of xanthan gum, and HO^- end-groups from PEG chains. When chitosan, xanthan gum, and polyethylene glycol are mixed together in solution, they can form a complex through electrostatic interactions among the positively charged chitosan and the negatively charged xanthan gum and PEG. The complexation process can have as a result the changes in the properties of each component from the mixture (hydrogel matrix). The interaction among chitosan, xanthan gum, and PEG can help to strengthen the hydrogel structure, leads to an increase in viscosity due to the formation of a three-dimensional polymer network, and improves its chemical and mechanical stability, as well as its adhesive properties.

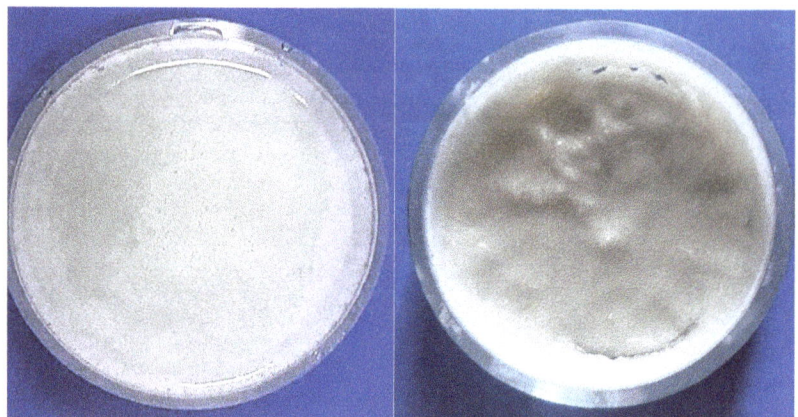

Figure 1. Optical images of the CPX hydrogel: wet (**left**) and dry (**right**).

2.2. Infrared Spectroscopy Measurements

FTIR spectra of chitosan, xanthan gum, PEG4000, and the developed hydrogel CPX are shown in Figure 2.

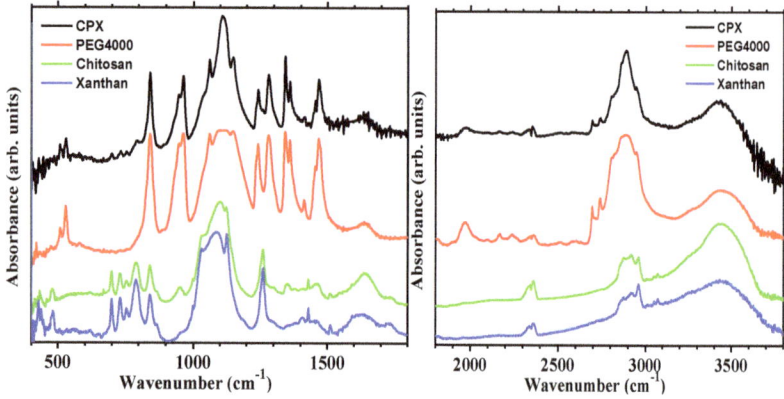

Figure 2. FTIR spectra of chitosan, xanthan gum, PEG4000, and the developed CPX hydrogel.

The main vibrational bands observed in the FTIR spectra of each individual component include O–H stretching vibrations observed by the broad band at around 3500–3200 cm^{-1}, corresponding to the stretching of the hydroxyl groups (–OH) present in the polymer chain for all compounds (red, green, and blue lines). Similar bands were noticed in the spectra of the polyethylene glycol 4000, xanthan gum, and chitosan previously reported [50–57]. C–H stretching vibrations were noticed by the band at around 3000–2800 cm^{-1}, corresponding to the asymmetric and symmetric stretching of the carbon–hydrogen (C–H) bonds present in the (–CH$_3$) groups, in agreement with previous reported bands observed in chitosan, xanthan gum, and PEG4000 [50–53]. C=O stretching vibrations registered by the band at around 1650 cm^{-1} in the chitosan spectrum (green line) corresponds to the stretching of the carbonyl group (C=O) and is relatively weak compared to the other peaks in the spectra, as also observed by Zajac et al. [50] and de Morais et al. [53] in chitosan. C–C and C–O–C bending vibrations are represented by the bands at 1450–1470 cm^{-1}, which correspond to the bending of the carbon–carbon (C–C) and carbon–oxygen–carbon (C–O–C) linkages. Weak bands in the same region were also noticed by Zajac et al. [50] and Nirmala [55]. C–H bending vibrations are assigned to the weak bands at around 1410 cm^{-1}

and 1340–1360 cm^{-1}, and they correspond to the bending of the carbon–hydrogen (C–H) bonds. Bands corresponding to C–H bending vibrations were observed at 1422 cm^{-1} and 1325–1377 cm^{-1} by Zajac et al. [50]. The band observed at 1280 cm^{-1} is typically assigned to the bending vibrations of the carbon–hydrogen (C–H) bonds present in the methylene (–CH$_2$–) groups in the PEG backbone (red line). This band was observed at 1262 cm^{-1} by Zajac et al. [50]. C–O–C asymmetric stretching vibrations: the peak at around 1240 cm^{-1} is characteristic of the asymmetric stretching of the ether linkages (–O–) present in the PEG molecule. The C–O–C and C–O stretching vibrations are represented by the intense peak at around 1100 cm^{-1}, corresponding to the stretching vibrations of the carbon–oxygen (C–O) bond present in the ether linkages (–O–). The 1150 cm^{-1} peak corresponds to the asymmetric stretching of the C–O bond and is often referred to as the "C–O stretching" peak. This band position correlates with other published spectra of the same polymers [56,57]. The intense peaks indicate the presence of multiple ether linkages in the polymer chain. The peak at around 1040 cm^{-1} corresponds to the stretching of the carbonyl groups (C=O) present and is relatively weak compared to the C–O–C stretching band. C–O–C rocking vibrations: the intense peak at 850 cm^{-1} corresponds to the rocking of the ether linkages (–O–) present in the PEG molecule (red line) and chitosan (green line), observed at 896 cm^{-1} by Zajac et al. [50] and Dey et al. [52]. In summary, the IR spectrum of PEG4000 (red line) is characterized by intense and broad O–H stretching vibrations, intense C–O–C stretching vibrations, and C–H and C=O stretching and bending vibrations.

The synthesized hydrogel spectrum (black line) exhibited some bands with slight shifting and lower intensities compared with the PEG4000 spectrum, especially in the main bands region in 2800–3000 cm^{-1}, 1000–1200 cm^{-1}, and 600–800 cm^{-1}, overlapping (in the 500–600 cm^{-1}, 800–1000 cm^{-1} regions), or the appearance of some characteristic peaks of pure components in the developed polymeric network (in 800–1000 cm^{-1} and 1200–1350 cm^{-1}, 2700–2800 cm^{-1}), which is a clear indication of interaction among the components through intramolecular re-arrangement, hydrogen bonding, or alteration in the positions of functional groups in the final hydrogel structure.

2.3. Raman Spectroscopy Results

As shown in Figure 3, the Raman spectra of the CPX hydrogel (Figure 3a) present well-defined bands in the 800–1700 cm^{-1} spectral region that correspond to the vibration modes of the gel components, shown in Figure 3b. The majority of the bands in Figure 3a correspond to PEG4000, with a stronger Raman signal. The presence of chitosan or xanthan gum is revealed by the wide Raman band centered at 1600 cm^{-1}; the G band from C–C bonds in the backbone of the polymeric structure, which is not present in the Raman spectra of PEG4000; and a shoulder at 1082 cm^{-1} from xanthan gum. The peak at 851 cm^{-1} is related to the superposition of a number of vibrations with the main contributions of the CH$_2$ rocking, C–O stretching, and C–C stretching vibrations. Therefore, the peak positions of the Raman bands at about 851, 1140, and 1477 cm^{-1} can serve as a quantitative measure of the molecular weight distribution for short PEG molecules. Their positions confirm molecular weights over 2000 [58]. According to Matsuura [59,60], the bands at 868, 936, 1070, 1133, and 1150 cm^{-1} are also assigned to the superposition of a number of vibrations with the main contributions of the CH$_2$ rocking, C–O stretching, and C–C stretching of terminal groups, while the bands observed at higher Raman shifts are assigned to CH$_2$ modes: in-plane twisting at 1289 cm^{-1}, out-of-plane wagging at 1392 cm^{-1}, and scissoring vibrations at 1441 and 1477 cm^{-1} [59–61].

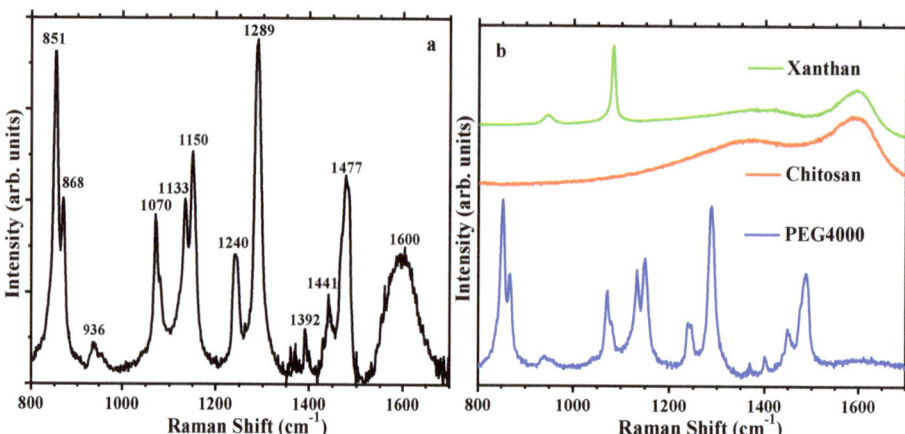

Figure 3. The Raman spectra of the CPX hydrogel (**a**) and the Raman vibration modes of the gel components (**b**).

2.4. XRD Results

The X-ray diffraction method was used to examine the structure of CPX hydrogel. The XRD patterns of the raw materials (chitosan, PEG4000, and xanthan gum) and the obtained CPX hydrogel are shown in Figure 4.

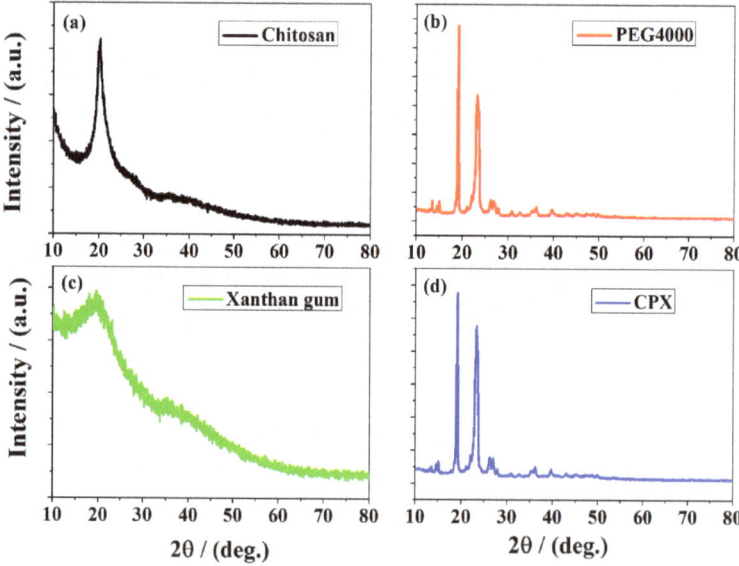

Figure 4. XRD diffractograms of (**a**) chitosan, (**b**) PEG4000, (**c**) xanthan gum, and (**d**) the developed CPX hydrogel.

The X-ray pattern of chitosan (Figure 4a) shows an intense reflection from (200) planes at $2\theta = 20.2°$, revealing a crystalline structure [62,63]. The pattern of PEG4000 (Figure 4b) displays two sharp and strong diffraction peaks at approximately 19.6° and 23.4° [64,65]. Xanthan gum (Figure 4c) presents no sharp peaks, indicating its amorphous structure, in agreement with the earlier reports [21,66]. The X-ray pattern of the CPX hydrogel (Figure 4d) is dominated by the intense diffraction peaks of PEG4000; however, the peaks are slightly shifted and with modified intensities, as a result of the interaction among the three components in the hydrogel network. Therefore, the observed XRD pattern is in good agreement with the findings observed in the FTIR analysis.

2.5. Differential Scanning Calorimetry (DSC) Analysis

Figure 5 shows the DSC thermogram of the CPX hydrogel.

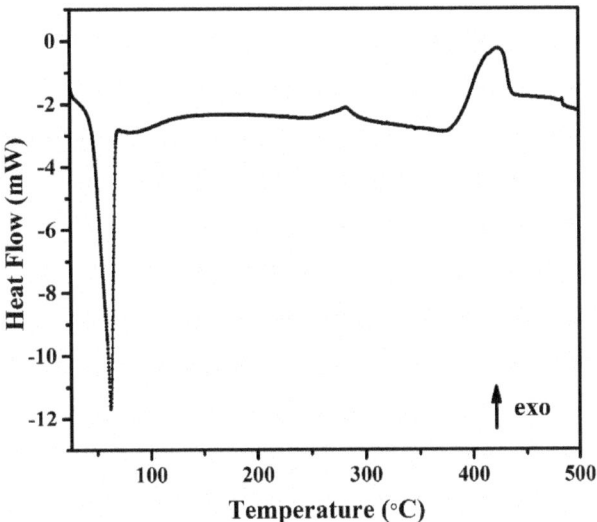

Figure 5. The DSC thermogram of the CPX sample.

The first thermal event from Figure 5 is a wide endothermic peak centered between 51 and 68 °C, with an onset at 51 °C. The values for the temperature and associated enthalpy of the peak minimum are 62 °C and 577.08 mJ. The endothermic peak is associated to dehydration, the loss of water associated with the hydrophilic groups of chitosan [67,68]. In the solid state, chitosan-based polysaccharides have disordered structures and have a strong affinity for water and, as a result, can be easily hydrated [69]. The hydration properties of these polysaccharides depend on the primary and supramolecular structures [70,71]. This peak indicates that the sample was not completely dried, and there was some bound water, which was not removed during drying. Appelqvist et al. [72] and Gidley and Robinson [73] reported an endothermic dehydration peak in a similar range of 60 ± 10 °C for a range of polysaccharides at a moisture level between 5 and 25%, which they attributed to the enthalpic association between water and carbohydrate.

The DSC thermogram of the sample also showed a small exothermic event between 270 and 300 °C, which can be attributed to scissions in the polymeric network of polysaccharides [68]. The wide exothermic peak between 370 and 430 °C can be attributed to the overall decomposition of highly substituted regions in polysaccharides. The exothermic peak is assigned to the thermal degradation of the composite polymeric material (monomer dehydration, glycoside bond cleavage, and decomposition of the acetyl and deacetylated units) [74,75].

Thus, the observed DSC events are in good agreement with the thermal decomposition profiles observed by TGA.

2.6. Thermogravimetric and Differential Thermal Gravimetric Analysis

Figure 6 shows the TG/DTG curves of the CPX hydrogel.

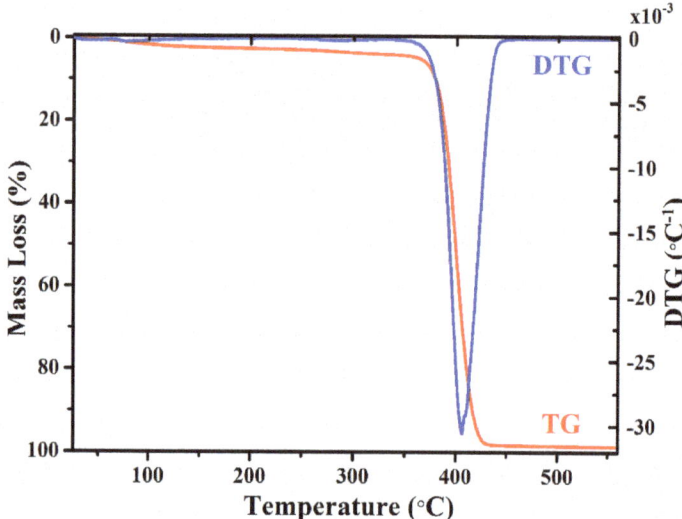

Figure 6. Thermogravimetric and differential thermal gravimetric analysis of the CPX hydrogel.

A small (<5%) weight loss is observed in the TG curve between 20 and 100 °C, associated to the evaporation of water (moisture loss), being a continuous, progressive weight loss, confirmed by the DTG curve. After dehydration, no further weight loss is observed up to 370 °C, indicating a good thermal stability of the polymeric network. A substantial weight loss occurs between 370 °C and 430 °C, indicating the complete pyrolysis of the polymeric composite (chitosan, xanthan gum, and PEG4000) in a single thermal event. TG measurements reveal a total mass loss of 99.37% at 550 °C, which confirms the full decomposition of the composite polymer. Similar results were also reported [53,76,77].

A plateau of thermal stability was observed after the thermal decomposition of the organic material of the components, whose weight loss started before finishing the dehydroxylation step. These results are compatible with the process of thermal degradation of the polymeric network [78,79]. During the pyrolysis, a random split of the glycosidic bonds occurs in the chitosan and PEG network, which is further followed by decomposition forming acetic, butyric, and fatty acids [52].

2.7. SEM Analysis

SEM images of the dried CPX hydrogel show a fibrillar morphology (Figure 7), consisting of bundles of two-dimensional stacked sheet-like structures (Figure 7a,b), with lengths of tens of microns, as shown in more detail in Figure 7c–d, and thicknesses in the nanorange (below 100 nm), illustrated in the inset of Figure 7d, a magnified image (300,000×) at the edge of a bundle of stacked nanosheets, at the position marked with an 'x'. The combination of negatively charged (PEG with OH^- groups and xanthan gum with COO^-) and positively charged (chitosan with NH^{3+} groups) electrolytes is known to induce intimate mixing in ordered morphologies. Chitosan–xanthan gum composite hydrogels have previously been shown to produce two-dimensional structures [80,81]. These two-dimensional ordered structures are highly advantageous for enhancing the swelling capacity of hydrogels by filling and expanding the cavities between individual sheets, while maintaining the structural integrity of the polymer network.

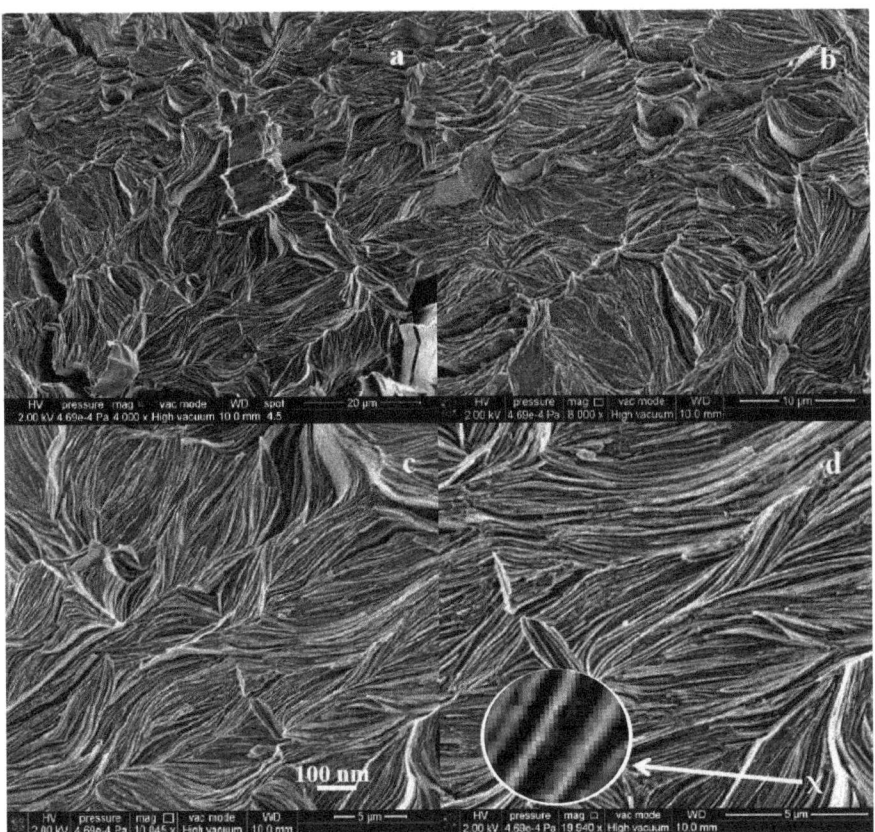

Figure 7. SEM micrographs at different magnifications showing the microstructure of the CPX hydrogel. (**a**,**b**) bundles of two-dimensional stacked sheet-like structures with lengths of tens of microns, (**c**,**d**) magnified images with thicknesses in the nanorange (below 100 nm); a magnified image (300,000×) at the edge of a bundle of stacked nanosheets, at the position marked with an 'x'.

2.8. Rheology

The gel nature of the obtained hydrogel was confirmed using rheology analysis. The dynamic rheological behavior of the CPX hydrogel was investigated, and the results are presented in Figure 8.

The CPX hydrogel (Figure 8a) showed a linear viscoelastic response at a shear strain less than 2%, with G' and G" being independent of applied strain, and the elastic modulus being 7 times greater than the viscous one. After a 2% strain stress, the non-linear viscoelastic behavior occurs, although G' remains the dominant part. After that, the elastic modulus considerably decreases, and the viscous component takes control of the nonlinear behavior at a strain amplitude of about 80% (the crossover point). The dependences of rheological moduli on applied frequency at 0.5% stress strain are illustrated in Figure 8b. The CPX hydrogel responded to frequency measurements with a gel-like response, with G' higher than G", and both rheological moduli appear to be almost frequency-independent. The same behavior was observed for other hydrogels used for adsorption of pharmaceuticals from water [31]. Further, for the CPX hydrogel, the shear viscosity decreases as the shear rate increases (Figure 8c), pointing to a non-Newtonian pseudoplastic fluid with shear-thinning properties.

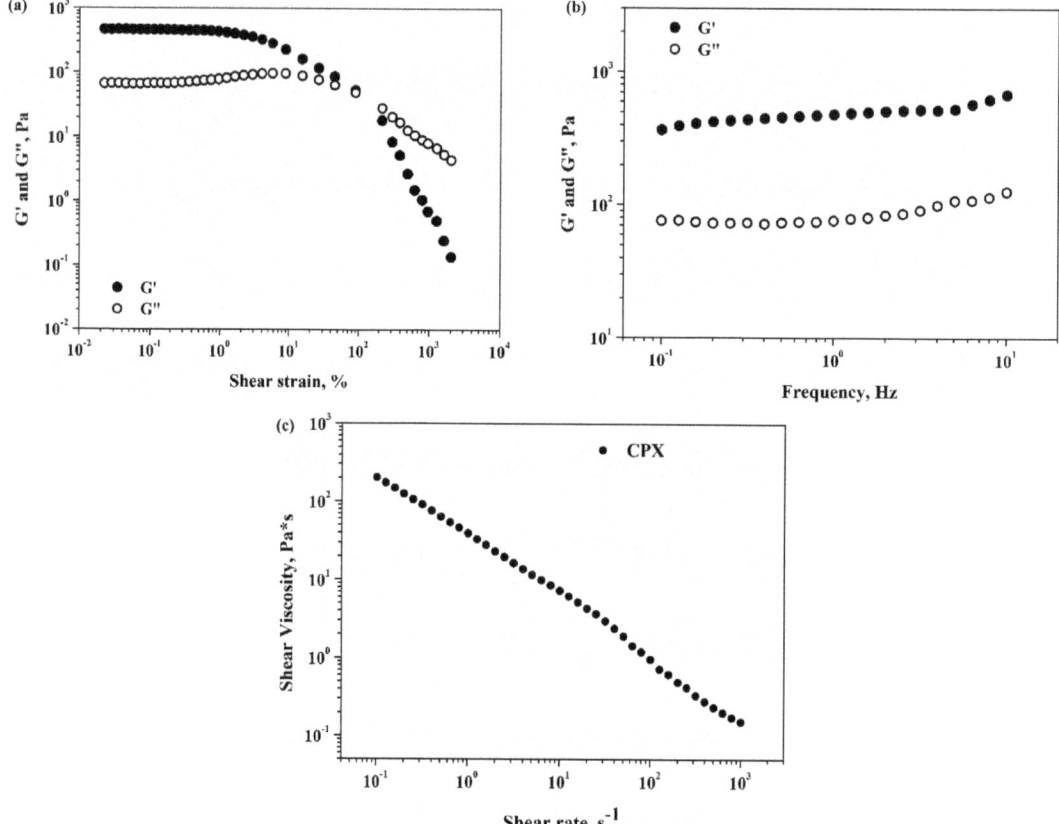

Figure 8. Rheological characterization of CPX hydrogel: (**a**) Storage and loss moduli vs. shear strain; (**b**) Storage and loss moduli vs. frequency; and (**c**) Shear viscosity vs. shear rate.

2.9. Pharmacotehnical Characterization

The dry CPX hydrogel showed a tensile strength of 0.84 ± 0.09 kg/mm^2 and 17.04 ± 0.12% elongation. The low moisture content, 5.22 ± 0.43%, explains the reduced hardness and flexibility, as expected for dehydrated hydrogels that contain little or no water.

The CPX hydrogel swelling degree over 6 h in 3 different media is shown in Figure 9. Swelling performance is affected by different parameters, such as the hydrophilicity and hydrophobicity of the polymers type, crosslinking density of the hydrogel network, and pH conditions [82].

The CPX hydrogel swelling behavior is almost similar at different pH values, proving that it is not pH-sensitive, and has the same performance in any medium. Similar results were reported in the literature [43,83,84]. These studies described that the hydrogel network, which was designed through the ionic interactions among the amino groups from chitosan and the carboxyl groups from xanthan gum in the hydrogel matrix, which facilitates the controlled adsorption of various molecules.

The swelling degree was demonstrated to be developed linearly for the first 4 hours (83%), after which the increase slowed down, with the differences between 240 and 360 min being essentially unimportant. The findings show that independent of pH conditions, the swelling ability is the highest in the first 30 min.

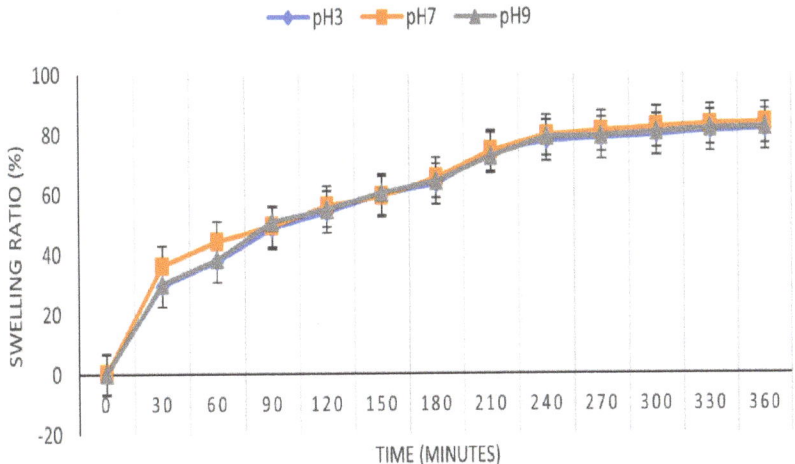

Figure 9. The CPX hydrogel swelling degree over 6 h at pH 3, pH 7, and pH 9.

2.10. Preliminary Adsorption Studies

Preliminary investigation of the adsorption tests was made using various dyes (Figure 10a; gentian violet, methyl orange, and eosin), and the results are depicted in Figure 10. A color change from the initial colors (Figure 10b) to lighter shades (Figure 10c) after 24 h of contact time was visually observed, indicating that the dye was easily adsorbed by the hydrogel.

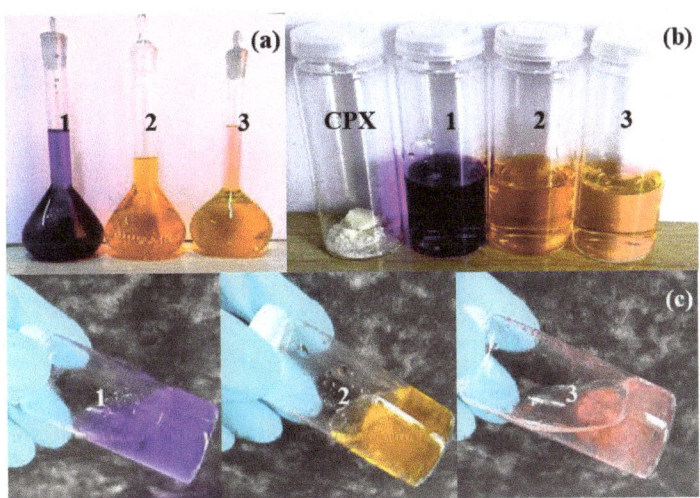

Figure 10. Visual dye preliminary adsorption tests of CPX hydrogel. (1) Gentian violet; (2) methyl orange; and (3) eosin; (**a**) stock solution; (**b**) initial time; (**c**) after 24 h.

2.11. Batch Adsorption Study

The effect of contact time on the adsorption of DCF on the CPX hydrogel is presented in Figure 11.

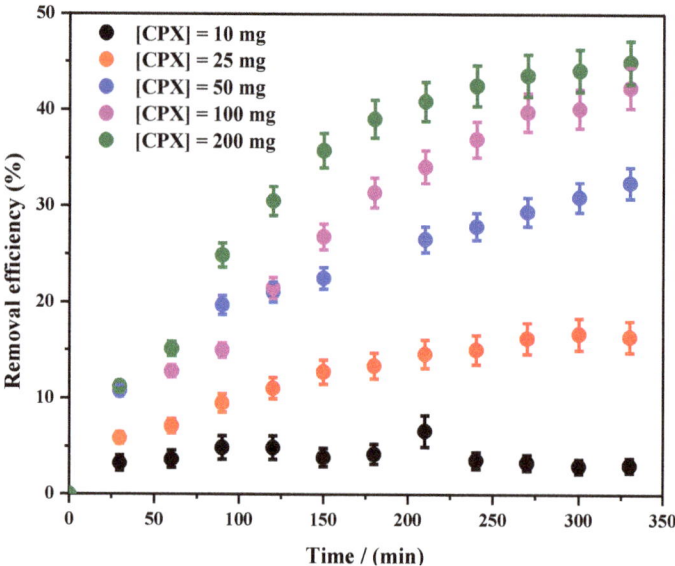

Figure 11. Effect of contact time on DCF adsorption on CPX hydrogel over time.

It was observed that the DCF is slowly adsorbed into the CPX hydrogel after 350 min, reaching 45% in the highest adsorbent amount (200 mg) (Figure 11). During the adsorption study, no blue or red shifting of the characteristic adsorption maximum of DCF was observed. Since for all tested CPX hydrogel amounts, the behavior was quite similar, Figure 12 shows an exemplification of the adsorbance variation in the considered time frame.

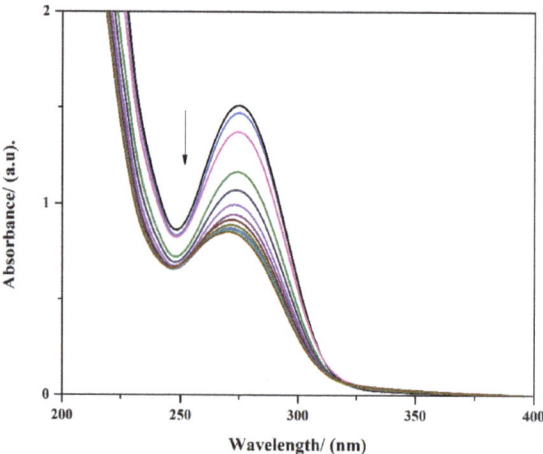

Figure 12. UV–Vis spectra of adsorption of DCF in presence of 50 mg CPX hydrogel during 330 min.

When chitosan, xanthan gum, and polyethylene glycol are mixed together in solution, they can form a complex through electrostatic interactions among the positively charged chitosan and the negatively charged xanthan gum and PEG. PEG4000 is considered to be hydrophilic and is soluble in aqueous solutions. This is because PEG is a polar molecule, with hydroxyl groups (–OH) along its polymer chain that can interact with water molecules through hydrogen bonding. The hydrophilic properties of PEG can vary, depending on the length of the polymer chain; shorter chains of PEG are more hydrophilic, while longer chains can become more hydrophobic due to the increased number of non-polar carbon atoms in the polymer backbone. Nevertheless, PEG4000 is still generally considered to be a hydrophilic compound. Xanthan gum contains numerous hydrophilic functional groups, such as hydroxyl (–OH) and carboxyl (–COOH) groups. As a result, xanthan gum is highly soluble in water and widely used as a thickener, stabilizer, and emulsifier. Chitosan contains hydroxyl and amino groups that are hydrophilic and can interact with water molecules through hydrogen bonding, as well as acetyl and amino groups that are hydrophobic and can interact with non-polar molecules through van der Waals interactions. DCF is a polar compound that contains both hydrophilic and hydrophobic functional groups. It has a carboxylic acid (–COOH) group and a phenolic (–OH) group that are both hydrophilic and can interact with water molecules through hydrogen bonding. However, it also has two chloro (–Cl) substituents that are hydrophobic and can interact with non-polar solvents through van der Waals interactions. Therefore, both electrostatic and hydrophobic interactions may contribute to the adsorption of DCF.

Subsequently, the pseudo first-order model was employed to fit the experimental data in order to study the DCF adsorption process (Figure 13).

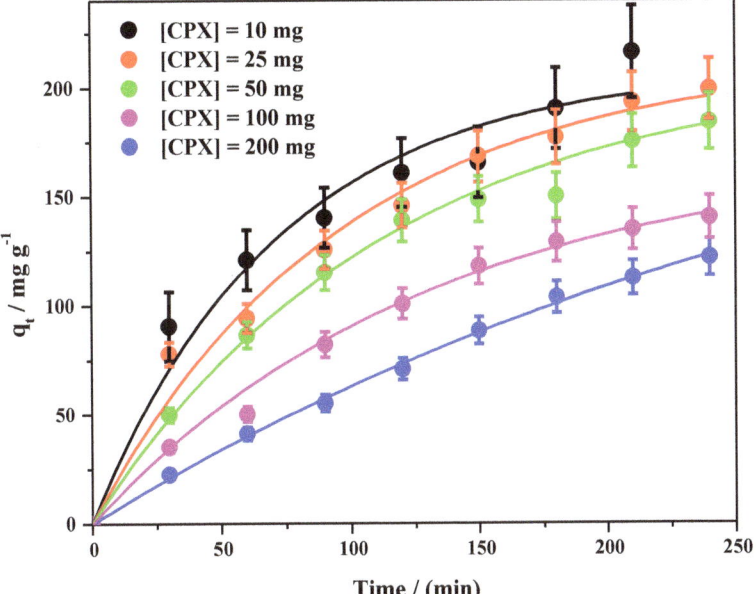

Figure 13. The plot of the non-linear form of the pseudo first-order model for different amounts of CPX hydrogel and 0.33 mg DCF [Solid lines represent the best fit of the experimental data points using Equation (10)].

Meanwhile, Table 1 contains the corresponding model parameters of the fits for a pseudo first-order kinetics. The pseudo first-order model assumes that the adsorption rate is limited mainly by the diffusion step [85,86].

The pseudo second-order model for the adsorption of DCF in different amounts of CPX hydrogel was also analyzed. The model was fitted using a linear equation. The results, together with the errors, are shown in Figure S1 and Table S1 from the Supplementary Materials. The adjusted R square is much lower than 1, suggesting that the adequate model for the investigated systems containing DCF follows a pseudo first-order model.

Since the adsorption takes place at equilibrium, the experimental data were analyzed by Langmuir [86] and Freundlich [87] models, and the type of adsorption that takes place between the aqueous DCF species and adsorption sites was determined (Figure 13). Table 2 provides the linear correlation coefficients (R^2) and the isotherm constants (q_{max}, b, n, and K_F) for the Langmuir and Freundlich models. In addition, the equilibrium parameter (R_L) was calculated. The Langmuir and Freundlich isotherm graphical representation is shown in Figures S2 and S3 from the Supplementary Materials.

Table 1. Pseudo first-order kinetic parameters of DCF adsorption on different masses of CPX hydrogel.

Pseudo First-Order Parameters	[CPX]				
	10 mg	25 mg	50 mg	100 mg	200 mg
$k_1/(min^{-1})$	0.018	0.012	0.013	0.004	0.006
$q_e /(mg \times g^{-1})$	208.67	202.44	189.93	187.06	175.01
R^2	0.99	0.99	0.98	0.98	0.99

Table 2. Langmuir and Freundlich isotherm parameters.

Langmuir Isotherm Parameters				Freundlich Isotherm Parameters		
q_{max}	b	R_L	R^2	K_F	n	R^2
172.41	0.0014	0.80	0.99	292.33	16.28	0.93

Obviously, the R^2 value of the Langmuir model (0.99) is substantially higher than that of the Freundlich model (0.93) and close to 1.0. Furthermore, the theoretical value of q_{max} (172.41 mg/g) is in agreement with the experimental data. The R_L parameter is lower than 1, which suggests that DCF adsorption is a favorable phenomenon. Similar results were reported in the literature using different hydrogel-based adsorbent materials for DCF adsorption studies [88,89]. The Langmuir model showed that the DCF adsorption process on the CPX hydrogel matrix was close to monolayer adsorption [81]; the adsorption sites were uniform and independent [81], with no interaction involving adsorption molecules [90].

The kinetic analysis indicates that the DCF adsorption process tends to follow a Langmuir-type adsorption. Therefore, it is likely that the adsorption of DCF and dyes involves mainly electrostatic interactions, governed by hydrogen bonds, with some contribution of Van der Waals forces.

The data obtained in the kinetic study reveal that the used adsorbent may have a high potential to be used in environmental applications as a water cleaning agent.

3. Conclusions

A green adsorbent hydrogel based on chitosan, polyethylene glycol 4000, and xanthan gum was synthesized using a simple and easy aqueous solution method. The FTIR, Raman, XRD, DSC, and TGA analyses confirmed the occurrence of interactions among the three components, through the intramolecular rearrangement of hydrogen bonds and in the modification of the positions of the functional groups, in the final hydrogel assembly. The SEM analysis evidenced a two-dimensional ordered structure that has a great advantage in improving the swelling capacity of CPX by approximately 83% after 360 min and does not depend on the pH medium.

Preliminary adsorption tests showed that CPX has the capacity to adsorb different dyes (gentian violet, methyl orange, and eosin) from water after 24 h of contact time, evaluated by visual observation. The adsorption ability of the CPX hydrogel towards the removal of pharmaceuticals from water was tested for DCF as the tested drug. CPX hydrogel exhibited a remarkable adsorption capacity (172.41 mg/g) for DCF. For the tested drug, the adsorption kinetics were found to follow the pseudo first-order model, while the adsorption mechanism was explained by the Langmuir and Freundlich models. Taken together as a first report, these results allow us to conclude that CPX-based adsorbent hydrogel can be used as a promising facile, ecological, and cost-effective adsorbent for the removal of pharmaceuticals from wastewater.

4. Materials and Methods

4.1. Chemicals and Reagents

All materials were of analytical grade. Chitosan (C) (deacetylated chitin, poly(D-glucosamine), medium molecular weight M.W. = 190,000–310,000 (MMWCH) and viscosity = 200–800 cps in 1% acetic acid, degree of deacetylation 75–85%), acetic acid ($\geq$99.7%), and polyethylene glycol of molecular weight 4000 (PEG4000) were purchased from Merck, Germany. Elemental SRL, Romania, supplied xanthan gum (X). The experiments were performed with deionized (DI) water (resistivity of 18.2 $\Omega\cdot$cm at 25 °C). All reagents were used as received, without further purification.

4.2. CPX Hydrogel Preparation

The hydrogel was prepared by a simple and easy mechanical mixing method in aqueous solution and with a final light heat treatment, by using the minimum quantities of materials able to promote the formation of hydrogel matrix. Firstly, 100 mL of an aqueous solution containing 1% chitosan (w/v) in 1% acetic acid (w/v) was prepared and stirred overnight until the chitosan was completely solubilized. Subsequently, 40% PEG4000 was added to this solution under vigorous continuous stirring, at 800 rpm, for approximately 1 h for homogeneous mixing. Next, 4% xanthan gum was added to the solution, under stirring, and slightly warmed below 40 °C for approximately 30 min, until gelation occurred. Finally, the obtained hydrogel, (notation: CPX, for its chitosan (1%)—PEG4000 (40%)—xanthan gum (4%) composition) was left to rest at room temperature until the next day. After being organoleptically examined, the synthesized CPX hydrogel was poured into Petri dishes and kept until completely dry. Figure 14 presents a schematic illustrating the synthesis and characterization of the hydrogel-based adsorbent material.

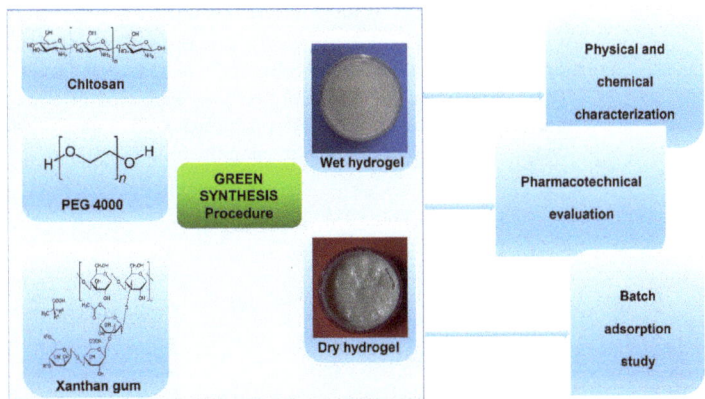

Figure 14. Schematic illustration of the synthesis procedure and the evaluation of the hydrogel-based adsorbent material.

4.3. Methods

4.3.1. Visual Examination

The obtained hydrogel was examined regarding its specific properties, such as color, appearance, homogeneity, consistency, and phase separation, or the presence of agglomerations [21].

4.3.2. Physical and Chemical Analysis

Fourier Transform Infrared (FTIR) analysis was carried out using a Nicolet 6700 apparatus in the range of 4000–400 cm^{-1}. Potassium bromide KBr of spectroscopic grade was used for mixing the samples. The measurements were obtained in absorbance mode, with a sensitivity of 4 cm^{-1}. For further qualitative analysis, the normalized spectra were used.

A Horiba Jobin Yvon LabRam HR spectrometer (Horiba, Ltd., Kyoto, Japan) was used for recording the Raman spectra. An excitation laser at 325 nm and a NUV 40× objective, using an integration time of 60 s, was utilized.

XRD spectra were obtained using a PANalytical Empyrean diffractometer with a Cu X-ray tube (λ Cu Kα1 = 1.541874 Å), at room temperature. X-ray diffractograms were collected using a 0.02° scan step, in the range of 20°–80°, in a Bragg–Brentano geometry.

Differential scanning calorimetry (DSC) analysis was recorded with a Mettler Toledo DSC 3 calorimeter and carried out to obtain the DSC curves, in a nitrogen atmosphere with a gas flow of 80 mL min^{-1}. The samples were sealed in crimped Al pans with a pinhole in the lid.

Thermogravimetric (TG/DTG) analysis was achieved with a Mettler Toledo TGA/SDTA851^e instrument, in a synthetic airflow atmosphere with a flow of 80 mL min^{-1}, using 70 µL open alumina pans. The used heating rates for DSC and TG/DTG analyses were 10 °C/min.

Scanning electron microscopy (SEM) was carried out to study the morphology of dry hydrogel in a Quanta 3D field emission microscope in secondary electron images, which operate in high vacuum mode, at an accelerating voltage of 2 kV.

The rheological measurements were performed on a Kinexus Pro rheometer at 25 °C, with a 0.8 mm gap using parallel plate geometry. The linear viscoelastic domain of storage and loss moduli (G' and G'') was determined at a constant frequency (1 Hz), as a function of shear strain. The G' and G'' were evaluated as a function of frequency (0.1–10 Hz) in the linear viscoelastic regime at a shear strain of 0.5%. Moreover, the shear viscosity was investigated at applied shear rates ranging from 0.1 to 1000 s^{-1}. Rheological data were presented using a logarithmic scale.

4.3.3. Pharmacotechnical Characterization and Elongation Ability and Tensile Strength

The mechanical performance was tested with a digital tensile force tester used for universal materials, produced by Lloyd Instruments Ltd., LR 10K Plus (West Sussex, UK). Between the two plates positioned at a distance of 30 mm, the dry hydrogel was placed vertically, and the breaking force was measured at a speed of 30 mm/min. The mechanical characteristics were calculated using the following formulas:

$$Tensile\ strength\ (kg/mm^2) = \frac{Force\ at\ breakage\ (kg)}{Film\ thickness\ (mm) \times Film\ width\ (mm)} \quad (1)$$

$$Elongation\ (\%) = \frac{Increased\ film\ length}{Inital\ film\ length} \times 100 \quad (2)$$

Moisture Content

The moisture content was evaluated using an HR 73 Mettler Toledo halogen humidity analyzer, produced by Mettler-Toledo GmbH (Greifensee, Switzerland), using the thermogravimetric technique [91]. It was calculated as the drying loss (%). The measurements were performed in triplicate.

Swelling Behavior

The swelling ability of hydrogel was evaluated by varying pH conditions, from an acidic to a basic value. A total of 3 different mediums were prepared and used in the analysis: HCl/H$_2$O solution (pH 3), purified water (pH 7), and NaOH/H$_2$O solution (pH 9). In each solution, 0.1 g of hydrogel was placed and kept at room temperature (22 °C). At every hour, the samples were withdrawn and weighed. The swelling behavior was determined at 6 h. The percentages of water absorption were calculated according to Equation (3):

$$Swelling\ ratio\ (\%) = \frac{Wt}{Wi} \times 100 \quad (3)$$

where W_t is the sample weight at time t after the incubation, and W_i is the initial weight [92–94]. The experiments were performed in triplicates.

4.4. Preliminary Adsorption Studies

Initially, dye adsorption test studies were conducted to preliminarily assess the capacity of adsorption of hydrogel-based material as an adsorbent. As a model for adsorption assays, the following dyes were used: gentian violet, methyl orange, and eosin (purchased from Merck, Germany). For adsorption experiments, 0.5 g of dry hydrogel was added to 10 mL of each dye solution, of concentration 20 mg/L, at room temperature, at pH 7. The evaluation of the visual adsorption was determined after 24 h of contact time.

4.5. Batch Adsorption Study

A Carry Varian X100 spectrophotometer was used for the batch adsorption study. The stock solution of DCF sodium salt of 1 mg per 1 mL was prepared with distilled water and stored refrigerated. For adsorption experiments, certain amounts of CPX were weighed out and left for 24 h in 10 mL of water to swell. After that, 0.33 mL of DCF stock solution was added to each sample. At the desired time interval, the solutions containing DCF were scanned in the 200–400 nm UV range using a 1 cm quartz cuvette. The residual concentration of DCF from the aqueous solution was quantified by following the decrease of the adsorption maximum of DCF at $\lambda = 276$ nm ($\varepsilon = 9580$ M^{-1} × cm^{-1}).

The DCF removal efficiency and CPX adsorption capacity were estimated using the following equations:

$$Ad(\%) = \frac{(C_0 - C_t) \times 100}{C_0} \quad (4)$$

$$q_t = \frac{(C_0 - C_t) \times V}{W} \quad (5)$$

$$q_e = \frac{(C_0 - C_e) \times V}{W} \quad (6)$$

where Ad (%) is the DCF removal efficiency of CPX; q_e and q_t are the adsorption capacity expressed in (mg × g^{-1}) at equilibrium and at time t (min), respectively; C_0, C_t, and C_e are the initial DCF concentration, the concentration of DCF that is still in the solution at a time t, and the equilibrium DCF concentration (mg × L^{-1}), respectively; V is the volume of the aqueous solution (L); and W is the mass of the CPX adsorbent (g).

The curves obtained by plotting the q_t function of t were fitted by nonlinear regression using an equation that describes a kinetic model specific for liquid–solid phase adsorption, a process of pseudo first order:

$$q_t = q_e\left(1 - e^{-k_1 t}\right) \quad (7)$$

The equation allowed the obtaining of k_1, which is the rate constant of the pseudo first-order sorption (min^{-1}) and also the q_e adsorption capacity at equilibrium (mg × g^{-1}).

Furthermore, the Langmuir and Freundlich adsorption models were both employed to correlate the obtained isotherm information.

The linearized Langmuir equation was expressed as:

$$\frac{C_e}{q_e} = \frac{1}{q_{max}}C_e + \frac{1}{bq_{max}} \quad (8)$$

where q_{max} was the maximum monolayer adsorption capacity (mg × g^{-1}), and b defines the Langmuir adsorption constant (L × mg^{-1}).

Using the parameters obtained with Equation (8), the equilibrium parameter (R_L) by means of Equation (9) was obtained [95]:

$$R_L = \frac{1}{1 + bC_0} \quad (9)$$

The Freundlich equation was also used to investigate the adsorption process on CPX at the equilibrium condition. The theoretical Freundlich isotherm was used in the linearized form as:

$$ln q_e = ln K_F + \frac{1}{n} ln C_e \qquad (10)$$

where K_F and n are the Freundlich parameters that are obtained from the graphical representation of $ln q_e$ versus $ln C_e$.

Supplementary Materials: The following supporting information can be downloaded at: https://www.mdpi.com/article/10.3390/gels9060454/s1, Figure S1: The plot of the non-linear form of the pseudo second-order model for different amounts of CPX hydrogel and 0.33 mg DCF; Table S1: Pseudo second-order kinetic parameters of DCF adsorption on different amounts of CPX hydrogel; Figure S2: Langmuir isotherm representation of DCF adsorption in presence of different CPX amounts; Figure S3: Freundlich isotherm representation of DCF adsorption in presence of different CPX amounts.

Author Contributions: Conceptualization, M.C. and A.M.M.; methodology, M.C., M.P. and A.M.M.; data curation, M.C., M.P. and A.M.M.; resources, M.C.; formal analysis, M.P., M.C., J.C.M., J.P.C., V.A.S., A.R.L., E.A.O., L.A. and A.M.M.; investigation, M.C., M.P., J.C.M., J.P.C., V.A.S., A.R.L., E.A.O., L.A. and A.M.M.; writing—original draft preparation, M.C., M.P., J.C.M. and A.M.M.; writing—review and editing, M.C., M.P. and A.M.M.; visualization, M.C. and A.M.M.; supervision A.M.M. All authors have read and agreed to the published version of the manuscript.

Funding: This research received no external funding.

Institutional Review Board Statement: Not applicable.

Informed Consent Statement: Not applicable.

Data Availability Statement: Not applicable.

Conflicts of Interest: The authors declare no conflict of interest.

References

1. Du Plessis, A. Persistent degradation: Global water quality challenges and required actions. *One Earth* **2022**, *5*, 129–131. [CrossRef]
2. Estêvão, M.D. Aquatic Pollutants: Risks, Consequences, Possible Solutions and Novel Testing Approaches. *Fishes* **2023**, *8*, 97. [CrossRef]
3. Dulsat-Masvidal, M.; Ciudad, C.; Infante, O.; Mateo, R.; Lacorte, S. Water pollution threats in important bird and biodiversity areas from Spain. *J. Hazard. Mater.* **2023**, *448*, 130938. [CrossRef] [PubMed]
4. Bhuyan, A.; Ahmaruzzaman, M. Recent advances in new generation nanocomposite materials for adsorption of pharmaceuticals from aqueous environment. *Environ. Sci. Pollut. Res.* **2023**, *30*, 39377–39417. [CrossRef]
5. Kumar, M.; Sridharan, S.; Sawarkar, A.D.; Shakeel, A.; Anerao, P.; Mannina, G.; Sharma, P.; Pandey, A. Current research trends on emerging contaminants pharmaceutical and personal care products (PPCPs): A comprehensive review. *Sci. Total Environ.* **2023**, *859*, 160031. [CrossRef]
6. Alessandretti, I.; Rigueto, C.V.T.; Nazari, M.T.; Rosseto, M.; Dettmer, A. Removal of diclofenac from wastewater: A comprehensive review of detection, characteristics and tertiary treatment techniques. *J. Environ. Chem. Eng.* **2021**, *9*, 106743. [CrossRef]
7. Matamoros, V.; Rodríguez, Y.; Albaig'es, J. A comparative assessment of intensive and extensive wastewater treatment technologies for removing emerging contaminants in small communities. *Water Res.* **2016**, *88*, 777–785. [CrossRef]
8. Sousa, J.C.G.; Ribeiro, A.R.; Barbosa, M.O.; Pereira, M.F.R.; Silva, A.M.T. A review on environmental monitoring of water organic pollutants identified by EU guidelines. *J. Hazard. Mater.* **2018**, *344*, 146–162. [CrossRef]
9. Loganathan, P.; Vigneswaran, S.; Kandasamy, J.; Cuprys, A.K.; Maletskyi, Z.; Ratnaweera, H. Treatment Trends and Combined Methods in Removing Pharmaceuticals and Personal Care Products from Wastewater—A Review. *Membranes* **2023**, *13*, 158. [CrossRef]
10. Rahman, T.U.; Roy, H.; Islam, M.R.; Tahmid, M.; Fariha, A.; Mazumder, A.; Tasnim, N.; Pervez, M.N.; Cai, Y.; Naddeo, V.; et al. The Advancement in Membrane Bioreactor (MBR) Technology toward Sustainable Industrial Wastewater Management. *Membranes* **2023**, *13*, 181. [CrossRef]
11. Rikabi, A.A.K.K.; Chelu, B.; Mariana; Harabor, I.; Albu, P.C.; Segarceanu, M.; Nechifor, G. Iono-molecular Separation with Composite Membranes I. Preparation and characterization of membranes with polysulfone matrix. *Rev. Chimie* **2016**, *67*, 1658–1665.
12. Diaconu, I.; Gardea, R.; Cristea, C.; Nechifor, G.; Ruse, E.; Eftimie Totu, E. Removal and recovery of some phenolic pollutants using liquid membranes. *Rom. Biotechnol. Lett.* **2010**, *15*, 2010.

13. Nechifor, A.C.; Goran, A.; Grosu, V.-A.; Pîrtac, A.; Albu, P.C.; Oprea, O.; Grosu, A.R.; Pascu, D.; Pancescu, F.M.; Nechifor, G.; et al. Reactional Processes on Osmium–Polymeric Membranes for 5–Nitrobenzimidazole Reduction. *Membranes* **2021**, *11*, 633. [CrossRef] [PubMed]
14. Diaconu, I.; Aboul-Enein, H.Y.; Bunaciu, A.A.; Ruse, E.; Mirea, C.; Nechifor, G. Selective separation of acetaminophene from pharmaceutical formulations through membrane techniques. *Rev. Roum. Chim.* **2015**, *60*, 521–525.
15. Serban, E.A.; Diaconu, I.; Ruse, E.; Ghe, B.; Nechifor, G.; Lazar, M.N. Bulk Liquid Membranes for Separation and Recovery of Pharmaceutical Products. *Rev. Chim.* **2018**, *69*, 3257–3260. [CrossRef]
16. Renault, F.; Sancey, B.; Badot, P.M.; Crini, G. Chitosan for coagulation/flocculation processes—An eco-friendly approach. *Eur. Polym. J.* **2009**, *45*, 1337–1348. [CrossRef]
17. Bhatt, P.; Joshi, S.; Urper Bayram, G.M.; Khati, P.; Simsek, H. Developments and application of chitosan-based adsorbents for wastewater treatments. *Environ. Res.* **2023**, *226*, 115530. [CrossRef]
18. Bezerra de Araujo, C.M.; Ghislandi, M.G.; Gonçalves Rios, A.; Bezerra da Costa, G.R.; do Nascimento, B.F.; Ferreira, A.F.P.; da Motta Sobrinho, M.A.; Rodrigues, A.E. Wastewater treatment using recyclable agar-graphene oxide biocomposite hydrogel in batch and fixed-bed adsorption column: Bench experiments and modeling for the selective removal of organics. *Colloids Surf. A Physicochem. Eng. Asp.* **2022**, *639*, 128357. [CrossRef]
19. Resende, J.F.; Paulino, I.M.R.; Bergamasco, R.; Vieira, M.F.; Vieira, A.M.S. Hydrogels produced from natural polymers: A review on its use and employment in water treatment. *Braz. J. Chem. Eng.* **2023**, *40*, 23–38. [CrossRef]
20. Chelu, M.; Musuc, A.M. Polymer Gels: Classification and Recent Developments in Biomedical Applications. *Gels* **2023**, *9*, 161. [CrossRef]
21. Chelu, M.; Popa, M.; Ozon, E.A.; Pandele Cusu, J.; Anastasescu, M.; Surdu, V.A.; Calderon Moreno, J.; Musuc, A.M. High-Content Aloe vera Based Hydrogels: Physicochemical and Pharmaceutical Properties. *Polymers* **2023**, *15*, 1312. [CrossRef] [PubMed]
22. Thakur, S.; Sharma, B.; Verma, A.; Chaudhary, J.; Tamulevicius, S.; Thakur, V.K. Recent progress in sodium alginate based sustainable hydrogels for environmental applications. *J. Clean. Prod.* **2018**, *198*, 143–159. [CrossRef]
23. Khan, S.A.; Shah, L.A.; Shah, M.; Jamil, I. Engineering of 3D polymer network hydrogels for biomedical applications: A review. *Polym. Bull.* **2022**, *79*, 2685–2705. [CrossRef]
24. Bezerra de Araujo, C.M.; Wernke, G.; Ghislandi, M.G.; Diório, A.; Vieira, M.F.; Bergamasco, R.; da Motta Sobrinho, M.A.; Rodrigues, A.E. Continuous removal of pharmaceutical drug chloroquine and Safranin-O dye from water using agar-graphene oxide hydrogel: Selective adsorption in batch and fixed-bed experiments. *Environ. Res.* **2023**, *216*, 114425. [CrossRef] [PubMed]
25. Machado, T.S.; Crestani, L.; Marchezi, G.; Melara, F.; Rafael de Mello, J.; Dotto, G.L.; Piccin, J.S. Synthesis of glutaraldehyde-modified silica/chitosan composites for the removal of water-soluble diclofenac sodium. *Carbohydr. Polym.* **2022**, *277*, 118868. [CrossRef]
26. Mottaghi, H.; Mohammadi, Z.; Abbasi, M.; Tahouni, N.; Panjeshahi, M.H. Experimental investigation of crude oil removal from water using polymer adsorbent. *J. Water Process Eng.* **2021**, *40*, 101959. [CrossRef]
27. Gkika, D.A.; Mitropoulos, A.C.; Kokkinos, P.; Lambropoulou, D.A.; Kalavrouziotis, I.K.; Bikiaris, D.N.; Kyzas, G.Z. Modified chitosan adsorbents in pharmaceutical simulated wastewaters: A review of the last updates. *Carbohydr. Polym. Technol. Appl.* **2023**, *5*, 100313. [CrossRef]
28. Petri, D.F.S. Xanthan gum: A versatile biopolymer for biomedical and technological applications. *Appl. Polym. Sci.* **2015**, *132*, 42035. [CrossRef]
29. Deng, S.; Wang, R.; Xu, H.; Jiang, X.; Yin, J. Hybrid hydrogels of hyperbranched poly(ether amine)s (hPEAs) for selective adsorption of guest molecules and separation of dyes. *J. Mater. Chem.* **2012**, *22*, 10055–10061. [CrossRef]
30. Zare, E.N.; Fallah, Z.; Le, V.T.; Doan, V.-D.; Mudhoo, A.; Joo, S.-W.; Vasseghian, Y.; Tajbakhsh, M.; Moradi, O.; Sillanpaa, M.; et al. Remediation of pharmaceuticals from contaminated water by molecularly imprinted polymers: A review. *Environ. Chem. Lett.* **2022**, *20*, 2629–2664. [CrossRef]
31. Fortunato, A.; Mba, M. A Peptide-Based Hydrogel for Adsorption of Dyes and Pharmaceuticals in Water Remediation. *Gels* **2022**, *8*, 672. [CrossRef] [PubMed]
32. Das, B.K.; Samanta, R.; Ahmed, S.; Pramanik, B. Disulphide Cross-Linked Ultrashort Peptide Hydrogelator for Water Remediation. *Chem. Eur. J.* **2023**, e202300312. [CrossRef]
33. Grohs, L.; Cheng, L.; Cönen, S.; Haddad, B.G.; Bülow, A.; Toklucu, I.; Ernst, L.; Körner, J.; Schmalzing, G.; Lampert, A.; et al. Diclofenac and other non-steroidal anti-inflammatory drugs (NSAIDs) are competitive antagonists of the human P2X3 receptor. *Front. Pharmacol.* **2023**, *14*, 1120360. [CrossRef] [PubMed]
34. Bonnefille, B.; Gomez, E.; Courant, F.; Escande, A.; Fenet, H. Diclofenac in the marine environment: A review of its occurrence and effects. *Mar. Pollut. Bull.* **2018**, *131*, 496–506. [CrossRef] [PubMed]
35. de Carvalho Filho, J.A.A.; da Cruz, H.M.; Fernandes, B.S.; Motteran, F.; de Paiva, A.L.R.; da Silva Pereira Cabral, J.J. Efficiency of the bank filtration technique for diclofenac removal: A review. *Environ. Pollut.* **2022**, *300*, 118916. [CrossRef]
36. Fahimi, A.; Zanoletti, A.; Federici, S.; Assi, A.; Bilo, F.; Depero, L.E.; Bontempi, E. New eco-materials derived from waste for emerging pollutants adsorption: The case of diclofenac. *Materials* **2020**, *13*, 3964. [CrossRef]
37. Godiya, C.B.; Kumar, S.; Xiao, Y. Amine functionalized egg albumin hydrogel with enhanced adsorption potential for diclofenac sodium in water. *J. Hazard. Mater.* **2020**, *393*, 122417. [CrossRef]

38. Khan, S.A.; Siddiqui, M.F.; Khan, T.A. Synthesis of Poly(methacrylic acid)/Montmorillonite Hydrogel Nanocomposite for Efficient Adsorption of Amoxicillin and Diclofenac from Aqueous Environment: Kinetic, Isotherm, Reusability, and Thermodynamic Investigations. *ACS Omega* **2020**, *5*, 2843–2855. [CrossRef]
39. Feng, Z.; Simeone, A.; Odelius, K.; Hakkarainen, M. Biobased Nanographene Oxide Creates Stronger Chitosan Hydrogels with Improved Adsorption Capacity for Trace Pharmaceuticals. *ACS Sustain. Chem. Eng.* **2017**, *5*, 11525–11535. [CrossRef]
40. Umbreen, N.; Sohni, S.; Ahmad, I.; Khattak, N.U.; Gul, K. Self-assembled three-dimensional reduced graphene oxide-based hydrogel for highly efficient and facile removal of pharmaceutical compounds from aqueous solution. *J. Colloid Interface Sci.* **2018**, *527*, 356–367. [CrossRef]
41. Mahmoodi, H.; Fattahi, M.; Motevassel, M. Graphene oxide–chitosan hydrogel for adsorptive removal of diclofenac from aqueous solution: Preparation, characterization, kinetic and thermodynamic modelling. *RSC Adv.* **2021**, *11*, 36289–36304. [CrossRef] [PubMed]
42. Zaka, A.; Ibrahim, T.H.; Khamis, M. Removal of selected non-steroidal anti-inflammatory drugs from wastewater using reduced graphene oxide magnetite. *Desalination Water Treat.* **2021**, *212*, 401–414. [CrossRef]
43. Argin-Soysal, S.; Kofinas, P.; Lo, Y.M. Effect of complexation conditions on xanthan–chitosan polyelectrolyte complex gels. *Food Hydrocoll.* **2009**, *23*, 202–209. [CrossRef]
44. Corrias, F.; Dolz, M.; Herraez, M.; Diez-Sales, O. Rheological properties of progesterone microemulsions: Influence of xanthan and chitosan biopolymer concentration. *J. Appl. Polym. Sci.* **2008**, *110*, 1225–1235. [CrossRef]
45. Phaechamud, T.; Ritthidej, G.C. Formulation variables influencing drug release from layered matrix system comprising chitosan and xanthan gum. *AAPS PharmSciTech* **2008**, *9*, 870–877. [CrossRef] [PubMed]
46. Popa, N.; Novac, O.; Profire, L.; Lupusoru, C.E.; Popa, M.I. Hydrogels based on chitosan–xanthan for controlled release of theophylline. *J. Mater. Sci. Mater. Med.* **2010**, *21*, 1241–1248. [CrossRef]
47. Bellini, M.Z.; Pires, A.L.R.; Vasconcelos, M.O.; Moraes, A.M. Comparison of the properties of compacted and porous lamellar chitosan–xanthan membranes as dressings and scaffolds for the treatment of skin lesions. *J. Appl. Polym. Sci.* **2012**, *125*, 421–431. [CrossRef]
48. Luo, Y.; Wang, Q. Recent development of chitosan-based polyelectrolyte complexes with natural polysaccharides for a drug delivery. *Int. J. Biol. Macromol.* **2014**, *64*, 353–367. [CrossRef]
49. Chelu, M.; Moreno, J.C.; Atkinson, I.; Cusu, J.P.; Rusu, A.; Bratan, V.; Aricov, L.; Anastasescu, M.; Seciu-Grama, A.-M.; Musuc, A.M. Green synthesis of bioinspired chitosan-ZnO-based polysaccharide gums hydrogels with propolis extract as novel functional natural biomaterials. *Int. J. Biol. Macromol.* **2022**, *211*, 410–424. [CrossRef]
50. Zajac, A.; Hanuza, J.; Wandas, M.; Dyminska, L. Determination of N-acetylation degree in chitosan using Raman spectroscopy. *Spectrochim. Acta Part A Mol. Biomol. Spectrosc.* **2015**, *134*, 114–120. [CrossRef]
51. Malik, N.S.; Ahmad, M.; Minhas, M.U.; Tulain, R.; Barkat, K.; Khalid, I.; Khalid, Q. Chitosan/Xanthan Gum Based Hydrogels as Potential Carrier for an Antiviral Drug: Fabrication, Characterization, and Safety Evaluation. *Front Chem.* **2020**, *8*, 1–16. [CrossRef]
52. Dey, S.C.; Al-Amin, M.; Rashid, T.U.; Sultan, M.Z.; Ashaduzzaman, M.; Sarker, M.; Shamsuddin, S.M. Reparation, characterization and performance evaluation of chitosan as an adsorbent for remazol red. *Int. J. Latest Res. Eng. Technol.* **2016**, *2*, 52–62.
53. De Morais Lima, M.; Carneiro, L.C.; Bianchini, D.; Dias, R.G.A.; da Rosa Zavareze, E.; Prentice, C.; da Silveira Moreira, A. Structural, Thermal, Physical, Mechanical and Barrier Properties of Chitosan Films with the Addition of Xanthan Gum. *J. Food Sci.* **2017**, *82*, 698–705. [CrossRef] [PubMed]
54. Horn, M.M.; Martins, V.C.A.; de Guzzi Plepis, A.M. Influence of collagen addition on the thermal and morphological properties of chitosan/xanthan hydrogels. *Int. J. Biol. Macromol.* **2015**, *80*, 225–230. [CrossRef] [PubMed]
55. Nirmala, R.; Il, B.W.; Navamathavan, R.; El-Newehy, M.H.; Kim, H.Y. Preparation and characterizations of anisotropic chitosan nanofibers via electrospinning. *Macromol. Res.* **2011**, *19*, 345–350. [CrossRef]
56. Hussien, M.A.; Ebtessam, A.E.; El, G.S.A. Investigation of the effect of formulation additives on telmisartan dissolution rate: Development of oral disintegrating tablets. *Eur. J. Biomed. Pharm. Sci.* **2019**, *6*, 12–20.
57. Essa, E.; Elmarakby, A.; Donia, A.; El Maghraby, G.M. Controlled precipitation for enhanced dissolution rate of flurbiprofen: Development of rapidly disintegrating tablets. *Drug Dev. Ind. Pharm.* **2017**, *24*, 1–10. [CrossRef]
58. Kuzmin, V.V.; Novikov, V.S.; Ustynyuk, L.Y.; Prokhorov, K.A.; Sagitova, E.A.; Nikolaeva, G.Y. Raman spectra of polyethylene glycols: Comparative experimental and DFT study. *J. Mol. Struct.* **2020**, *1217*, 128331. [CrossRef]
59. Matsuura, H.; Fukuhara, K. Conformational analysis of poly(oxyethylene) chain in aqueous solution as a hydrophilic moiety of nonionic surfactants. *J. Mol. Struct.* **1985**, *126*, 251–260. [CrossRef]
60. Matsuura, H.; Fukuhara, K. Vibrational spectroscopic studies of conformation of poly(oxyethylene). II. Conformation–spectrum correlations. *J. Polym. Sci. Part B Polym. Phys.* **1986**, *24*, 1383–1400. [CrossRef]
61. Takahashi, Y.; Tadokoro, H. Structural studies of polyethers, (-(CH2)m-O-)n. X. crystal structure of poly(ethylene oxide). *Macromolecules* **1973**, *6*, 672–675. [CrossRef]
62. Kumar, S.; Dutta, P.K.; Koh, J. A physiocochemical and biological study of novel chitosan-chloroquinoline derivative for biomedical applications. *Int. J. Biol. Macromol.* **2011**, *49*, 356–361. [CrossRef] [PubMed]
63. Podgorbunskikh, E.; Kuskov, T.; Rychkov, D.; Lomovskii, O.; Bychkov, A. Mechanical Amorphization of Chitosan with Different Molecular Weights. *Polymers* **2022**, *14*, 4438. [CrossRef]

64. Liu, Z.; Fu, X.; Jiang, L.; Wu, B.; Wang, J.; Lei, J. Solvent-free synthesis and properties of novel solid–solid phase change materials with biodegradable castor oil for thermal energy storage. *Sol. Energy Mater. Sol. Cells* **2016**, *147*, 177–184. [CrossRef]
65. Fița, A.C.; Secăreanu, A.A.; Musuc, A.M.; Ozon, E.A.; Sarbu, I.; Atkinson, I.; Rusu, A.; Mati, E.; Anuta, V.; Pop, A.L. The Influence of the Polymer Type on the Quality of Newly Developed Oral Immediate-Release Tablets Containing Amiodarone Solid Dispersions Obtained by Hot-Melt Extrusion. *Molecules* **2022**, *27*, 6600. [CrossRef]
66. Kang, Y.; Li, P.; Zeng, X.; Chen, X.; Xie, Y.; Zeng, Y.; Zhang, Y.; Xie, T. Biosynthesis, structure and antioxidant activities of xanthan gum from Xanthomonas campestris with additional furfural. *Carbohydr. Polym.* **2019**, *216*, 369–375. [CrossRef] [PubMed]
67. Cheung, M.K.; Wan, K.P.; Yu, P.H. Miscibility and morphology of chiral semicrystalline poly-(R)-(3-hydroxybutyrate)/chitosan and poly-(R)-(3-hydroxybutyrate-co-3-hydroxyvalerate)/chitosan blends studied with DSC,1HT1 andT1? CRAMPS. *J. Appl. Polym. Sci.* **2002**, *86*, 1253–1258. [CrossRef]
68. Kittur, F.; Prashanth, K.H.; Sankar, K.U.; Tharanathan, R. Characterization of chitin, chitosan and their carboxymethyl derivatives by differential scanning calorimetry. *Carbohydr. Polym.* **2002**, *49*, 185–193. [CrossRef]
69. Cardenas, G.; Miranda, S.P. FTIR and TGA studies of chitosan composite films. *J. Chil. Chem. Soc.* **2004**, *49*, 291–295. [CrossRef]
70. Kacurakova, M.; Belton, P.S.; Hirsch, J.; Ebringerova, A. Hydration Properties of Xylan-Type Structures: An Study of Xylo-oligosaccharides FTIR. *J. Sci. Food Agric.* **1998**, *77*, 38–44. [CrossRef]
71. Phillips, G.O.; Takigami, S.; Takigami, M. Hydration characteristics of the gum exudate from Acacia Senegal. *Food Hydrocoll.* **1996**, *10*, 11–19. [CrossRef]
72. Appelquist, I.A.M.; Cooke, D.; Gidley, M.J.; Lane, S.J. Effect of Heat Treatment on the Pectins of Tomatoes during Tomato Paste Manufacturing. *Carbohydr. Polym.* **1993**, *20*, 291–299. [CrossRef]
73. Gidley, M.J.; Robinson, G. 18—Techniques for Studying Interactions Between Polysaccharides. *Methods Plant Biochem.* **1990**, *2*, 607–642. [CrossRef]
74. Sreenivasan, K. Thermal stability studies of some chitosan metal ion complexes using differential scanning calorimetry. *Polym. Degrad. Stab.* **1996**, *52*, 85–87. [CrossRef]
75. Deng, L.; Qi, H.; Yao, C.; Feng, M.; Dong, A. Investigation on the properties of methoxy poly (ethylene glycol)/chitosan graft co-polymers. *J. Biomater. Sci. Polym. Ed.* **2007**, *18*, 1575–1589. [CrossRef] [PubMed]
76. Nunes, M.M.; Menezes, P.F.; Alves, R.M.; Gonçalves, R.J.; Junges, A.; Formentin, M.W.; Mendonça, D.F.; Ligabue, R.A.; Bueno, M.F.; Severino, P.; et al. Chitosan and chitosan/PEG nanoparticles loaded with in-dole-3-carbinol: Characterization, computational study and potential effect on human bladder cancer cells. *Mater. Sci. Eng. C* **2021**, *124*, 112089. [CrossRef]
77. Jayaramudu, T.; Raghavendra, G.M.; Varaprasad, K.; Reddy, G.V.S.; Reddy, A.B.; Sudhakar, K.; Sadiku, E.R. Preparation and characterization of poly(ethylene glycol) stabilized nano silver particles by a mechanochemical assisted ball mill process. *J. Appl. Polym. Sci.* **2016**, *133*, 43027. [CrossRef]
78. Yang, J.H.; Han, Y.S.; Park, M.; Park, T.; Hwang, S.J.; Choy, J.H. New inorganic-based drug delivery system of indole-3-acetic acid-layered metal hydroxide nanohybrids with controlled release rate. *Chem. Mater.* **2007**, *19*, 2679–2685. [CrossRef]
79. Chitra, G.; Franklin, D.S.; Guhanathan, S. Indole-3-acetic acid based tunable hydrogels for antibacterial, antifungal and antioxidant applications. *J. Macromol. Sci. Part A Pure Appl. Chem.* **2017**, *54*, 151–163. [CrossRef]
80. Yongmei, G.; Chengqun, Y.; Zhenzhong, Z.; Xinhao, W.; Abid, N.; Rui, Z.; Weifeng, Z. Chitosan/xanthan gum-based (Hydroxypropyl methylcellulose-co-2-Acrylamido-2-methylpropane sulfonic acid) interpenetrating hydrogels for con-trolled release of amorphous solid dispersion of bioactive constituents of Puerariae lobatae. *Int. J. Biol. Macromol.* **2023**, *224*, 380–395. [CrossRef]
81. Kulkarni, N.; Wakte, P.; Naik, J. Development of floating chitosan-xanthan beads for oral controlled release of glipizide. *Int. J. Pharm. Investig.* **2015**, *5*, 73–80. [CrossRef]
82. Nafisa, G.; Shahzad, M.K.; Osama, M.B.; Atif, I.; Attaullah, S.; Sehrish, J.; Saba, U.K.; Afrasyab, K.; Rafi, U.K.; Muhammad, T.Z.B. Inflammation targeted chitosan-based hydrogel for controlled release of diclofenac sodium. *Int. J. Biol. Macromol.* **2020**, *162*, 175–187. [CrossRef]
83. Lessa, E.F.; Nunes, M.L.; Fajardo, A.R. Chitosan/waste coffee-grounds composite: An efficient and eco-friendly adsorbent for removal of pharmaceutical contaminants from water. *Carbohydr. Polym.* **2018**, *189*, 257–266. [CrossRef]
84. Martinez-Ruvalcaba, A.; Chornet, E.; Rodrigue, D. Viscoelastic properties of dispersed chitosan/xanthan hydrogels. *Carbohydr. Polym.* **2007**, *67*, 586–595. [CrossRef]
85. Hu, D.; Huang, H.; Jiang, R.; Wang, N.; Xu, H.; Wang, Y.G.; Ouyang, X.K. Adsorption of diclofenac sodium on bilayer amino-functionalized cellulose nanocrystals/chitosan composite. *J. Hazard. Mater.* **2019**, *369*, 483–493. [CrossRef] [PubMed]
86. Liang, X.X.; Omer, A.M.; Hu, Z.-H.; Wang, Y.G.; Di, Y.; Ouyang, X.-K. Efficient adsorption of diclofenac sodium from aqueous solutions using magnetic amine-functionalized chitosan. *Chemosphere* **2019**, *217*, 270–278. [CrossRef] [PubMed]
87. Li, S.; Cui, J.; Wu, X.; Zhang, X.; Hu, Q.; Hou, X. Rapid in situ microwave synthesis of Fe3O4@MIL-100(Fe) for aqueous diclofenac sodium removal through integrated adsorption and photodegradation. *J. Hazard. Mater.* **2019**, *373*, 408–416. [CrossRef]
88. Zhuang, S.; Cheng, R.; Wang, J. Adsorption of diclofenac from aqueous solution using UiO-66-type metal-organic frameworks. *Chem. Eng. J.* **2019**, *359*, 354–362. [CrossRef]
89. Xiong, T.; Yuan, X.; Wang, H.; Wu, Z.; Jiang, L.; Leng, L.; Xi, K.; Cao, X.; Zeng, G. Highly efficient removal of diclofenac sodium from medical wastewater by Mg/Al layered double hydroxide-poly(m-phenylenediamine) composite. *Chem. Eng. J.* **2019**, *366*, 83–91. [CrossRef]

90. Zhao, R.; Zheng, H.; Zhong, Z.; Zhao, C.; Sun, Y.; Huang, Y.; Zheng, X. Efficient removal of diclofenac from surface water by the functionalized multilayer magnetic adsorbent: Kinetics and mechanism. *Sci. Total Environ.* **2021**, *760*, 144307. [CrossRef]
91. Balaci, T.; Velescu, B.; Karampelas, O.; Musuc, A.M.; Nitulescu, G.M.; Ozon, E.A.; Nitulescu, G.; Gird, C.E.; Fita, C.; Lupuliasa, D. Physico-Chemical and Pharmaco-Technical Characterization of Inclusion Complexes Formed by Rutoside with beta-Cyclodextrin and Hydroxypropyl-beta-Cyclodextrin Used to Develop Solid Dosage Forms. *Processes* **2021**, *9*, 26. [CrossRef]
92. Nafee, N.A.; Ismail, F.A.; Boraie, N.A.; Mortada, L.M. Mucoadhesive buccal patches of miconazole nitrate: In vitro/in vivo performance and effect of ageing. *Int. J. Pharm.* **2003**, *264*, 1–14. [CrossRef] [PubMed]
93. Don, T.M.; Huang, M.L.; Chiu, A.C. Preparation of thermo-responsive acrylic hydrogels useful for the application in transdermal drug delivery systems. *Mater Chem Phys* **2008**, *107*, 266–273. [CrossRef]
94. Wang, K.; Fu, Q.; Chen, X.; Gao, Y.; Dong, K. Preparation and characterization of pH-sensitive hydrogel for drug delivery system. *RSC Adv.* **2012**, *2*, 7772–7780. [CrossRef]
95. Yi, Z.; Yao, J.; Zhu, M.; Chen, H.; Wang, F.; Liu, X. Kinetics, equilibrium, and thermodynamics investigation on the adsorption of lead (II) by coal-based activated carbon. *SpringerPlus* **2016**, *5*, 1160. [CrossRef] [PubMed]

Disclaimer/Publisher's Note: The statements, opinions and data contained in all publications are solely those of the individual author(s) and contributor(s) and not of MDPI and/or the editor(s). MDPI and/or the editor(s) disclaim responsibility for any injury to people or property resulting from any ideas, methods, instructions or products referred to in the content.

Article

Effect of Chitosan as Active Bio-colloidal Constituent on the Diffusion of Dyes in Agarose Hydrogel

Martina Klučáková

Faculty of Chemistry, Brno University of Technology, Purkyňova 464/118, 612 00 Brno, Czech Republic; klucakova@fch.vutbr.cz

Abstract: Agarose hydrogel was enriched by chitosan as an active substance for the interactions with dyes. Direct blue 1, Sirius red F3B, and Reactive blue 49 were chosen as representative dyes for the study of the effect of their interaction with chitosan on their diffusion in hydrogel. Effective diffusion coefficients were determined and compared with the value obtained for pure agarose hydrogel. Simultaneously, sorption experiments were realized. The sorption ability of enriched hydrogel was several times higher in comparison with pure agarose hydrogel. Determined diffusion coefficients decreased with the addition of chitosan. Their values included the effects of hydrogel pore structure and interactions between chitosan and dyes. Diffusion experiments were realized at pH 3, 7, and 11. The effect of pH on the diffusivity of dyes in pure agarose hydrogel was negligible. Effective diffusion coefficients obtained for hydrogels enriched by chitosan increased gradually with increasing pH value. Electrostatic interactions between amino group of chitosan and sulfonic group of dyes resulted in the formation of zones with a sharp boundary between coloured and transparent hydrogel (mainly at lower pH values). A concentration jump was observed at a given distance from the interface between hydrogel and the donor dye solution.

Keywords: chitosan; agarose; dyes; diffusion; sorption

1. Introduction

Chitosan is a crystalline polysaccharide obtained by the deacetylation of chitin, a by-product of the seafood industry [1,2]. As a result of the unique chemical structure, chitosan and its derivatives have been paid close and extensive attention as a potential bio-functional material [3] and they have prospective applications in many fields such as biomedicine, wastewater treatment, functional membranes, and flocculation [4]. Most of the commercial or practical applications of chitosan are confined to its unmodified forms [5]. However, synthesis of modified chitosan via N-substitution, O-substitution, free radical graft copolymerization, and other modification methods are developed to improve the application potential of this material [4–8]. Chitosan belongs to polyelectrolytes which can be found anywhere around us. In the form of charged biopolymers, such as nucleic acids and some polysaccharides and proteins, they form vital structural and functional constituents of living organisms. Additionally, they represent the crucial component of many non-living parts of nature, such as soils, waters, and sediments, where they—in the form of humus—regulate environmental and biological uptake and transport of essential nutrients as well as harmful pollutants. Similarly, chitosan can be used as an active substance able to interact with many pollutants, immobilize them, and affect their migration [9]. Therefore, it was chosen for this study as a representant of bio-polyelectrolytes able to interact with different constituents and affect their migration ability. It can be applied in natural systems as well as in artificial hydrogels. Agarose (a linear polysaccharide of red algae, made up of the repeating monomeric unit of agarobiose) is proposed as material-of-choice for the preparation of the hydrogel which can be enriched by an active substance for the investigation of the interactions during the transport [10,11]. The network of agarose chains can be interpenetrated by chitosan at higher temperatures where both compounds are dissolved, and the mixture is then easily gelled by cooling to normal temperature. The mechanical and textural properties of agarose hydrogels as well as the gelation mechanism are well understood [12–14]. The diffusion in agarose hydrogels has already been subject to vast concern [15–20]. Golmohamadi et al. [15] studied self- and mutual diffusion of Cd^{2+} and charged rhodamine derivatives. Lead et al. [16] determined diffusion coefficients of humic acids in agarose hydrogel and in water. They obtained values between 0.9 and 2.5×10^{-10} m^2 s^{-1} which were generally 10–20% lower than in water. Gutenwik et al. [17] measured the effective diffusion coefficients of lysozyme and bovine serum albumin. They demonstrated the influence of pH and ionic strength on their diffusive properties. The same proteins were studied by Liang et al. [18]. At the considered range of agarose concentration (0.5–3.0 wt.%), the diffusion coefficients range from 4.98 to 8.21×10^{-11} m^2/s for BSA and 1.15 to 1.56×10^{-10} m^2/s for lysozyme, respectively. Tan et. Al. [19] applied a real-time electronic speckle pattern interferometry method to study the diffusion behavior of levofloxacin mesylate. Their results confirmed that the diffusivity of solute decreased with the increase of concentration of agarose. Its value extrapolated to infinite dilution was equal to 5.3×10^{-10} m^2/s. Labille et al. [20] used fluorescence correlation spectroscopy to study the diffusion of nanometric solutes in agarose hydrogel and determined values of diffusion coefficients between 0.5 and 2.8×10^{-10} m^2 s^{-1}. Their results showed that, at the liquid/gel interface, a thin hydrogel layer is formed with characteristics significantly different from those of the bulk gel. In particular, in this layer, the porosity of agarose fiber network is significantly lower than in the bulk gel. The diffusion coefficient of solutes in this layer is consequently decreased for steric reasons. The diffusion characteristics are the crucial parameters reflecting the migration ability of diffusing particles which can be affected not only by the hydrogel structure but also their interactions with the active substance incorporated in the hydrogel. Chitosan, as the bio-functional material with high affinity to many harmful substances, was used in this study for the functionalizing of inert agarose hydrogel. The addition of chitosan as an active substance allowed to investigate the interactions directly in the motion of diffusing particles. Thus, the interactions can be included directly in the parameters obtained on the basis of diffusion experiments.

Studies dealing with the interactions of dyes with chitosan are often focused on traditional batch experiments (e.g., [21–27]). Some of them deal with hydrogels containing chitosan as an active substance in combination with other materials such as gelatin [28], pectin, DNA [29], activated carbon [30], Fe(III) [31], cellulose [32], and tri-polyphosphate [33]. Concepts for developing physical gels of chitosan and of chitosan derivatives are summarized in the review in [34]. As described in our previous study [9], the adsorption properties of chitosan are attributed to high hydrophilicity (due to OH groups), primary amino groups with high activity, and the flexible structure of polymer chains. This means that chitosan is a material with good affinity to different substances, including dyes, which are the subject matter of this study. Adsorbents based on chitosan have very good adsorption capacities and relatively low cost.

The studies on the transport of dyes and diffusion processes in chitosan materials are relatively scarce. Barron-Zambrano et al. [35] investigated the dynamic sorption of Reactive Black 5 onto chitosan in fixed-bed column. The obtained breakthrough curves were typical of systems that do not reach equilibrium which indicated that adsorption was affected by mass transfer limitations, probably due to intraparticle diffusion. It had a significant impact on column performance strongly affected by particle size. A smaller particle size resulted in a faster pore diffusion rate because the diffusion path was shorter and the resistance to diffusion was lower. Lazaridis and Keenan [36] used chitosan beads as barriers to the transport of azo dye in soil column. The used non-equilibrium transport models were divided into three parts: physical, chemical, and physical and chemical non-equilibrium transport. The application of a chitosan barrier resulted in a strong increase in the retardation factor of soil. García-Aparicio et al. [37] studied the diffusion of three small molecules, caffeine, theophylline and caprolactam, in chitosan gels with different concentrations of water by means of proton-localized NMR spectroscopy. The measured concentration profiles were in agreement with the Fickian law. The values of the diffusion coefficients ranged from 6.1×10^{-10} to 3.4×10^{-10} m^2 s^{-1}, depending on chitosan concentration and type of diffusant molecule. Cheung et al. [38] realized batch adsorption experiments with Orange 10, Acid Orange 12, Acid Red 18, Acid Red 73, and Acid Green 25. They concluded that the adsorption mechanism was predominantly intraparticle diffusion, but there was also a dependence on pore size as the dye diffuses through macropore, mesopore, and micropore, respectively. Similarly, two distinct linear parts were observed in plots of data obtained for the adsorption of Reactive Blue 4 dye onto Chitosan 10B. The initial linear portion may be attributed to the macropore diffusion and the second linear part to the micropore diffusion [21]. The intraparticle diffusion was presented as the rate-limiting process in the adsorption of dyes on double network gelatin/chitosan hydrogel [28], Reactive Black 5 on quartzite/chitosan composite [39], Rhodamine-6G on chitosan, nanoclay and chitosan–nanoclay composite [40], malachite green on chitosan beads [25,41], and indigo carmine on functional chitosan and β-cyclodextrin/chitosan beads [26]. Bilal et al. [42] developed Agarose-chitosan hydrogel-immobilized horseradish peroxidase and studied its bio-catalytic activity and effectivity in degradation of dye (Reactive Blue 19). Except for the main goals, the study provided detailed characteristics of hydrogel properties such as morphology and thermal stability.

Other studies are focused on the diffusion through chitosan membranes and films [15,43–47]. Hartig et al. [43] studied the diffusion of fructose in precipitated chitosan membranes using diffusion cells. They determined a diffusion coefficient which was dependent on the concentration and ranged between 6.2×10^{-10} and 2.1×10^{-10} m^2 s^{-1}. Yang et al. [44] performed permeation studies of model drug through preswollen chitosan/PVA blended hydrogel membranes using side-by-side diffusion cells. Similarly, Yang and Su [15] investigated the diffusion of 5-Fluorouracil through four kinds of chitosan membranes. They determined the permeability coefficient which was indirectly proportional to chitosan content. Waluga and Scholl [45] determined diffusion coefficients of different sugars and the sugar alcohol sorbitol in chitosan membranes and beads. Obtained diffusion coefficients ranged from 1.1×10^{-10} to 2.3×10^{-10} m^2 s^{-1} for chitosan membranes, and from 1.4×10^{-10} to 2.4×10^{-10} m^2 s^{-1} for chitosan beads. Xu et al. [46] incorporated four types of polyhedral oligosilsesquioxanes into chitosan by solution blending to fabricate composite membranes, and permeation studies were conducted for riboflavin. Their diffusion coefficients varied between 1.0×10^{-12} m^2 s^{-1} and 2.7×10^{-12} m^2 s^{-1}. Carlough et al. [47] produced chitosan films and determined diffusion coefficients for Direct Red 81, Green 26, Blue 75, and Black 22. Their values (for 60 °C and pH 9) differed in magnitude and ranged from 4.5×10^{-15} m^2 s^{-1} (Green 26) to 4.5×10^{-14} m^2 s^{-1} (Black 22).

Our study is focused on the reactivity mapping of chitosan distributed in hydrogel during the dye migration. Since chitosan is considered a material with high potential to immobilize dyes, it is desirable to investigate the interactions in detail. In order to distinguish between the diffusivity through reactive and non-reactive medium, agarose hydrogel was chosen as the basic non-reactive material. Agarose hydrogel proved to be the suitable medium for the investigation of diffusion of different substances [10,11,47–49]. It can be enriched by an active substance which can interact with diffusing particles and (partially) immobilize them. Therefore, the interactions of diffusing dyes with active substance incorporated in the hydrogel can be studied in their motion. The main aim of this study is thus the detailed analysis of these two parallel processes.

2. Results and Discussion

In Figure 1, the time development of a concentration profile in pure agarose hydrogel for Direct blue 1 is shown. Experimental data are fitted by Equation (1) derived on the basis of on Fick's laws [9,50–53] and initial and boundary conditions listed in Table 1.

$$c = c_s \operatorname{erfc} \frac{x}{2\sqrt{D_h t}}, \tag{1}$$

where t is time, x is distance from interface, c is concentration of dye, c_s is concentration at interface, and D_h is the diffusion coefficient of dye in pure agarose hydrogel. If Equation (1) is applied for the data obtained for the diffusion of dyes in hydrogels enriched by chitosan, the diffusion coefficient D_h in Equation (1) and (following) Equation (2) should be replaced by effective diffusion coefficient D_{ef} (including interactions between dyes and chitosan). Both diffusion coefficients can be determined from the slope of the dependence of the total diffusion flux m_t on the square root of time [9,50–53]:

$$m_t = 2c_s\sqrt{\frac{D_h t}{\pi}} \qquad (2)$$

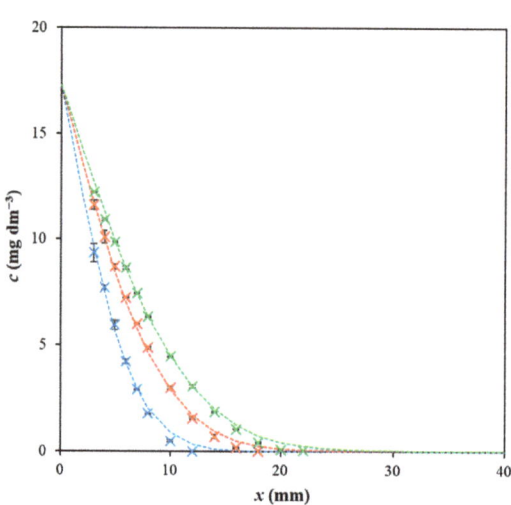

Figure 1. Concentration profiles of Direct blue 1 in agarose hydrogel after 24 h (blue), 48 h (red), and 72 h (green). Experimental data are fitted by Equation (1).

Table 1. Initial and boundary conditions of diffusion experiments.

Time t	Distance x	Concentration c
$t = 0$	$x > 0$	$c = 0$
$t > 0$	$x = 0$	$c = c_s$
$t > 0$	$x \to \infty$	$c = 0$

As can be seen, the concentration between the donor solution and hydrogel remained constant during the whole experiment as it agrees with the boundary condition (Table 1). Similarly, the hydrogel can be considered as the semi-infinite medium. This means that the hydrogel closer to the bottom of the cuvette remained free of dye during the whole diffusion experiment. The initial condition was that the initial concentration of dye in hydrogel was equal to zero. The conditions listed in Table 1 were valid for all realized experiments (for both types of hydrogels and all three dyes).

The comparison of concentration profiles obtained for pure agarose hydrogel and the enriched one is shown in Figure 2. As can be seen, the profiles differ mainly in the distances close to the interface and the surface concentration c_s is higher for the hydrogel enriched by chitosan. It was found that the surface concentrations are higher for hydrogel enriched by chitosan for all studied dyes and pH values.

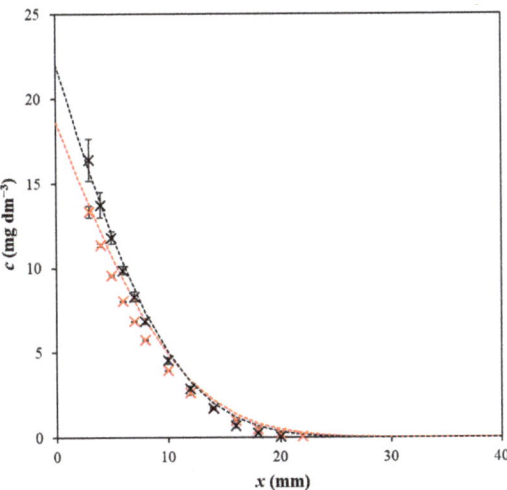

Figure 2. Concentration profiles of Direct blue 1 in agarose hydrogel (red), and hydrogel enriched by chitosan (black) after 72 h at pH 11. Experimental data are fitted by Equation (1).

The ratio between surface concentration in hydrogel enriched by chitosan and pure agarose hydrogel is strongly affected by character of dye and pH value (see Figure 3a). The ratios decreased with increasing pH values for all dyes, and they are not dependent on the duration of the experiment. It means that the surface concentration is constant for the given dye and given pH value. The ratios obtained for Reactive blue 49 at pH 7 and 11 are practically the same. The highest values were achieved for Sirius red F3B, the lowest ones for Reactive blue 49. However, the differences between dyes are negligible at pH 11. The results showed that the increase in diffusion rates and concentrations of dyes in the hydrogel enriched by chitosan (in comparison with pure agarose hydrogel) was caused mainly by the increase in surface concentration, which is a crucial factor determining the concentrations in hydrogels, as can be deduced from Equation (1). Another crucial factor is the (effective) diffusion coefficient which is lower in the hydrogel enriched by chitosan than in the pure agarose hydrogel (see Tables 2 and 3). Therefore, the effect of the increase in the surface concentration preponderated over the effect of the decrease in the diffusivities of dyes.

Table 2. Values of diffusion coefficient (D_h) determined for pure agarose hydrogel.

Dye	D_h (pH 3) (m^2 s^{-1})	D_h (pH 7) (m^2 s^{-1})	D_h (pH 11) (m^2 s^{-1})
Direct blue 1	$(1.51 \pm 0.05) \times 10^{-10}$	$(1.55 \pm 0.07) \times 10^{-10}$	$(1.54 \pm 0.10) \times 10^{-10}$
Sirius red F3B	$(2.05 \pm 0.12) \times 10^{-10}$	$(2.04 \pm 0.10) \times 10^{-10}$	$(2.03 \pm 0.12) \times 10^{-10}$
Reactive blue 49	$(2.98 \pm 0.09) \times 10^{-10}$	$(2.90 \pm 0.13) \times 10^{-10}$	$(3.02 \pm 0.07) \times 10^{-10}$

Table 3. Values of effective diffusion coefficient (D_{ef}) determined for hydrogel enriched by chitosan.

Dye	D_{ef} (pH 3) (m^2 s^{-1})	D_{ef} (pH 7) (m^2 s^{-1})	D_{ef} (pH 11) (m^2 s^{-1})
Direct blue 1	$(6.32 \pm 0.37) \times 10^{-11}$	$(1.00 \pm 0.05) \times 10^{-10}$	$(1.34 \pm 0.11) \times 10^{-10}$
Sirius red F3B	$(4.64 \pm 0.22) \times 10^{-11}$	$(7.42 \pm 0.37) \times 10^{-11}$	$(1.72 \pm 0.13) \times 10^{-10}$
Reactive blue 49	$(2.36 \pm 0.13) \times 10^{-10}$	$(2.52 \pm 0.05) \times 10^{-10}$	$(2.67 \pm 0.08) \times 10^{-10}$

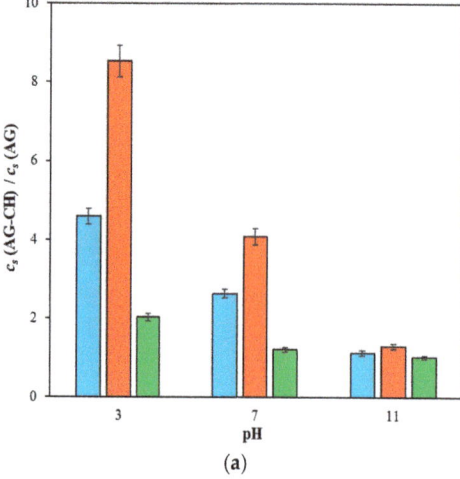
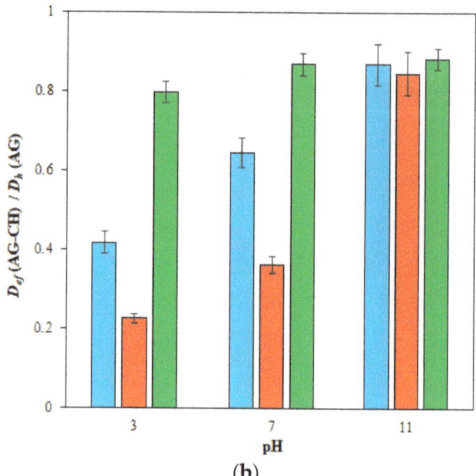

Figure 3. (a) The ratio between surface concentrations in hydrogels enriched by chitosan c_s (AG-CH) and pure agarose hydrogels c_s (AG); (b) the ratio between effective diffusion coefficient D_{ef} (for hydrogel enriched by chitosan) and diffusion coefficients D_h for pure agarose hydrogel; Direct blue 1 (blue), Sirius red F3B (red), and Reactive blue 49 (green).

The values of diffusion coefficients D_h in Table 2 agree with values published for dyes in hydrogels [10,11,48,49,54–59]. Values of D_{ef} obtained for hydrogels enriched by chitosan are lower as mentioned above (see Table 3 and Figure 3b). The highest diffusivity was determined for Reactive blue 49, the lowest for Direct blue 1. As can be seen, the diffusion coefficients D_h changed with pH only slightly. They are practically independent from pH because errors of their determination are higher than the differences between values obtained for different pH values. The effective diffusion coefficients include the effect of interactions of dyes with chitosan. If we assume that the pore structure of hydrogel did not change with the addition of chitosan, the differences between D_h and D_{ef} should be caused mainly by the interactions. Rheological behaviour of agarose hydrogels and the hydrogels enriched by chitosan was investigated in detail in previous work [9]. Its results showed that the rheological behaviour of hydrogels was changed by the addition of chitosan. The changes were influenced by two contrary effects. The storage modulus was higher than the loss one and elastic character predominated for all studied hydrogels. However, the addition of chitosan caused the hydrogels to become more liquid and therefore, more permeable for diffusing particles. This effect can slightly suppress the decrease in the diffusivity of dyes in hydrogel containing chitosan. On the other hand, the decrease in D_{ef} values (in comparison with D_h ones) should be caused mainly by the dye–chitosan interactions.

While the ratio between surface concentrations in enriched and pure hydrogel decreased with increasing pH, the effect of pH on the mobility of dyes is the opposite. The common feature is that the diffusivities of studied dyes are comparable at pH 11. In neutral and acidic pH values, the decrease in diffusivity was stronger for Sirius red F3B and Reactive blue 49 and Direct blue 1 (in comparison with Reactive blue 49). The addition of chitosan into inert agarose hydrogel thus resulted in the increase in surface concentration and the decrease in the diffusion coefficient. Both parameters are mostly affected by the chitosan addition for Sirius red F3B. In this case, the biggest increase in surface concentration and the biggest decrease in diffusion coefficient were observed. The diffusivity of Sirius red F3B as well as its surface concentration are strongly influenced by pH value. In contrast, the effect of pH on the parameters determined for Reactive blue 49 was much weaker.

Both discussed parameters (surface concentration and diffusion coefficient) influenced the distribution of dyes in hydrogel in the diffusion. In Figure 4, the concentration profiles of Sirius red F3B in pure agarose hydrogel and hydrogel enriched by chitosan are compared. We can see that the profiles differed only slightly with pH changes when dyes were diffused in inert agarose hydrogel. In contrast, the changes in hydrogel containing chitosan as active substance are dramatical. The surface concentration at pH 3 is really high, which influenced the distribution of concentration in whole hydrogel. The difference between distribution of the dye in inert agarose hydrogel and hydrogel enriched by chitosan is shown in Figure 5. We can see that the dye particles diffuse in hydrogel containing chitosan as a layer with a sharp interface between hydrogel containing diffusing dye particles and hydrogel without them. We suppose that the reason for the formation of zones with a sharp boundary between coloured and transparent hydrogel are the electrostatic interactions between the amino group of chitosan and the sulfonic group of the dye. The amino groups are protonated at lower pH values [5,6,24,60], which resulted in their higher reactivity with dye and the formation of a concentration jump at a given distance from the interface between hydrogel and the donor dye solution. A similar sharp interface was observed for Direct blue 1.

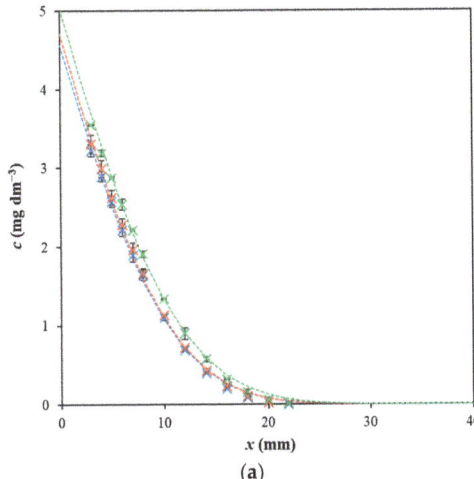
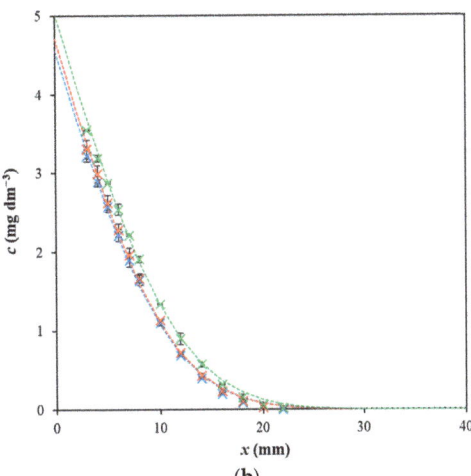

Figure 4. (a) The concentration profiles of Sirius red F3B in pure agarose hydrogel after 48 h; (b) the concentration profiles of Sirius red F3B in hydrogel enriched by chitosan after 48 h; pH 3 (blue), pH 7 (red), and pH 11 (green). Experimental data are fitted following Equation (1).

On the basis of the obtained results, mainly having observed sharp interfaces between hydrogel containing dye and hydrogel without that, it was decided to realize additional diffusion experiments with Direct blue 1 and hydrogels with different contents of chitosan (similarly to previous work [9]). The obtained results are listed in Table 4. We can see that the diffusion coefficient gradually decreased with increasing content of chitosan. Simultaneously, the distribution of dye in hydrogel changed gradually into the sharply bordered dye layer as chitosan content gradually increased (see Figure 6). This effect is probably caused by electrostatic interactions between the amino group of chitosan and the sulfonic group of dyes. Since pure agarose hydrogel does not contain an active substance, the formation of zones with a sharp boundary between coloured and transparent hydrogel was not observed. The sharp interface was observed for all chitosan additions and this phenomenon was more pronounced for its larger amounts. Simultaneously, a deceleration of diffusion (related to a decrease in effective diffusion coefficient) was observed.

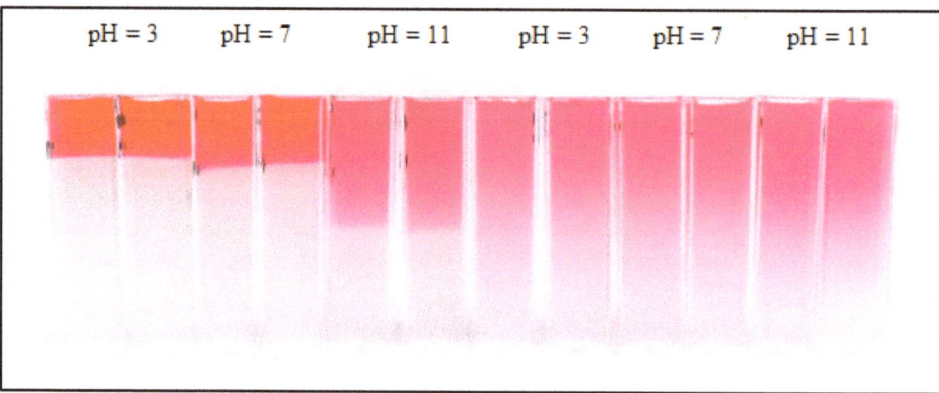

Figure 5. The concentration distribution of Sirius red F3B in hydrogel enriched by chitosan (**left**) and pure agarose hydrogel (**right**) after 72 h.

Table 4. Values of diffusion coefficient (D_h) and effective diffusion coefficient (D_{ef}) determined for Direct blue 1 and pH 3 for different contents of chitosan.

D_h (m² s⁻¹)	D_{ef} (m² s⁻¹)		
Without Chitosan	0.2 mg g⁻¹	0.5 mg g⁻¹	1 mg g⁻¹
$(1.51 \pm 0.05) \times 10^{-10}$	$(1.32 \pm 0.03) \times 10^{-10}$	$(9.83 \pm 0.16) \times 10^{-11}$	$(6.32 \pm 0.37) \times 10^{-11}$

Since the effective diffusion coefficient is strongly affected by the interactions between dyes and chitosan as the active substance in the enriched hydrogel, we can analyse their relationship on the basis of the mathematical model published in previous work [9] and considering Fickian laws and a simple equilibrium between immobilized and free movable dye particles [50–53]. The relationship between the diffusion coefficient of dyes in pure agarose hydrogel (D_h) and effective diffusion coefficient of dyes in hydrogels enriched by chitosan (D_{ef}) can be expressed by the following equation [9,50–53]:

$$D_{ef} = \frac{D_h}{\frac{c_{im}}{c_{free}} + 1} = \frac{D_h}{K+1} = \frac{\mu D}{K+1}. \qquad (3)$$

In Equation (3), the apparent equilibrium constant K represents the ratio between immobilized c_{im} and free movable c_{free} dye particles. It is supposed that the immobilization is caused by the interactions between dyes and chitosan. This simple equilibrium can be included into Fickian laws [9,50–53] and the value of K can be determined on the basis of Equation (4):

$$K = \frac{D_h}{D_{ef}} - 1. \qquad (4)$$

The values of apparent equilibrium constants K are listed in Table 5. As can be seen, the values of K depended on the type of dye and pH value. In some cases, the values of K are greater than 1, which means that immobilized dye particles predominate over free mobile ones. In the opposite case ($K < 1$), most of the dye remains free and can migrate in hydrogel. Sirius red F3B can be strongly immobilized at acidic and neutral pH values, although the fraction of immobilized particles decreased, and free mobile particles predominated in alkaline environment. In contrast, free mobile particles predominated in the case of Reactive blue 49 and the decrease of K with increasing pH is only gentle. Direct blue 1 can be strongly immobilized in acidic conditions, but free mobile particles predominate in neutral and alkaline environment. Nevertheless, the local equilibrium between free movable and immobilized dyes cannot be considered as a stable state. It is dynamic, therefore, the values obtained here can be considered as average and effective. Another aspect is that it is assumed that chitosan particles are incorporated in hydrogel and trapped in their positions due to its non-diffusive dynamic state being strongly influenced by thermodynamic parameters (entropic traps) [61]. Chitosan, because of its structure, can fluctuate in its conformational arrangement, which can affect its reactivity.

Figure 6. The concentration distribution of Direct blue 1 in hydrogels with different contents of chitosan after 72 h at pH 3.

Table 5. Values of effective diffusion coefficient (D_{ef}) determined for hydrogel enriched by chitosan.

Dye	K (pH 3)	K (pH 7)	K (pH 11)
Direct blue 1	1.39 ± 0.08	0.55 ± 0.03	0.15 ± 0.01
Sirius red F3B	3.42 ± 0.23	1.75 ± 0.12	0.18 ± 0.01
Reactive blue 49	0.25 ± 0.01	0.15 ± 0.01	0.13 ± 0.01

Portions of free mobile fraction and immobilized dye fraction in hydrogel enriched by chitosan are shown in Figure 7. The calculation was based on the K values. We can see graphically the predomination of free mobile fraction for Reactive blue 49 as well as the strong immobilization in the case of Sirius red F3B (pH 3 and 7) and Direct blue 1 (pH 3).

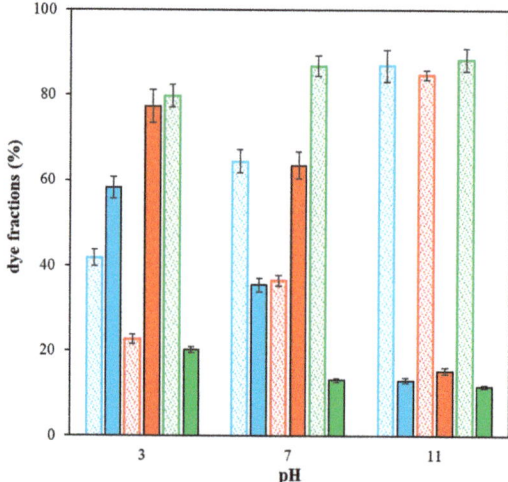

Figure 7. The portions of free mobile fraction (empty dotted columns) and immobilized dye fraction (full columns) in hydrogel enriched by chitosan; Direct blue 1 (blue), Sirius red F3B (red), and Reactive blue 49 (green).

Additional sorption experiments were conducted with both types of hydrogels in order to determine the efficiency of chitosan in agarose hydrogel for studied dyes. The results are listed in Table 6.

Table 6. Adsorption efficiency of pure agarose hydrogels and hydrogels enriched by chitosan.

Dye	Pure Agarose Hydrogels	Enriched Hydrogels
Direct blue 1	$(0.84 \pm 0.02)\%$	$(21.26 \pm 0.61)\%$
Sirius red F3B	$(1.14 \pm 0.03)\%$	$(24.48 \pm 0.44)\%$
Reactive blue 49	$(3.37 \pm 0.06)\%$	$(12.70 \pm 0.17)\%$

Experiments were conducted with aqueous solutions and no pH values were adjusted. Their aim was to compare adsorption efficiency of individual dyes (as such) without the presence of other substances and ions. As expected, the efficiency of enriched hydrogel was higher for Sirius red F3B and Direct blue 1 in comparison with Reactive blue 49. The results obtained for these dyes were comparable. Reactive blue 49 differed in the efficiency of pure agarose hydrogel as well as enriched one. The obtained values are in agreement with the results of diffusion experiments.

3. Conclusions

In this work, the transport properties of Direct blue 1, Sirius red F3B, and Reactive blue 49 in hydrogels were studied. Inert agarose hydrogel was enriched by chitosan as an active substance for the interactions with dyes. It was found that the presence of chitosan strongly affected the diffusion of dyes, mainly in the cases of Sirius red F3B and Reactive blue 49. Electrostatic interactions between the amino group of chitosan and the sulfonic group of dyes resulted in the formation of dye layers with a sharp interface between coloured hydrogel containing dye and transparent hydrogel without it. This effect was better observed in an acidic environment. The specific interaction between chitosan and dyes resulted in an increase in surface concentration and decrease in diffusivity. The decrease in diffusion coefficient caused by the interactions provided information about apparent equilibrium constant defined as the ratio between immobilized dye and free movable dye particles as well as their portions in enriched hydrogel. Immobilized particles predominated over dyes able to migrate in the cases of Sirius red F3B in acidic and neutral conditions and Direct blue 1 at pH 3.

The results obtained in this study provided information on reactivity mapping of dyes in hydrogel enriched by chitosan as an active substance. The advantage of this approach is the possibility to investigate the interactions of dyes with chitosan directly in their diffusion and characterize their transport affected by the interactions by means of a relatively simple mathematical model. The model is also usable for different bio-functional materials containing active sites for the immobilization of diffusing particles. Concentrations of free movable dyes were measured directly in hydrogels in defined distances from interfaces between hydrogel and donor solution. The experimental concentration profiles of dyes in hydrogels provided data for the determination of effective diffusion coefficients in which the effect of chemical interactions is included. Comparing with results obtained for inert agarose hydrogel, the fractions of free movable and immobilized particles can be calculated. This method is universal, its main requirements are the formation of hydrogel with defined size and shape, the possibility to determine a concentration profile in hydrogel, and the experimental arrangement corresponding with initial and boundary conditions given for the mathematical model.

4. Materials and Methods

4.1. Chemicals

Chitosan (medium molecular weight), agarose (routine use class), and Direct blue 1 were purchased from Sigma Aldrich (St. Luis, MO, USA). Sirius red F3B and Reactive blue 49 were purchased from Synthesia (Pardubice, Czech Republic). Acetic acid for the preparation of chitosan solution was purchased from Lachner (Neratovice, Czech Republic). Disodium hydrogen phosphate, sodium dihydrogen phosphate, citric acid, and sodium hydroxide for the preparation of buffer solutions were purchased from Penta (Chrudim, Czech Republic).

The exact molecular weights of chitosan and agarose were determined by means of size exclusion chromatography coupled with multiangle static light scattering, differential refractive index, and UV/VIS detection (SEC chromatographic system from Agilent Technologies, detectors from Wyatt Technology). The exact molecular weights were 251 ± 4 kDa for chitosan and 146 ± 3 kDa for agarose.

The deacetylation degree of chitosan was determined by potentiometric titration as described by Garcia et al. [62]. The degree was determined as $83.8 \pm 0.2\%$ mol.

4.2. Preparation of Hydrogels

The preparation of hydrogels was based on the thermo-reversible gelation of agarose solution described in previous works [9,50–53]. Agarose hydrogel gelatinized from the solution of agarose in water. The agarose content in hydrogel was 10 mg g^{-1}. The mixture was slowly heated with continuous stirring up to 80 °C, stirred at this temperature in order to obtain a transparent solution, and finally sonicated (1 min) to remove gasses. Afterwards, the mixture was slowly poured into the PMMA spectrophotometric cuvette (inner dimensions: $10 \times 10 \times 42$ mm). The cuvette orifice was immediately covered with a pre-heated plate of glass to prevent drying and shrinking of gel. The flat surface of the boundary of resulting hydrogels was provided by wiping an excess solution away. Gentle cooling of cuvettes at the laboratory temperature led to the gradual gelation of the mixture.

Agarose–chitosan hydrogels were prepared from agarose solution mixed with the solution of chitosan. An accurately weighed amount of chitosan was dissolved in 50 cm^3 of acetic acid (5% wt.) The solution was titrated by 1M NaOH up to pH equal to 7 and diluted by distilled water (the final volume was 100 cm^3). The agarose content in hydrogel was 10 mg g^{-1}, the content of chitosan was 1 mg g^{-1}.

4.3. Diffusion Experiments

Two cuvettes (with both types of hydrogels) were placed into 250 cm^3 of dye solution. Dye solutions were prepared in buffers with pH equal to 3, 7, and 10. Buffers were composed of disodium hydrogen phosphate, sodium dihydrogen phosphate, citric acid, and sodium hydroxide in appropriate ratios. The bulk concentration of dyes was 50 mg dm^{-3}. The solution was stirred continuously by the magnetic stirrer and the dye were left to diffuse from the solution into the hydrogels through the square orifices of the cuvettes. Diffusion experiments were triplicated, it means that three different vessels for the same type of dye were used. The durations of the diffusion experiments were 24, 48 and 72 h. In these time intervals, the cuvettes were taken out of the solution and the UV-VIS spectra were measured in dependence on distances from the interface between hydrogel and donor solution. Varian Cary 50 UV–VIS spectrophotometer (Agilent Technologies, Palo Alto, CA, USA) equipped with the special accessory providing controlled fine vertical movement of the cuvette in the spectrophotometer was used for this purpose [21,28,40]. The concentration of dyes was determined at different positions in the hydrogels by means of a calibration line. The spectra were calibrated for the hydrogels with the known concentration, homogeneously distributed in the whole volume of the hydrogel.

The experiments were performed at laboratory temperature (25 ± 1 °C). Data are presented as average values with standard deviation bars.

4.4. Sorption Experiments

Glass tubes (length and diameter, 1 cm) were filled by hydrogels and placed separately into vessels with 20 cm^3 of dye solution. Vessels were closed and covered by parafilm to prevent evaporation. Diffusion experiments were triplicated, which means that three different vessels for the same type of dye and the same type of hydrogel were used. Hydrogels were taken out after 6 days and the UV-VIS spectra of solutions were collected. The decrease in concentration and the sorption efficiency were determined on the basis of calibration line.

Experiments were performed at laboratory temperature (25 ± 1 °C). Data are presented as average values with standard deviation bars.

Funding: This research was funded by the National Program for Sustainability I (Ministry of Education, Youth and Sports), grant number REG LO1211, Materials Research Centre at FCH BUT Sustainability and Development.

Institutional Review Board Statement: Not applicable.

Informed Consent Statement: Not applicable.

Data Availability Statement: Data will be available on request.

Conflicts of Interest: The authors declare no conflict of interest.

References

1. Falk, B.; Garramone, S.; Shivkumar, S. Diffusion coefficient of paracetamol in a chitosan hydrogel. *Mater. Lett.* **2004**, *58*, 3261–3265. [CrossRef]
2. Molinaro, G.; Leroux, J.; Damas, J.; Adam, A. Biocompatibility of thermosensitive chitosan-based hydrogels: An in vivo experimental approach to injectable biomaterials. *Biomaterials* **2002**, *23*, 2717–2722. [CrossRef] [PubMed]
3. Wang, J.; Wang, L.; Yu, H.; Zain-Ul-Abdin; Chen, Y.; Chen, Q.; Zhou, W.; Zhang, H.; Chen, X. Recent progress on synthesis, property and application of modified chitosan: An overview. *Int. J. Biol. Macromol.* **2016**, *88*, 333–344. [CrossRef] [PubMed]
4. Jayakumar, R.; Prabaharan, M.; Reis, R.L.; Mano, J.F. Graft copolymerized chitosan—Present status and applications. *Carbohydr. Polym.* **2005**, *62*, 142–158. [CrossRef]
5. Ji, J.; Wang, L.; Yu, H.; Chen, Y.; Zhao, Y.; Zhang, H.; Amer, W.A.; Sun, Y.; Huang, L.; Saleem, M. Chemical modifications of chitosan and its applications. *Polym.-Plast. Technol. Eng.* **2014**, *53*, 1494–1505. [CrossRef]
6. An-Chong, C.; Shin-Shing, S.; Yu-Chuang, L.; Fwu-Long., M. Enzymatic grafting of carboxyl groups on to chitosan—To confer on chitosan the property of a cationic dye adsorbent. *Bioresour. Technol.* **2004**, *91*, 157–162.
7. Kyzas, G.Z.; Bikiaris, D.N. Recent modifications of chitosan for adsorption applications: A critical and systematic review. *Mar. Drugs* **2015**, *13*, 312–337. [CrossRef]

8. Kausar, A. Scientific potential of chitosan blending with different polymeric materials: A review. *J. Plast. Film Sheeting* **2017**, *33*, 384–412. [CrossRef]
9. Klučáková, M. How the addition of chitosan affects the transport and rheological properties of agarose hydrogels. *Gels* **2023**, *9*, 99. [CrossRef]
10. Sedláček, P.; Smilek, J.; Klučáková, M. How interactions with polyelectrolytes affect mobility of low molecular ions—Results from diffusion cells. *React. Funct. Polym.* **2013**, *73*, 1500–1509. [CrossRef]
11. Sedláček, P.; Smilek, J.; Klučáková, M. How the interactions with humic acids affect the mobility of ionic dyes in hydrogels—2. Non-stationary diffusion experiments. *React. Funct. Polym.* **2014**, *75*, 41–50. [CrossRef]
12. Xiong, J.Y.; Narayanan, J.; Liu, X.Z.; Chong, T.K.; Chen, S.B.; Chung, T.S. Topology evolution and gelation mechanism of agarose gel. *J. Phys. Chem B* **2005**, *109*, 5638–5643. [CrossRef] [PubMed]
13. Barrangou, L.M.; Daubert, C.R.; Foegeding, E.A. Textural properties of agarose gels. I. Rheological and fracture properties. *Food Hydrocoll.* **2006**, *20*, 184–195. [CrossRef]
14. Barrangou, L.M.; Drake, N.A.; Daubert, C.R.; Foegeding, E.A. Textural properties of agarose gels. II. Relationships between rheological properties and sensory texture. *Food Hydrocoll.* **2006**, *20*, 196–203. [CrossRef]
15. Golmohamadi, M.; Davis, T.A.; Wilkinson, K.J. Diffusion and partitioning of cations in an agarose hydrogel. *J. Phys. Chem. A* **2012**, *116*, 6505–6510. [CrossRef] [PubMed]
16. Lead, J.R.; Starchev, K.; Wilkinson, K.J. Diffusion coefficients of humic substances in agarose gel and in water. *Environ. Sci. Technol.* **2003**, *37*, 482–487. [CrossRef]
17. Gutenwik, J.; Nilson, B.; Axxelson, A. Determination of protein diffusion coefficients in agarose gel with a diffusion cell. *Biochem. Eng. J.* **2004**, *19*, 1–7. [CrossRef]
18. Liang, S.M.; Xu, J.; Weng, L.; Dai, H.; Zhang, X.; Zhang, L. Protein diffusion in agarose hydrogel in situ measured by improved refractive index method. *J. Control. Release* **2006**, *115*, 189–196. [CrossRef]
19. Tan, S.X.; Dai, H.J.; Wu, J.; Zhao, N.; Zhang, X.; Xu, J. Optical investigation of diffusion of levofloxacin mesylate in agarose hydrogel. *J. Biomed. Opt.* **2009**, *14*, 050503. [CrossRef] [PubMed]
20. Labille, J.; Fatin-Rouge, N.; Buffle, J. Local and average diffusion of nanosolutes in agarose gel: The effect of the gel/solution interface structure. *Langmuir* **2007**, *23*, 2083–2090. [CrossRef]
21. Karmaker, S.; Nag, A.J.; Saha, T.K. Adsorption of reactive blue 4 dye onto Chitosan 10B in aqueous solution: Kinetic modeling and isotherm analysis. *Russ. J. Phys. Chem.* **2020**, *94*, 2349–2359. [CrossRef]
22. Liu, D.; Cheng, W.; Yu, J.; Ding, Y. Polyamine chitosan adsorbent for the enhanced adsorption of anionic dyes from water. *J. Dispers. Sci. Technol.* **2017**, *38*, 1832–1841.
23. Qin, Y.; Cai, L.; Feng, D.; Shi, J.; Liu, J.; Zhang, W.; Shen, Y. Combined use of chitosan and alginate in the treatment of wastewater. *J. Appl. Polym. Sci.* **2007**, *104*, 3581–3587. [CrossRef]
24. Pietrelli, L.; Francolini, I.; Piozzi, A. Dyes adsorption from aqueous solutions by chitosan. *Sep. Sci. Technol.* **2015**, *50*, 1101–1107. [CrossRef]
25. Bekci, Z.; Ozveri, C.; Seki, Y.; Yurdakoc, K. Sorption of malachite green on chitosan bead. *J. Hazard. Mater.* **2008**, *154*, 254–261. [CrossRef]
26. Kekes, T.; Tzia, C. Adsorption of indigo carmine on functional chitosan and β-cyclodextrin/chitosan beads: Equilibrium, kinetics and mechanism studies. *J. Environ. Manag.* **2020**, *262*, 110372. [CrossRef]
27. Sutirman, Z.A.; Sanagi, M.M.; Karim, K.J.A.; Naim, A.A.; Ibrahim, W.A.W. Enhanced removal of Orange G from aqueous solutions by modified chitosan beads: Performance and mechanism. *Int. J. Biol. Macromol.* **2019**, *133*, 1260–1267. [CrossRef] [PubMed]
28. Ren, J.; Wang, X.; Zhao, L.; Li, M.; Yang, W. Double network gelatin/chitosan hydrogel effective removal of dyes from aqueous solutions. *J. Polym. Environ.* **2022**, *30*, 2007–2021. [CrossRef]
29. Cesco, C.T.; Valente, A.J.M.; Paulino, A.T. Methylene blue release from chitosan/pectin and chitosan/DNA blend hydrogels. *Pharmaceutics* **2021**, *13*, 842. [CrossRef]
30. Gonçalves, J.O.; da Silva, K.A.; Rios, E.C.; Crispim, M.M.; Dotto, G.L.; de Almeida Pinto, L.A. Single and binary adsorption of food dyes on chitosan/activated carbon hydrogels. *Chem. Eng. Technol.* **2019**, *42*, 454–464. [CrossRef]
31. Shen, C.; Shen, Y.; Wen, Y.; Wang, H.; Liu, W. Fast and highly efficient removal of dyes under alkaline conditions using magnetic chitosan-Fe(III) hydrogel. *Water Res.* **2011**, *45*, 5200–5210. [CrossRef]
32. Kim, U.J.; Kimura, S.; Wada, M. Characterization of cellulose–chitosan gels prepared using a LiOH/urea aqueous solution. *Cellulose* **2019**, *26*, 6189–6199. [CrossRef]
33. Le, H.Q.; Sekiguchi, Y.; Ardiyanta, D.; Shimoyama, Y. CO_2-activated adsorption: A new approach to dye removal by chitosan hydrogel. *ACS Omega* **2018**, *3*, 14103–14110. [CrossRef]
34. Sacco, P.; Furlani, F.; De Marzo, G.; Marsich, E.; Paoletti, S.; Donati, I. Concepts for developing physical gels of chitosan and of chitosan derivatives. *Gels* **2018**, *4*, 67. [CrossRef]
35. Barron-Zambrano, J.; Szygula, A.; Ruiz, M.; Sastre, A.M.; Guibal, E. Biosorption of Reactive Black 5 from aqueous solutions by chitosan: Column studies. *J. Environ. Manag.* **2010**, *91*, 2669–2675. [CrossRef]
36. Lazaridis, N.K.; Keenan, H. Chitosan beads as barriers to the transport of azo dye in soil column. *J. Hazard. Mater.* **2010**, *173*, 144–150. [CrossRef] [PubMed]

37. García-Aparicio, C.; Quijada-Garrido, I.; Garrido, L. Diffusion of small molecules in a chitosan/water gel determined by proton localized NMR spectroscopy. *J. Colloid Interface Sci.* **2012**, *268*, 14–20. [CrossRef]
38. Cheung, W.H.; Szeto, Y.S.; McKay, G. Intraparticle diffusion processes during acid dye adsorption onto chitosan. *Bioresour. Technol.* **2007**, *98*, 2897–2904. [CrossRef] [PubMed]
39. Coura, J.C.; Profeti, D.; Profeti, L.P.R. Eco-friendly chitosan/quartzite composite as adsorbent for dye removal. *Mater. Chem. Phys.* **2020**, *256*, 123711. [CrossRef]
40. Vanamudan, A.; Pamidimukkala, P. Chitosan, nanoclay and chitosan–nanoclay composite as adsorbents for Rhodamine-6G and the resulting optical properties. *Int. J. Biol. Macromol.* **2015**, *74*, 127–135. [CrossRef]
41. Sadiq, A.C.; Rahim, N.Y.; Suah, F.B.M. Adsorption and desorption of malachite green by using chitosan-deep eutectic solvents beads. *Int. J. Biol. Macromol.* **2020**, *164*, 3965–3973. [CrossRef] [PubMed]
42. Bilal, M.; Rasheed, T.; Zhao, Y.; Iqbal, H.M.N. Agarose-chitosan hydrogel-immobilized horseradish peroxidase with sustainable bio-catalytic and dye degradation properties. *Int. J. Biol. Macromol.* **2019**, *124*, 742–749. [CrossRef]
43. Hartig, D.; Hacke, S.; Scholl, S. Concentration-dependent diffusion coefficients for fructose in highly permeable chitosan polymers. *Chem. Eng. Technol.* **2018**, *41*, 454–460. [CrossRef]
44. Yang, J.M.; Su, W.Y.; Leu, T.L.; Yang, M.C. Evaluation of chitosan/PVA blended hydrogel membranes. *J. Membr. Sci.* **2004**, *236*, 39–51. [CrossRef]
45. Waluga, T.; Scholl, S. Diffusion of saccharides and the sugar alcohol sorbitol in chitosan membranes and beads. *Chem. Eng. Technol.* **2013**, *36*, 681–686. [CrossRef]
46. Xu, D.; Loo, L.S.; Wang, K. Characterization and diffusion behavior of chitosan–POSS composite membranes. *J. Appl. Polym. Sci.* **2011**, *122*, 427–435. [CrossRef]
47. Carlough, M.; Hudson, S.; Smith, B.; Spadgenske, D. Diffusion coefficients of direct dyes in chitosan. *J. Appl. Polym. Sci.* **1991**, *42*, 3035–3038. [CrossRef]
48. Klučáková, M.; Smilek, J.; Sedláček, P. How humic acids affect the rheological and transport properties of hydrogels. *Molecules* **2019**, *24*, 1545. [CrossRef]
49. Klučáková, M. Agarose hydrogels enriched by humic acids as complexation agent. *Polymers* **2020**, *12*, 687. [CrossRef]
50. Crank, J. *The Mathematics of Diffusion*, 1st ed.; Clarendon Press: Oxford, UK, 1956; pp. 26–41.
51. Cussler, E.L. *Diffusion: Mass Transfer in Fluid Systems*, 2nd ed.; Cambridge University Press: Cambridge, MA, USA, 1984; pp. 13–49.
52. Klučáková, M.; Pekař, M. Study of structure and properties of humic and fulvic acids. IV. Study of interactions of Cu^{2+} ions with humic gels and final comparison. *J. Polym. Mater.* **2003**, *20*, 155–162.
53. Klučáková, M.; Pekař, M. Study of diffusion of metal cations in humic gels. In *Humic Substances: Nature's Most Versatile Materials*, 1st ed.; Ghabbour, E.A., Davies, G., Eds.; Taylor & Francis: New York, NY, USA, 2004; pp. 263–273.
54. Maekawa, M.; Kamada, C. Mixture diffusion of sulfonated dyes into cellulose membrane: IV. Effects of complex formation between a couple of dyes. *Colloids Surf. A Physicochem. Eng. Asp.* **2003**, *216*, 83–90. [CrossRef]
55. Mansurov, R.R.; Zverev, V.S.; Safronov, A.P. Dynamics of diffusion-limited photocatalytic degradation of dye by polymeric hydrogel with embedded TiO_2 nanoparticles. *J. Catal.* **2022**, *406*, 9–18. [CrossRef]
56. Şolpan, D.; Duran, S.; Torun, M. Removal of cationic dyes by poly(acrylamide-co-acrylic acid) hydrogels in aqueous solutions. *Radiat. Phys. Chem.* **2008**, *77*, 447–452. [CrossRef]
57. Abdel-Aal, S.E. Synthesis of copolymeric hydrogels using gamma radiation and their utilization in the removal of some dyes in wastewater. *J. Appl. Polym. Sci.* **2006**, *102*, 3720–3731. [CrossRef]
58. Al-Mubaddel, F.S.; Haider, S.; Aijaz, M.O.; Haider, A.; Kamal, T.; Almasry, W.; Javid, M.; Khan, S.U.-D. Preparation of the chitosan/polyacrylonitrile semi-IPN hydrogel via glutaraldehyde vapors for the removal of Rhodamine B dye. *Polym. Bull.* **2017**, *74*, 1535–1551. [CrossRef]
59. Sandrin, D.; Wagner, D.; Sitta, C.E.; Thoma, R.; Felekyan, S.; Hermes, H.E.; Janiak, C.; de Sousa Amadeu, N.; Kühnemuth, R.; Löwen, H.; et al. Diffusion of macromolecules in a polymer hydrogel: From microscopic to macroscopic scales. *Phys. Chem. Chem. Phys.* **2016**, *18*, 12860–12876. [CrossRef] [PubMed]
60. Roussy, J.; Van Vooren, M.; Dempsey, B.A.; Guibal, E. Influence of chitosan characteristics on the coagulation and the flocculation of bentonite suspensions. *Water Res.* **2005**, *39*, 3247–3258. [CrossRef]
61. Chen, K.; Muthukumar, M. Entropic barrier of topologically immobilized DNA in hydrogels. *Proc. Natl. Acad. Sci. USA* **2021**, *118*, e210638011. [CrossRef]
62. Garcia, L.G.S.; de Melo Guedes, G.M.; da Silva, M.L.Q.; Castelo-Branco, D.S.C.M.; Sidrim, J.J.C.; de Aguiar Cordeiro, R.; Rocha, M.F.G.; Vieira, R.S.; Brilhante, R.S.N. Effect of the molecular weight of chitosan on its antifungal activity against *Candida* spp. in planktonic cells and biofilm. *Carbohydr. Polym.* **2018**, *195*, 662–669. [CrossRef]

Disclaimer/Publisher's Note: The statements, opinions and data contained in all publications are solely those of the individual author(s) and contributor(s) and not of MDPI and/or the editor(s). MDPI and/or the editor(s) disclaim responsibility for any injury to people or property resulting from any ideas, methods, instructions or products referred to in the content.

Article

Poly(ethylene glycol) Diacrylate Hydrogel with Silver Nanoclusters for Water Pb(II) Ions Filtering

Luca Burratti [1,*], Marco Zannotti [2,*], Valentin Maranges [1], Rita Giovannetti [2], Leonardo Duranti [3], Fabio De Matteis [1], Roberto Francini [1] and Paolo Prosposito [1]

1. Department of Industrial Engineering, University of Rome Tor Vergata, Via del Politecnico 1, 00133 Rome, Italy
2. Department School of Science and Technology, Chemistry Division, ChIP Research Center, University of Camerino, Via Madonna delle Ceneri, 62032 Camerino, Italy
3. Department of Chemical Science and Technologies, University of Rome Tor Vergata, Via Della Ricerca Scientifica 1, 00133 Rome, Italy
* Correspondence: luca.burratti@uniroma2.it (L.B.); marco.zannotti@unicam.it (M.Z.)

Abstract: Poly(ethylene glycol) diacrylate (PEGDA) hydrogels modified with luminescent silver nanoclusters (AgNCs) are synthesized by a photo-crosslinking process. The hybrid material thus obtained is employed to filter Pb(II) polluted water. Under the best conditions, the nanocomposite is able to remove up to 80–90% of lead contaminant, depending on the filter composition. The experimental results indicate that the adsorption process of Pb(II) onto the modified filter can be well modeled using the Freundlich isotherm, thus revealing that the chemisorption is the driving process of Pb(II) adsorption. In addition, the parameter n in the Freundlich model suggests that the adsorption process of Pb(II) ions in the modified hydrogel is favored. Based on the obtained remarkable contaminant uptake capacity and the overall low cost, this hybrid system appears to be a promising sorbent material for the removal of Pb(II) ions from aqueous media.

Keywords: silver nanoclusters; fluorescent nanoclusters; poly(ethylene glycol) diacrylate; PEGDA hydrogel; hybrid material; water remediation; heavy metal ions filtering; Pb(II) ions

1. Introduction

Heavy metals pollution has become one of the most serious environmental issues nowadays. Pb(II) ions are commonly present in many types of industrial effluents and are responsible for environmental contamination. The toxic effect and bioconcentration of lead ions represent an important issue for the health of many ecosystems [1,2]. Lead is a particularly hazardous heavy metal because once it enters the human body, it disperses immediately and causes harmful effects wherever it lands. For instance, Pb(II) can accumulate in many organs, such as the kidneys, adversely affecting the nervous, immune, reproductive, and cardiovascular systems [3,4]. Lead toxicity also impacts vegetation health as it reduces leaf growth by decreasing the level of photosynthetic pigments, alters the chloroplast structure, and decreases the enzymatic activity of CO_2 assimilation [5,6]. Many industrial processes, such as battery manufacturing, pigments and printing, metal plating and soldering materials, ceramic and glass industries, and iron and steel manufacturing units, are the main sources of lead contamination in wastewater [7,8]. Limiting lead pollution is of paramount importance for human health, as well as environmental and economic considerations.

The most commonly used techniques to eliminate heavy metal ions (HMIs) from water are represented by adsorption, membrane-based filtration and separation, chemical separation, and electric-based separation [9].

In the adsorption process, the adsorbent material, through chemical interaction with the functional groups (such as carboxyl, phenyl, etc.) on its surface, can adsorb HMIs. The materials principally used as adsorbents are carbon-based materials [9,10], which

sometimes are very expensive. Mineral adsorbents, such as zeolite, silica, and clay, show a high cation exchange capacity [11], and synthetic materials, such as MOFs (metal-organic frameworks), which show stability in water and toxicity, can be a serious problem for the depuration of polluted water [12].

Other techniques involve the use of membranes for the filtration and the extraction of HMIs, such as Pb(II), using membrane pores larger than the HM ions; in this case, additives bonded to metal ions to enlarge the size can be used, while ultrafiltration and polymer enhanced ultrafiltration are developed [13–15]. Chemical methods involve the use of hydroxides and sulfides that allow for HMI precipitation; in this case, the main problem is the great dosage of precipitating agent and the large volume of sludge [16]. Coagulation and flocculation methods are also used, followed by a sedimentation process in order to obtain clean water [17,18]. Finally, electrochemical methods such as electrochemical reduction, electrocoagulation, electroflotation, and electrooxidation can also be used [19,20].

Among all methods, adsorption is the most used and studied due to its ease of operation, low cost, and high adsorption capacity. The adsorption method can be divided into two types depending on the type of interactions between the adsorbent and adsorbate (pollutant): chemisorption and physisorption. Chemisorption is an irreversible process where the driving force is the strong chemical interaction between the adsorbate and adsorbent surface [21]. Conversely, physisorption is a reversible process where weak intermolecular physical forces are involved between the adsorbate and adsorbent, such as π-π and dipole-dipole interactions, van der Waals forces, hydrogen bonding, etc. [22]. Several factors, such as the surface area of the adsorbent, temperature, pH, type of interactions between adsorbent and adsorbate, contact time, adsorbent dosage, etc., significantly influence the adsorption efficiency and affect the removal of pollutants from effluent [23].

Adsorbent materials can be divided into several categories [24]: gels (hydrogels, aero-gels), layered nanomaterials (nanoclays, layered double hydroxides (LDH)), carbon nanomaterials (fullerene, graphene), nanoparticles (metal nanoparticles, magnetic nanoparticles), polymer-based nanomaterials (biopolymers, conjugated polymers), and conventional materials (activated carbon, silica gels), and it is also possible to combine two materials to increase the adsorption performances [25–27]. Noble metal nanomaterials such as nanoparticles [28,29] or fluorescent nanoclusters [30–32] are already employed in many fields of science and engineering [33–37] thanks to their different properties from those of bulk metal [38,39], but only recently they have been combined with adsorbent matrices to boost the adsorption of water pollutants [40–43].

To synthesize the hydrogel adsorbent matrix, many materials are available; among these, cellulose is available, which is the most abundant natural polymer on Earth. Cellulose has different useful features, such as biocompatibility, biodegradability, non-toxicity, good mechanical properties, etc. [44,45]. The synthesis of the cellulose-based hydrogel consists of several steps and chemical processes as described in the literature [46,47], including copper(I) as a catalyst [48]. Nonetheless, the employment of poly(ethylene glycol) derivatives such as poly(ethylene glycol) diacrylate (PEGDA) represents a valid alternative to the classic cellulose-based hydrogels. PEGDA is a synthetic biopolymer widely used in different fields of research [49–51], and its low cost and ease of hydrogel synthesis by photo-polymerization techniques (one-step process) make it suitable for a variety of practical applications. Moreover, these techniques can be easily scaled up to synthesize 3D scaffold hydrogels with different dimensions and shapes [52,53].

In the present work, we synthesized adsorbent materials based on poly(ethylene glycol) diacrylate with silver nanoclusters capped with poly(methacrylic acid) [AgNCs-PMAA], and we compared their properties with those of the unmodified matrix. For matrix synthesis, a photo-polymerization process was exploited using a UV source to activate the photoinitiator (Irgacure® 184), which reacts with the acrylate moieties of PEGDA and cross-linking the polymeric chains. The modified matrix was obtained by mixing the previously synthesized AgNCs-PMAA with PEGDA and Irgacure® 184 before the UV exposition. The two types of hydrogels (with and without AgNCs) were characterized by UV-Vis absorption, emission spectroscopy, and scanning electron microscopy (SEM).

The main aim of this study was to improve the removal efficiency (RE) towards Pb(II) ions of unmodified PEGDA hydrogel by adding the AgNCs-PMAA. The adsorption capacities for both types of hydrogels have been evaluated and compared. In addition, we performed equilibrium studies using different isotherm models, which show an improvement in the RE for the AgNCs-modified filters.

2. Materials and Methods

2.1. Chemicals

Silver nitrate ($AgNO_3$), poly(methacrylic acid) sodium salt water solution (PMAA, MW = 9500, 30% in wt), poly(ethylene glycol) diacrylate (PEGDA, Mn = 700), ethanol (>99.8%), nitric acid HNO_3 (70%), and lead nitrate [$Pb(NO_3)_2$, > 99.0%] were purchased from Sigma-Aldrich. The photoinitiator Irgacure® 184 (Irg.184) was purchased from Ciba Specialty Chemicals. All reagents were used without any further purification. All the water solutions were prepared with deionized water with resistivity equal to 18.2 MΩ·cm (Semplicity® UV, MERCK, Darmstadt, Germany).

2.2. Synthesis of Silver Nanoclusters (AgNCs-PMAA)

Silver nanoclusters were synthesized according to the previous papers [54,55], where a fresh water solution of $AgNO_3$ was prepared and mixed with a solution of PMAA. The pH was adjusted to reach the value of 4 by adding HNO_3. A volume equal to 3.5 mL was poured into a Petri dish and exposed to strong UV radiation (300 W, NEWPORT, Oriel Instruments, Irvine, CA, USA.) for 6 min to promote the reduction reaction of silver ions to silver metal [56,57]. To hamper the AgNCs surface oxidation, UV exposition took place under a flux of nitrogen gas. The colloidal solution was then purified by centrifugation for 20 min at 10,000 rpm (Thermo Scientific, Heraeus Megafuge 8, Waltham, MA, USA) and was applied in order to obtain, as far as possible, a monodispersed solution of AgCNs. Only the supernatant solution was collected and subsequently used for the synthesis of PEGDA/AgNCs-PMAA filters. The final solution was kept in the dark at T = 4 °C before use.

2.3. Synthesis of PEGDA/AgNCs-PMAA Filters

Filters in the final form were obtained via a photo-crosslinking process under UV irradiation (λ = 366 nm, 1.2 mW/cm^2 at 10 cm of distance, MinUVIS, DESAGA company, Germany). In a typical synthesis, the Irg.184 was dissolved in ethanol and PEGDA and stirred for 5 min in the dark. Subsequently, the AgNCs-PMAA solution was added dropwise and mixed with the previous precursor solution for 5 min. A volume of 2 mL of the photoactive solution was poured into a square box of 2.7 × 2.7 × 1.0 cm^3 until complete photo-reticulation. For a given PEGDA to Irg.184 amount ratio, the photo-crosslinking process was optimized by varying the UV exposure time in the range of 2–6 min. Other filters were prepared as references using deionized water instead of the AgNCs solution. In this case, the filters were fully formed after 2 min of UV exposure time. The details of the investigated compositions and exposure times can be found in Table 1.

Before the filtration tests, all hydrogels were cleaned in ethanol for 24 h to remove unreacted Irgacure and PEGDA, followed by 5 days in H_2O, changing the deionized water every 24 h. Figure S1 of the Supporting Information shows a digital picture of both types (unmodified and modified) of dried filters.

Table 1. Composition of the investigated samples and the UV exposure time.

Sample	PEGDA (% in wt.)	AgNCs (in mg)	UV Exposure Time (min)
1–0	14	0	2
1-A	14	180	6
1-B	14	225	6
1-C	14	255	6
2–0	19	0	2
2-A	19	180	4
2-B	19	225	4
2-C	19	240	4
3–0	24	0	2
3-A	24	180	2
3-B	24	225	2

2.4. Filtering Tests

Pb(II) filtering tests were carried out in static conditions: a single filter was immersed in a beaker containing 10 mL of Pb(II) polluted water for 24 h and at T = 25 °C. To select the best composition of filter, the quantities of PEGDA (14%, 19%, and 24% in wt.) and AgNCs (from 0 to 255 mg) were varied. These hydrogels were kept in contact with a solution of 1500 ppm of Pb(II) ions, and subsequently, the adsorption percentage was evaluated to select the best composition of sorbent material. The RE percentage was calculated by applying the following formula:

$$RE\ (\%) = \left(\frac{C_i - C_e}{C_i}\right) * 100, \quad (1)$$

where C_i and C_e represent the initial and equilibrium concentrations of metal ion (mg/L), respectively, measured by inductively coupled plasma–mass spectrometry (ICP-MS).

The performance of the filters with the best composition was further investigated by varying the Pb(II) concentration from 1500 ppm to 50 ppm. The same investigation was replicated for the filters without AgNCs for comparison. The number of metal ions adsorbed per unit mass of the sorbent material (mg/g) was evaluated with the following equation:

$$q_e = \left(\frac{C_i - C_e}{m}\right) * V, \quad (2)$$

where q_e (mg/g) is the adsorption capacity at equilibrium; C_i and C_e have the same meaning as in Equation (1), while V is the volume of adsorbate solution in liters, and m is the mass of the filter in grams. The error analysis is reported in the Supporting Information.

2.5. Instrumentation

All filters were dried in an oven at 40 °C overnight before being characterized. Optical absorption in the range of 300–700 nm was measured with a spectrophotometer (Perkin-Elmer, Lambda 750), and the photoluminescence (PL) spectra were obtained with a custom-made apparatus equipped with a LED (Light-Emitting Diode) source peaked at 430 nm with a 17 nm bandwidth (Thorlabs, M430L4). The excitation light was focalized onto the sample surface, and the PL was collected and analyzed by a grating monochromator (ARC, SpectraPro-300i). A long pass filter at 500 nm (Melles Griot, 03LWP003) at the entrance slit prevented the excitation wavelength from entering the monochromator. A photomultiplier tube (Hamamatsu Photonic Corp., R2949) was employed for detection. The output signal was analyzed by a lock-in amplifier (Stanford Research System, model SR830 DSP) in the spectral range of 500–800 nm. A computer running a LabView program controlled the whole setup. A full description of the apparatus can be found in the reference [58].

Morphological analyses were performed using a Zeiss Leo SUPRA 35 field emission scanning electron microscope (FE-SEM), while elemental analysis was performed by energy dispersive X-ray spectroscopy (EDX, INCAx-sight, Model: 7426, Oxford Instruments, Abington, UK). Hydrogels were treated in an oven at 120 °C for 24 h. Once completely dried, samples were cut into 1 mm-thick slices and gold-sputtered prior to the FE-SEM and EDX analyses.

The concentrations of Pb in the filtered water were measured with ICP-MS analysis. All the samples were diluted with an appropriate amount of aqueous HNO_3 (Supra Pure) 1% v/v. The measurement was carried out by Agilent Technologies 7500cx Series ICP-MS. The analysis was performed under the following conditions: power 1550 W; carrier gas 1.11 L/min; sample depth 7.5 mm; nebulizer pump 0.1 rps; spray chamber temperature 2 °C. The concentration of Pb was calculated as the sum of ^{208}Pb, ^{206}Pb, and ^{207}Pb.

3. Results and Discussion

3.1. Optical Characterization

Figure 1 shows the optical characterizations: Figure 1a represents the UV-Vis absorption spectra of AgNCs-PMAA (black curve), while the red and blue curves refer to the hydrogels with and without AgNCs, respectively. The colloidal solution presents an orange color after synthesis, but its spectrum does not show peculiar features in the analyzed range. The hydrogel with only PEGDA is colorless in the visible range; indeed, there are no absorption bands in this range. The spectrum of the hydrogel with AgNCs-PMAA shows an increasing optical absorption as the wavelength decreases. Figure S1 shows the difference in the color of both filters. In Figure 1b, we report the normalized photoluminescence emission spectra of the three samples in the range of 500–800 nm by exciting at λ = 430 nm. The colloidal solution exhibits an emission band centered at approximately 660 nm with an FWHM of roughly 150 nm (black curve in the graph). The same emission belongs to the NCs-based hydrogel (red curve in the figure), underlining that the insertion of the silver nanomaterial was successful and without modification of the AgNCs' optical properties. Finally, the hydrogel (blue curve in the graph) shows a null PL. The spectra shown in Figure 1 refer to the hydrogels samples 1–0 and 1-C (see Table 1), and the overall behavior is the same for all the investigated compositions.

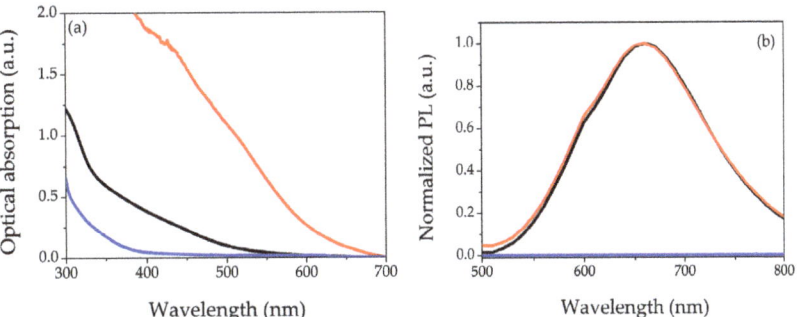

Figure 1. Optical characterizations: (**a**) UV-Vis optical absorption of AgNCs-PMAA colloidal solution (black curve), dried hydrogels with (red line) and without (blue curve) AgNCs-PMAA; (**b**) normalized photoemission spectra excited at λ = 430 nm of AgNCs-PMAA colloidal solution (black curve), dried hydrogels with (red line) and without (blue curve) AgNCs-PMAA.

3.2. Optimization of PEGDA/AgNCs-PMAA Composition

In Table 1, we list the explored compositions and the time needed to polymerize the entire volume poured in the square box.

Figure 2 shows the removal efficiency for 1500 ppm of Pb(II) solution as a function of filter composition as listed in Table 1; the red bars represent the filters with AgNCs-PMAA and the blue ones those without nanoclusters. For all PEGDA compositions, the presence of AgNCs inside the filter increases the removal percentage of Pb(II) and the efficiency increases by lowering the PEGDA content. Indeed, the best removal efficiency (83%)

corresponds to the filter 1-C. Within the same percentage of PEGDA, by raising the AgNCs content, the Pb(II) adsorption increases. In hydrogels without AgNCs, RE is low and almost constant (on average 26%) for all compositions. The hydrogel with composition 1-C was then selected for the adsorption capacity study.

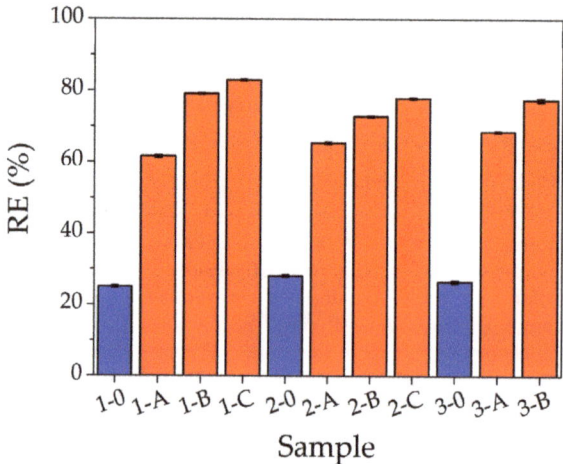

Figure 2. Removal efficiency as a function of filter composition after the filtration of a 1500 ppm of Pb(II) solution. Red bars represent the filters with AgNCs-PMAA, while blue bars correspond to those without AgNCs.

3.3. Morphological and Elemental Characterizations

The morphology and the elemental composition of samples with and without AgNCs-PMAA were investigated. The SEM images of modified and unmodified hydrogels are similar and independent of PEGDA percentage, as shown in Figure 3. SEM images at higher magnifications are reported in Figures S2 and S3 of the Supporting Information file. The first two panels of Figure 3 refer to the samples with 14% wt. of PEGDA without silver nanoclusters (1–0, a) and with 255 mg of AgNCs-PMAA (1-C, b). Here, the structures of both hydrogels are similar. The last two panels of Figure 3 display the case of 24% wt. containing PEGDA unmodified (3–0, c) and modified with 180 mg of AgNCs (3-A, d) filters. A similar morphology is present in both cases.

According to SEM analyses, the morphology of the hydrogel does not affect the volumetric adsorption of lead ions, while the presence of AgNCs and especially of the carboxylic groups of the capping agents seems to play a crucial role.

EDX analyses were performed on the cross sections of the samples after the Pb(II) filtration process to investigate the element distributions in the hydrogel volume: both on the surface and in bulk. The micrographs of samples 1–0 and 1-C (without and with AgNCs), together with Pb distributions, are reported in Figure 4, while the weight percentages of the detected elements are shown in Table 2. In both hydrogels (without and with AgNCs), the main elements are C and O, as expected, due to the presence of PEGDA (C, O, and Ag EDX maps are reported in Figure S4 of the Supporting Information file). In the case of the unmodified filter, the Pb distribution underlines that the adsorption occurred mainly on the surface of the hydrogel, as shown in the Pb panel of Figure 4a, with a total Pb uptake of 0.97%. In the case of AgNCs-based hydrogel, Figure 4b, the Pb content (9.37%) is approximately one order of magnitude higher compared to that of the filter without AgNCs, as reported in Table 2. The amount of detected silver in the AgNCs-based hydrogel is approximately 1% on average, and Ag distribution appears homogeneous throughout the hydrogel, as shown in Figure S4 of the Supporting Information.

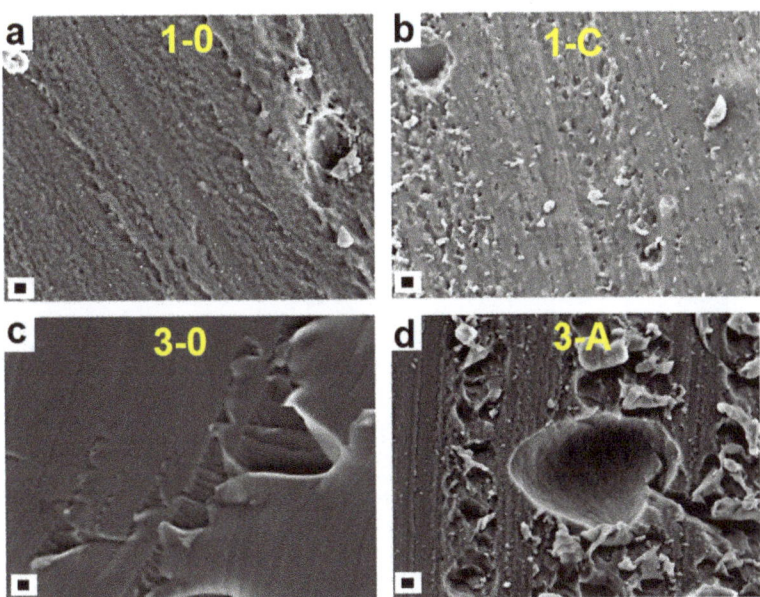

Figure 3. SEM images of (**a**) sample 1-0 with 14%wt of PEGDA and without AgNCs; (**b**) sample 1-C with 14%wt of PEGDA and 255 mg of AgNCs-PMAA; (**c**) sample 3-0 with 24%wt of PEGDA and without AgNCs; (**d**) sample 3-A with 24%wt of PEGDA and 180 mg of AgNCs-PMAA (scale bar 1 µm).

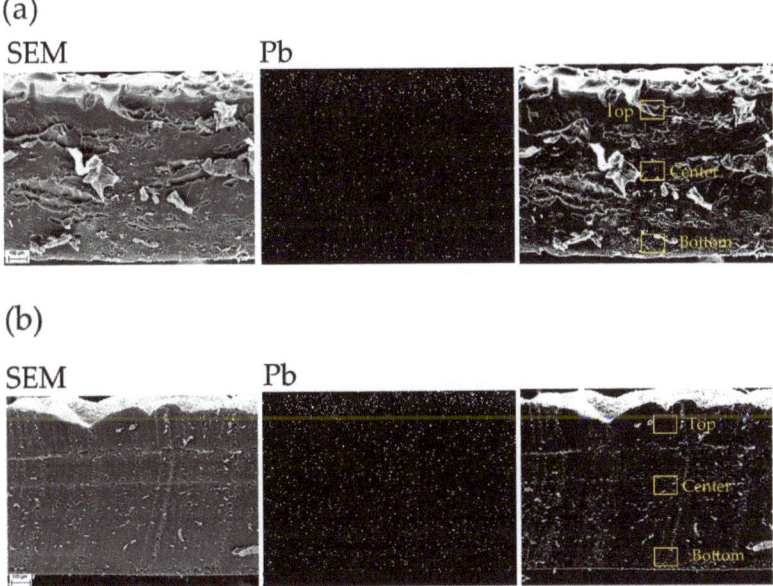

Figure 4. EDX analyses results of 14% PEGDA samples without (1-0, (**a**)) and with (1-C, (**b**)) after the filtration of 1500 ppm Pb(II). For each sample, the SEM image of the recorded area and the Pb distribution map are reported. The yellow boxes in the third panel underline the areas where the localized EDX spectra were acquired. The obtained elemental distributions of localized analyses are reported in Table S1.

Table 2. Elemental composition of the AgNCs-free and AgNCs hydrogels after the filtration of a 1500 ppm of Pb(II) solution.

Sample	Element	% in Weight
1–0	C	56.00 ± 1.63
	O	42.73 ± 1.71
	Au *	0.30 ± 0.06
	Pb	0.97 ± 0.40
	Tot.	100.00
1-C	C	57.34 ± 3.00
	O	29.92 ± 1.29
	Au *	1.51 ± 0.64
	Na †	0.81 ± 0.26
	Ag	1.04 ± 0.54
	Pb	9.37 ± 2.73
	Tot.	100.00

* The presence of Au is due to the metallization of the samples. † The Na element is the counter ion present in the PMAA solution; it results from the synthesis of AgNCs.

In the AgNCs-hydrogel, the Pb signal shows a high intensity not only on the surface but also inside the bulk of the filter. Thus, indicating that the presence of AgNCs-PMAA plays a key role in the enhancement of the Pb(II) uptake.

By focusing the analysis on three different areas of each sample (top, center, and bottom), the elemental composition was recorded in localized spots of 10 μm width. In the unmodified sample, the local amount of Pb is higher on the top surface (1.77%) than in the rest of the sample (only 0.57% for the center and bottom, see Table S1 of the Supporting Information file for details). This can be explained considering that the top surface was in direct contact with most of the volume of Pb(II) solution during the filtration test, while the bottom part was in contact with the glass of the beaker. On the other hand, in the modified sample, the Pb local distribution is higher at the center (14.50%), as compared to the top (8.43%) and bottom (5.18%). The same trend was observed for silver distribution in the filter (Table S1), showing that, indeed, silver nanoclusters are responsible for the higher Pb(II) uptake.

3.4. Adsorption Capacity and Equilibrium Studies

The adsorption capacity study was carried out using different aqueous concentrations of Pb(II) from 50 to 1500 ppm. Figure 5a shows the adsorption capacity as a function of the Pb(II) equilibrium concentration of the unmodified filter. In addition, the best fits of the experimental points are reported: the red line is the Henry isotherm; meanwhile, the green line represents the Freundlich model. Figure 5b shows the adsorption capacity as a function of the Pb(II) equilibrium concentration of the AgNCs-hydrogel filter. In this case, the best fits of the experimental points are reported: the red line is the Henry isotherm, the green line represents the Freundlich isotherm, and the blue curve refers to the Langmuir isotherm.

The parameters obtained by the best fits are listed in Table 3 for both filter types.

The Henry isotherm is defined as follows [59]:

$$q_e = K_{HE} * C_e, \qquad (3)$$

where K_{HE} represents Henry's partition coefficient calculated by the slope of the experimental fit. This constant is related to the interaction that occurs among adsorbate and adsorbent, or in other words, how the adsorbate is partitioned between the solution and the filter. The Henry model describes a weak interaction between adsorbate and adsorbent of the electrostatic or van der Waals kind or hydrophobic interactions [59].

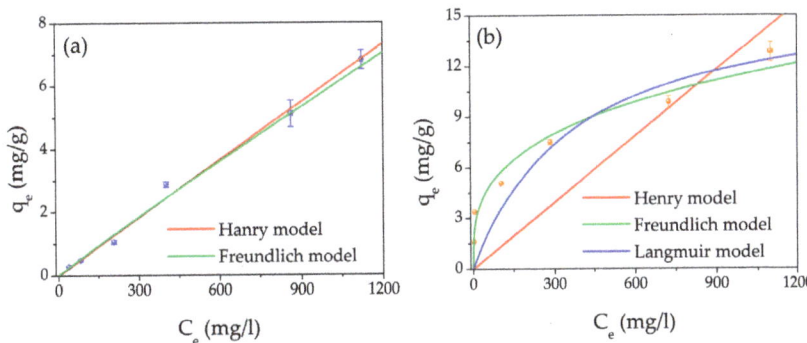

Figure 5. Plot of adsorption capacity (q_e) as a function of equilibrium concentration C_e for: (a) AgNCs-free filter, blue points and (b) AgNCs modified filter, orange points; the isotherm best fit are also shown in the graphs: red line Henry model, green line Freundlich model, and blue line Langmuir model.

Table 3. Isotherm parameters determined by fitting the experimental points.

Sample	Isotherm Model	Parameters	Sample	Isotherm Model	Parameters
1-0	Henry	K_{HE} = 6.1 ± 0.4 l/kg R^2 = 0.996	1-C	Henry	K_{HE} = 13 ± 9 l/kg R^2 = 0.494
	Freundlich	K_F = 0.008 ± 0.003 mg/g n = 1.04 ± 0.06 R^2 = 0.991		Freundlich	K_F = 1.6 ± 0.2 mg/g n = 3.4 ± 0.5 R^2 = 0.942
	Langmuir	Does not converge		Langmuir	K_L = 0.05 ± 0.02 l/mg q_m = 16 ± 3 mg/g R^2 = 0.899

The Freundlich isotherm is defined as follows [59]:

$$q_e = K_F * C_e^{\frac{1}{n}}, \qquad (4)$$

where K_F is Freundlich's coefficient and is related to the adsorption capacity, while n is related to the strength constant of the isotherm model. The Freundlich isotherm model describes a reversible and non-ideal adsorption process and defines the heterogeneity of the surface as well as the exponential distribution of the energies of the active site. Unlike the Langmuir isotherm model, the Freundlich one does not predict the saturation of the adsorbent material; hence, infinite surface coverage is mathematically predicted, indicating that multilayer adsorption on the surface is possible. The Freundlich model describes adsorption of a single layer when chemisorption is the fundamental adsorption mechanism; otherwise, it describes multilayer adsorption when physisorption is the main mechanism [59].

The values of $1/n$ $\frac{1}{n}$ depend on temperature and indicate adsorption conditions such as adsorption intensity or surface heterogeneity. The adsorption process is favored if $\frac{1}{n}$ is between 0 and 1, disadvantaged if it is greater than one, and irreversible if it equals one [60].

The Langmuir isotherm is defined as follows [59]:

$$q_e = \frac{q_m * K_L * C_e}{1 + K_L * C_e}, \qquad (5)$$

where q_m (mg/g) is the maximum adsorption capacity, and K_L (l/mg) is the Langmuir isotherm constant. The main point of the Langmuir isotherm assumes that the thickness of the adsorbed layer is one molecule (monolayer adsorption) in which the process of adsorption occurs at identical and equivalent definite localized sites, and the active sites are identical in terms of activation energy.

For the unmodified filter, by considering the R^2 factors of both models, the best fit is represented by the Henry isotherm (R^2 = 0.996) compared to the Freundlich one (R^2 = 0.991). By considering the chemical structure of PEGDA (see Figure S5.a of the Supporting Information), we can suppose that the electrostatic interaction takes place on the oxygen atoms present in the PEGDA molecule.

Regarding the filter with AgNCs-PMAA, from Figure 5b and Table 3, it appears that the best model is the Freundlich isotherm (R^2 = 0.942). From the experimental points, it is evident that the Henry model (R^2 = 0.494) does not describe what happens during the interaction between the AgNCs-based filter and Pb(II) ions. Moreover, the Langmuir model (R^2 = 0.899) does not interpolate the experimental points well, and the precision (defined as the relative error) in the estimation of K_L and q_m is lower than that of K_F and n.

Therefore, the best model that describes the adsorption process in the case of AgNCs-based filters is the Freundlich one. In fact, the process is favored ($0 < \frac{1}{n} < 1$), and the presence of the nanomaterial enhances the filtering effect since the carboxyl groups of PMAA (see Figure S5b of the Supporting Information), which are very close to each other, act as a chelating agent for the Pb ions. The interaction mechanism between Pb(II) and AgNCs-PMAA has already been investigated in previous work [54], where it was reported that the carboxylic groups exactly chelated the ions. In the literature, the chelating action is considered chemisorption [61,62]; thus, we can conclude that the main adsorption mechanism is chemisorption, which is assumed in the Freundlich isotherm. Finally, the values of K_F and n obtained from the fit are compared with those of the literature, and they are in good agreement with the values previously found [63–66].

4. Conclusions

In this work, the insertion of AgNCs-PMAA into a PEGDA matrix by a photopolymerization process was proven. Once inserted in the hydrogel, the AgNCs-PMAA showed the same optical properties as the original colloidal solution, demonstrating that the incorporation took place successfully. The optimization process of the composition revealed that the filters with 14% in weight of PEGDA and 255 mg of AgNCs-PMAA have the best performance by removing 90% of the Pb(II) ions present in the water solution. This value represents an efficiency 60% larger compared to that of the filters without active nanomaterial [RE ≈ 25%, for 1500 ppm of Pb(II) solution]. The study of the adsorption capacity revealed that the Pb(II) adsorption mechanisms are different in the case of unmodified and modified filters with AgNCs. The performed best fits for PEGDA hydrogels suggest an electrostatic interaction between Pb(II) ions and oxygen atoms of the polymeric chains (Henry model). Meanwhile, in the case of PEGDA/AgNCs-PMAA filters, the absorption model follows the Freundlich isotherm, demonstrating that the presence of the nanomaterial enhances the filter effect through the carboxyl groups (from PMAA), giving rise to a strong chelating action towards the Pb(II) ions (chemisorption). As a final remark, the employment of a light-curable polymer such as PEGDA opens new possibilities regarding the synthesis of three-dimensional structured materials to increase the surface interaction between water pollutants and filters.

Supplementary Materials: The following Supporting Information can be downloaded at: https://www.mdpi.com/article/10.3390/gels9020133/s1, Figure S1. Picture of dried filters: unmodified PEGDA hydrogel (left); modified with AgNCs-PMAA hydrogel (right). Figure S2. SEM images at different magnifications of samples with 14%wt of PEGDA and without NCs (a)–(d) (from left to right: 10kX scale bare 1 μm, 50kX scale bar 1 μm, 100kX scale bar 100 nm, 500kX scale bar 100 nm); and with 255 mg of AgNCs-PMAA (e)–(h) (from left to right: 10kX scale bare 1 μm, 50kX scale bar 1 μm, 100kX scale bar 100 nm, 500kX scale bar 100 nm). Figure S3. SEM images at different magnifications of samples with 24%wt of PEGDA and without AgNCs (a)–(d) (from left to right: 10kX scale bare 1 μm, 50kX scale bar 1 μm, 100kX scale bar 100 nm, 500kX scale bar 100 nm); and with 180 mg of

AgNCs-PMAA (e)–(h) (from left to right: 10kX scale bare 1 µm, 50kX scale bar 1 µm, 100kX scale bar 100 nm, 500kX scale bar 100 nm). Figure S4. Map distributions of chemical elements of hydrogel samples after the filtration of 1500 ppm of Pb(II): (a) unmodified and (b) modified filters. Table S1. Elemental composition of filters as a function of different areas of interest. Figure S5. Chemical structure of PEGDA molecule (a) and PMAA molecule (b).

Author Contributions: Conceptualization, L.B., V.M. and P.P.; methodology, L.B., P.P., R.G. and M.Z.; software, L.B. and V.M.; validation, R.G. and P.P.; formal analysis, L.B. and V.M. investigation, L.B., V.M., R.G. and M.Z.; data curation, L.B., P.P., R.G., M.Z. and L.D.; writing—original draft preparation, L.B.; writing—review and editing, L.D., R.G., M.Z., P.P., F.D.M. and R.F.; supervision, P.P.; project administration, P.P. All authors have read and agreed to the published version of the manuscript.

Funding: The project A0375-2020-36521 (CUP: E85F21002440002) FACS (Filtraggio di acque contaminate tramite sistemi nanostrutturati) received funding from the Regione Lazio (Italy) by "Gruppi di ricerca 2020"—POR FESR Lazio 2014–2020.

Institutional Review Board Statement: Not applicable.

Informed Consent Statement: Not applicable.

Data Availability Statement: Not applicable.

Conflicts of Interest: The authors declare no conflict of interest.

References

1. Özdemir, S.; Kilinc, E.; Poli, A.; Nicolaus, B.; Güven, K. Biosorption of Cd, Cu, Ni, Mn and Zn from aqueous solutions by thermophilic bacteria, *Geobacillus toebii* sub.sp. *decanicus and Geobacillus thermoleovorans sub.sp. stromboliensis: Equilibrium, kinetic and thermodynamic studies*. *Chem. Eng. J.* **2009**, *152*, 195–206. [CrossRef]
2. Kim, S.U.; Cheong, Y.H.; Seo, D.C.; Hur, J.S.; Heo, J.S.; Cho, J.S. Characterisation of heavy metal tolerance and biosorption capacity of bacterium strain CPB4 (*Bacillus* spp.). *Water Sci. Technol.* **2007**, *55*, 105–111. [CrossRef] [PubMed]
3. Gillis, B.; Arbieva, Z.; Gavin, I. Analysis of lead toxicity in human cells. *BMC Genom.* **2012**, *13*, 344. [CrossRef]
4. World Health Organization. *Lead in Drinking-Water Background Document for Development of WHO Guidelines for Drinking-Water Quality*; World Health Organization: Geneva, Switzerland, 2016.
5. Qufei, L.; Fashui, H. Effects of Pb2+ on the Structure and Function of Photosystem II of Spirodela polyrrhiza. *Biol. Trace Elem. Res.* **2009**, *129*, 251–260. [CrossRef]
6. Islam, E.; Liu, D.; Li, T.; Yang, X.; Jin, X.; Mahmood, Q.; Tian, S.; Li, J. Effect of Pb toxicity on leaf growth, physiology and ultrastructure in the two ecotypes of Elsholtzia argyi. *J. Hazard. Mater.* **2008**, *154*, 914–926. [CrossRef]
7. Jamali, M.K.; Kazi, T.G.; Arain, M.B.; Afridi, H.I.; Jalbani, N.; Memon, A.R. Heavy Metal Contents of Vegetables Grown in Soil, Irrigated with Mixtures of Wastewater and Sewage Sludge in Pakistan, using Ultrasonic-Assisted Pseudo-digestion. *J. Agron. Crop Sci.* **2007**, *193*, 218–228. [CrossRef]
8. Ghaedi, M.; Asadpour, E.; Vafaie, A. Simultaneous Preconcentration and Determination of Copper, Nickel, Cobalt, Lead, and Iron Content Using a Surfactant-Coated Alumina. *Bull. Chem. Soc. Jpn.* **2006**, *79*, 432–436. [CrossRef]
9. Qasem, N.A.A.; Mohammed, R.H.; Lawal, D.U. Removal of heavy metal ions from wastewater: A comprehensive and critical review. *npj Clean Water* **2021**, *4*, 36. [CrossRef]
10. Demiral, İ.; Samdan, C.; Demiral, H. Enrichment of the surface functional groups of activated carbon by modification method. *Surf. Interfaces* **2021**, *22*, 100873. [CrossRef]
11. Zhang, T.; Wang, W.; Zhao, Y.; Bai, H.; Wen, T.; Kang, S.; Song, G.; Song, S.; Komarneni, S. Removal of heavy metals and dyes by clay-based adsorbents. From natural clays to 1D and 2D nano-composites. *Chem. Eng. J.* **2021**, *420*, 127574. [CrossRef]
12. Wang, C.; Lin, G.; Xi, Y.; Li, X.; Huang, Z.; Wang, S.; Zhao, J.; Zhang, L. Development of mercaptosuccinic anchored MOF through one-step preparation to enhance adsorption capacity and selectivity for Hg(II) and Pb(II). *J. Mol. Liq.* **2020**, *317*, 113896. [CrossRef]
13. Abdullah, N.; Yusof, N.; Lau, W.J.; Jaafar, J.; Ismail, A.F. Recent trends of heavy metal removal from water/wastewater by membrane technologies. *J. Ind. Eng. Chem.* **2019**, *76*, 17–38. [CrossRef]
14. Sharma, P.R.; Sharma, S.K.; Lindström, T.; Hsiao, B.S. Nanocellulose-Enabled Membranes for Water Purification: Perspectives. *Adv. Sustain. Syst.* **2020**, *4*, 1900114. [CrossRef]
15. Singh, J.; Saharan, V.; Kumar, S.; Gulati, P.; Kapoor, R.K. Laccase grafted membranes for advanced water filtration systems: A green approach to water purification technology. *Crit. Rev. Biotechnol.* **2018**, *38*, 883–901. [CrossRef]
16. Park, J.-H.; Choi, G.-J.; Kim, S.-H. Effects of pH and slow mixing conditions on heavy metal hydroxide precipitation. *J. Korea Org. Resour. Recycl. Assoc.* **2014**, *22*, 55–56. [CrossRef]
17. Song, S.; Lopez-Valdivieso, A.; Hernandez-Campos, D.J.; Peng, C.; Monroy-Fernandez, M.G.; Razo-Soto, I. Arsenic removal from high-arsenic water by enhanced coagulation with ferric ions and coarse calcite. *Water Res.* **2006**, *40*, 364–372. [CrossRef] [PubMed]

18. Zouboulis, A.; Katsoyiannis, I. Removal of Arsenates from contaminated water by coagulation-direc filtrattion. *Sep. Sci. Technol.* **2002**, *37*, 2859–2873. [CrossRef]
19. Moussa, D.T.; El-Naas, M.H.; Nasser, M.; Al-Marri, M.J. A comprehensive review of electrocoagulation for water treatment: Potentials and challenges. *J. Environ. Manag.* **2017**, *186*, 24–41. [CrossRef] [PubMed]
20. Choumane, R.; Peulon, S. Development of an efficient electrochemical process for removing and separating soluble Pb(II) in aqueous solutions in presence of other heavy metals: Studies of key parameters. *Chem. Eng. J.* **2021**, *423*, 130161. [CrossRef]
21. Takijiri, K.; Morita, K.; Nakazono, T.; Sakai, K.; Ozawa, H. Highly stable chemisorption of dyes with pyridyl anchors over TiO_2: Application in dye-sensitized photoelectrochemical water reduction in aqueous media. *Chem. Commun.* **2017**, *53*, 3042–3045. [CrossRef] [PubMed]
22. Kumar, N.; Mittal, H.; Parashar, V.; Ray, S.S.; Ngila, J.C. Efficient removal of rhodamine 6G dye from aqueous solution using nickel sulphide incorporated polyacrylamide grafted gum karaya bionanocomposite hydrogel. *RSC Adv.* **2016**, *6*, 21929–21939. [CrossRef]
23. Gao, Z.; Bandosz, T.J.; Zhao, Z.; Han, M.; Qiu, J. Investigation of factors affecting adsorption of transition metals on oxidized carbon nanotubes. *J. Hazard. Mater.* **2009**, *167*, 357–365. [CrossRef]
24. Gusain, R.; Kumar, N.; Ray, S.S. Recent advances in carbon nanomaterial-based adsorbents for water purification. *Coord. Chem. Rev.* **2020**, *405*, 213111. [CrossRef]
25. Huang, S.; Jiang, S.; Pang, H.; Wen, T.; Asiri, A.M.; Alamry, K.A.; Alsaedi, A.; Wang, X.; Wang, S. Dual functional nanocomposites of magnetic MnFe2O4 and fluorescent carbon dots for efficient U(VI) removal. *Chem. Eng. J.* **2019**, *368*, 941–950. [CrossRef]
26. Xu, L.; Li, J.; Zhang, M. Adsorption Characteristics of a Novel Carbon-Nanotube-Based Composite Adsorbent toward Organic Pollutants. *Ind. Eng. Chem. Res.* **2015**, *54*, 2379–2384. [CrossRef]
27. Diel, J.C.; Franco, D.S.; Nunes, I.D.S.; Pereira, H.A.; Moreira, K.S.; Thiago, A.D.L.; Foletto, E.L.; Dotto, G.L. Carbon nanotubes impregnated with metallic nanoparticles and their application as an adsorbent for the glyphosate removal in an aqueous matrix. *J. Environ. Chem. Eng.* **2021**, *9*, 105178. [CrossRef]
28. Venditti, I. Engineered gold-based nanomaterials: Morphologies and functionalities in biomedical applications. a mini review. *Bioengineering* **2019**, *6*, 53. [CrossRef] [PubMed]
29. Vilela, D.; González, M.C.; Escarpa, A. Sensing colorimetric approaches based on gold and silver nanoparticles aggregation: Chemical creativity behind the assay. A review. *Anal. Chim. Acta* **2012**, *751*, 24–43. [CrossRef]
30. Burratti, L.; Bolli, E.; Casalboni, M.; de Matteis, F.; Mochi, F.; Francini, R.; Casciardi, S.; Prosposito, P. Synthesis of Fluorescent Ag Nanoclusters for Sensing and Imaging Applications. *Mater. Sci. Forum* **2018**, *941*, 2243–2248. [CrossRef]
31. Hu, X.; Liu, T.; Zhuang, Y.; Wang, W.; Li, Y.; Fan, W.; Huang, Y. Recent advances in the analytical applications of copper nanoclusters. *TrAC Trends Anal. Chem.* **2016**, *77*, 66–75. [CrossRef]
32. Zheng, J.; Nicovich, P.R.; Dickson, R.M. Highly Fluorescent Noble-Metal Quantum Dots. *Annu. Rev. Phys. Chem.* **2007**, *58*, 409–431. [CrossRef] [PubMed]
33. Burratti, L.; Ciotta, E.; De Matteis, F.; Prosposito, P. Metal Nanostructures for Environmental Pollutant Detection Based on Fluorescence. *Nanomaterials* **2021**, *11*, 276. [CrossRef] [PubMed]
34. Jarujamrus, P.; Meelapsom, R.; Pencharee, S.; Obma, A.; Amatatongchai, M.; Ditcharoen, N.; Chairam, S.; Tamuang, S. Use of a smartphone as a colorimetric analyzer in paper-based devices for sensitive and selective determination of mercury in water samples. *Anal. Sci.* **2018**, *34*, 75–81. [CrossRef] [PubMed]
35. Le Guevel, X. Recent advances on the synthesis of metal quantum nanoclusters and their application for bioimaging. *IEEE J. Sel. Top. Quantum Electron.* **2014**, *20*, 6801312. [CrossRef]
36. Wu, F.-N.; Zhu, J.; Weng, G.-J.; Li, J.-J.; Zhao, J.-W. Gold nanocluster composites: Preparation strategies, optical and catalytic properties, and applications. *J. Mater. Chem. C* **2022**, *10*, 14812–14833. [CrossRef]
37. Tong, Y.; Xue, G.; Wang, H.; Liu, M.; Wang, J.; Hao, C.; Zhang, X.; Wang, D.; Shi, X.; Liu, W.; et al. Interfacial coupling between noble metal nanoparticles and metal–organic frameworks for enhanced catalytic activity. *Nanoscale* **2018**, *10*, 16425–16430. [CrossRef]
38. Bolli, E.; Mezzi, A.; Burratti, L.; Prosposito, P.; Casciardi, S.; Kaciulis, S. X-ray and UV photoelectron spectroscopy of Ag nanoclusters. *Surf. Interface Anal.* **2020**, *52*, 1017–1022. [CrossRef]
39. Schiesaro, I.; Battocchio, C.; Venditti, I.; Prosposito, P.; Burratti, L.; Centomo, P.; Meneghini, C. Structural characterization of 3d metal adsorbed AgNPs. *Phys. E Low Dimens. Syst. Nanostruct.* **2020**, *123*, 114162. [CrossRef]
40. Vicente-Martínez, Y.; Caravaca, M.; Soto-Meca, A.; Solana-González, R. Magnetic core-modified silver nanoparticles for ibuprofen removal: An emerging pollutant in waters. *Sci. Rep.* **2020**, *10*, 18288. [CrossRef]
41. Babaladimath, G.; Badalamoole, V. Silver nanoparticles embedded pectin-based hydrogel: A novel adsorbent material for separation of cationic dyes. *Polym. Bull.* **2019**, *76*, 4215–4236. [CrossRef]
42. Pal, J.; Deb, M.K.; Deshmukh, D.K.; Verma, D. Removal of methyl orange by activated carbon modified by silver nanoparticles. *Appl. Water Sci.* **2013**, *3*, 367–374. [CrossRef]
43. Lei, X.; Li, H.; Luo, Y.; Sun, X.; Guo, X.; Hu, Y.; Wen, R. Novel fluorescent nanocellulose hydrogel based on gold nanoclusters for the effective adsorption and sensitive detection of mercury ions. *J. Taiwan Inst. Chem. Eng.* **2021**, *123*, 79–86. [CrossRef]
44. Sapuła, P.; Bialik-Wąs, K.; Malarz, K. Are Natural Compounds a Promising Alternative to Synthetic Cross-Linking Agents in the Preparation of Hydrogels? *Pharmaceutics* **2023**, *15*, 253. [CrossRef] [PubMed]

45. Jiang, G.; Wang, G.; Zhu, Y.; Cheng, W.; Cao, K.; Xu, G.; Zhao, D.; Yu, H. A Scalable Bacterial Cellulose Ionogel for Multisensory Electronic Skin. *Research* **2022**, *2022*, 9814767. [CrossRef]
46. Akter, M.; Bhattacharjee, M.; Dhar, A.K.; Rahman, F.B.A.; Haque, S.; Rashid, T.U.; Kabir, S.M.F. Cellulose-Based Hydrogels for Wastewater Treatment: A Concise Review. *Gels* **2021**, *7*, 30. [CrossRef]
47. Zainal, S.H.; Mohd, N.H.; Suhaili, N.; Anuar, F.H.; Lazim, A.M.; Othaman, R. Preparation of cellulose-based hydrogel: A review. *J. Mater. Res. Technol.* **2021**, *10*, 935–952. [CrossRef]
48. Koschella, A.; Hartlieb, M.; Heinze, T. A "click-chemistry" approach to cellulose-based hydrogels. *Carbohydr. Polym.* **2011**, *86*, 154–161. [CrossRef]
49. Guo, J.; Zhou, M.; Yang, C. Fluorescent hydrogel waveguide for on-site detection of heavy metal ions. *Sci. Rep.* **2017**, *7*, 7902. [CrossRef]
50. Choi, J.R.; Yong, K.W.; Choi, J.Y.; Cowie, A.C. Recent advances in photo-crosslinkable hydrogels for biomedical applications. *Biotechniques* **2019**, *66*, 40–53. [CrossRef]
51. Ju, H.; McCloskey, B.D.; Sagle, A.C.; Kusuma, V.A.; Freeman, B.D. Preparation and characterization of crosslinked poly(ethylene glycol) diacrylate hydrogels as fouling-resistant membrane coating materials. *J. Memb. Sci.* **2009**, *330*, 180–188. [CrossRef]
52. Kim, J.; Lee, K.-W.; Hefferan, T.E.; Currier, B.L.; Yaszemski, M.J.; Lu, L. Synthesis and Evaluation of Novel Biodegradable Hydrogels Based on Poly(ethylene glycol) and Sebacic Acid as Tissue Engineering Scaffolds. *Biomacromolecules* **2008**, *9*, 149–157. [CrossRef]
53. Yang, W.; Yu, H.; Liang, W.; Wang, Y.; Liu, L. Rapid Fabrication of Hydrogel Microstructures Using UV-Induced Projection Printing. *Micromachines* **2015**, *6*, 1903–1913. [CrossRef]
54. Burratti, L.; Ciotta, E.; Bolli, E.; Kaciulis, S.; Casalboni, M.; De Matteis, F.; Garzón-Manjón, A.; Scheu, C.; Pizzoferrato, R.; Prosposito, P. Fluorescence enhancement induced by the interaction of silver nanoclusters with lead ions in water. *Colloids Surfaces A Physicochem. Eng. Asp.* **2019**, *579*, 123634. [CrossRef]
55. Burratti, L.; Ciotta, E.; Bolli, E.; Casalboni, M.; De Matteis, F.; Francini, R.; Casciardi, S.; Prosposito, P. Synthesis of fluorescent silver nanoclusters with potential application for heavy metal ions detection in water. In *AIP Conference Proceedings*; AIP Publishing: Woodbury, NY, USA, 2019; Volume 2145, p. 020007.
56. Shang, L.; Dong, S. Facile preparation of water-soluble fluorescent silver nanoclusters using a polyelectrolyte template. *Chem. Commun.* **2008**, *9*, 1088–1090. [CrossRef] [PubMed]
57. Lu, F.; Zhou, S.; Zhu, J.J. Photochemical synthesis of fluorescent Ag nanoclusters and enhanced fluorescence by ionic liquid. *Int. J. Hydrog. Energy* **2013**, *38*, 13055–13061. [CrossRef]
58. Burratti, L.; Maranges, V.; Sisani, M.; Naryyev, E.; De Matteis, F.; Francini, R.; Prosposito, P. Determination of Pb(II) Ions in Water by Fluorescence Spectroscopy Based on Silver Nanoclusters. *Chemosensors* **2022**, *10*, 385. [CrossRef]
59. Chen, X.; Hossain, M.F.; Duan, C.; Lu, J.; Tsang, Y.F.; Islam, M.S.; Zhou, Y. Isotherm models for adsorption of heavy metals from water—A review. *Chemosphere* **2022**, *307*, 135545. [CrossRef]
60. Febrianto, J.; Kosasih, A.N.; Sunarso, J.; Ju, Y.-H.; Indraswati, N.; Ismadji, S. Equilibrium and kinetic studies in adsorption of heavy metals using biosorbent: A summary of recent studies. *J. Hazard. Mater.* **2009**, *162*, 616–645. [CrossRef]
61. Cui, G.; Li, Y.; Liu, J.; Wang, H.; Li, Z.; Wang, J. Tuning Environmentally Friendly Chelate-Based Ionic Liquids for Highly Efficient and Reversible SO 2 Chemisorption. *ACS Sustain. Chem. Eng.* **2018**, *6*, 15292–15300. [CrossRef]
62. Fuerstenau, D.; Herrera-Urbina, R.; McGlashan, D. Studies on the applicability of chelating agents as universal collectors for copper minerals. *Int. J. Miner. Process.* **2000**, *58*, 15–33. [CrossRef]
63. Shahrokhi-Shahraki, R.; Benally, C.; El-Din, M.G.; Park, J. High efficiency removal of heavy metals using tire-derived activated carbon vs commercial activated carbon: Insights into the adsorption mechanisms. *Chemosphere* **2021**, *264*, 128455. [CrossRef] [PubMed]
64. Tuomikoski, S.; Runtti, H.; Romar, H.; Lassi, U.; Kangas, T. Multiple heavy metal removal simultaneously by a biomass-based porous carbon. *Water Environ. Res.* **2021**, *93*, 1303–1314. [CrossRef]
65. Kim, W.-K.; Shim, T.; Kim, Y.-S.; Hyun, S.; Ryu, C.; Park, Y.-K.; Jung, J. Characterization of cadmium removal from aqueous solution by biochar produced from a giant Miscanthus at different pyrolytic temperatures. *Bioresour. Technol.* **2013**, *138*, 266–270. [CrossRef] [PubMed]
66. Bordoloi, N.; Goswami, R.; Kumar, M.; Kataki, R. Biosorption of Co (II) from aqueous solution using algal biochar: Kinetics and isotherm studies. *Bioresour. Technol.* **2017**, *244*, 1465–1469. [CrossRef] [PubMed]

Disclaimer/Publisher's Note: The statements, opinions and data contained in all publications are solely those of the individual author(s) and contributor(s) and not of MDPI and/or the editor(s). MDPI and/or the editor(s) disclaim responsibility for any injury to people or property resulting from any ideas, methods, instructions or products referred to in the content.

Article

How the Addition of Chitosan Affects the Transport and Rheological Properties of Agarose Hydrogels

Martina Klučáková

Faculty of Chemistry, Brno University of Technology, Purkyňova 118, 612 00 Brno, Czech Republic; klucakova@fch.vutbr.cz

Abstract: Agarose hydrogels enriched by chitosan were studied from a point of view diffusion and the immobilization of metal ions. Copper was used as a model metal with a high affinity to chitosan. The influence of interactions between copper and chitosan on transport properties was investigated. Effective diffusion coefficients were determined and compared with values obtained from pure agarose hydrogel. Their values increased with the amount of chitosan added to agarose hydrogel and the lowest addition caused the decrease in diffusivity in comparison with hydrogel without chitosan. Liesegang patterns were observed in the hydrogels with higher contents of chitosan. The patterns were more distinct if the chitosan content increased. The formation of Liesegang patterns caused a local decrease in the concentration of copper ions and concentration profiles were affected by this phenomenon. Thus, the values of effective diffusion coefficient covered the influences of pore structure of hydrogels and the interactions between chitosan and metal ions, including precipitation on observed Liesegang rings. From the point of view of rheology, the addition of chitosan resulted in changes in storage and loss moduli, which can show on a "more liquid" character of enriched hydrogels. It can contribute to the increase in the effective diffusion coefficients for hydrogels with higher content of chitosan.

Keywords: chitosan; copper; diffusion; Liesegang pattern; solubility product; immobilization

Citation: Klučáková, M. How the Addition of Chitosan Affects the Transport and Rheological Properties of Agarose Hydrogels. Gels 2023, 9, 99. https://doi.org/10.3390/gels9020099

Academic Editors: Luca Burratti, Paolo Prosposito and Iole Venditti

Received: 22 December 2022
Revised: 10 January 2023
Accepted: 18 January 2023
Published: 23 January 2023

Copyright: © 2023 by the author. Licensee MDPI, Basel, Switzerland. This article is an open access article distributed under the terms and conditions of the Creative Commons Attribution (CC BY) license (https://creativecommons.org/licenses/by/4.0/).

1. Introduction

Chitosan is a hydrophilic polyelectrolyte obtained by the deacetylation of chitin, which is widely spread among marine and terrestrial invertebrates and in lower forms of the plant kingdom. It has unique properties as biocompatibility, bioactivity, and biodegradability [1–4]. It has low toxicity, hemostatic potential, and a good ability to form films [5–8]. It can be used in many industrial and biomedical applications, e.g., [9–15]. Chitosan is one of the most efficient adsorbents, receiving considerable interest for heavy metals removal due to its very good adsorption capacities and relatively low cost [4,7,8,16]. Its adsorption properties are attributed to high hydrophilicity (due to OH groups), primary amino groups with high activity, and flexible structure of polymer chains [17]. Major sources of heavy metals in nature are mainly from industry, mining, metal plating, corrosion, and electronic device manufactures. Waste streams from these workings may contain considerable amount of toxic and polluting metal ions. Copper is one of the most widespread metal contaminants in nature [8]. Copper is extensively used in the electrical industry, production of fungicides, and anti-fouling paints. Additionally, it is an essential trace element for people, although it can be harmful when a large dosage is ingested [4,18,19]. It is not biodegradable and can accumulate in living organisms. It can cause various toxicological effects, e.g., gastrointestinal irritation, hypertension, sterility, and intellectual disability [19–21]. The adsorption of metal ions is a widely used method for removing metal ions from nature. The use of chitosan is also widely used for these purposes [16,19,21–25], although its adsorption capacity for copper shows significant differences among published studies [16]. Chitosan can be physically and chemically modified to improve their mechanical properties [4]. They can be used in the forms of hydrogel beads [4,16,18], resin [26], nanoparticles [12], and membranes [8]. Detail analysis of adsorption of copper (and other metal ions) on chitosan and chitosan-based materials can be found in [22–25,27].

As mentioned above, many authors deal with chitosan as the material has a good affinity with different substances, including metal ions. However, published papers are mainly focused on the adsorption and the use of chitosan as an important constituent of adsorbents. The studies on the mobility of metal ions and diffusion processes in chitosan materials are scarce. Some authors observed an influence of intraparticle diffusion on the adsorption kinetics of metal ions on chitosan [4,28–31]. In general, the adsorption can include additional steps of external and intraparticle diffusion: film or external diffusion, pore diffusion, and surface diffusion. The external diffusion is usually not the major limiting step in the adsorption, so the attention is mainly focused on simple intra-particular diffusion models [4,27–29]. All of the above-mentioned approaches relate to the adsorption processes. The diffusion in chitosan materials as the main topic is studied only in several works [32–34]. In ref. [32], chitosan membranes were prepared, then characterized from point of view of the permeation of KCl. Concentration changes in the donor and acceptor compartments were determined on the basis of conductivity measurements. It means that measured values of conductivity included influence of all present ions. No diffusion coefficient was calculated. It was stated that results can be influenced by changes in the membrane structure during experiments and the occurrence of osmosis. Krajewska et al. [33] studied diffusion of 15 metal ions through several types of hydrogel chitosan membranes. The diffusive permeability coefficients of metal ions were strongly influenced by the treatment of membrane. This value was determined as 1.8×10^{-11} $m^2\,s^{-1}$ for the diffusion through pure chitosan membrane, and decreased for the membrane treated by glutaraldehyde to 1.6×10^{-12} $m^2\,s^{-1}$. The permeability coefficient is a product of the solute solubility coefficient in the membrane (a dimensionless number expressed by ratio of solute content in membrane and in solution), and the solute diffusion coefficient. The salt diffusive permeability coefficients were obtained from the recorded steady states of salt diffusion through the membranes. It was found that, except for the diffusion of Cu(II) and Ni(II) ions through chitosan membranes, the steady states were established within a short time. For the diffusion of Cu(II) and Ni(II) through chitosan membranes, time lags between 10 and 20 min were observed. These results were attributed to high affinity of chitosan to these metal ions.

Yoshida et al. [34] dealt with the diffusion of potassium sorbate through chitosan and palmitic acid chitosan films. They obtained diffusion coefficient in the range from 1.22×10^{-13} to 1.16×10^{-12} $m^2\,s^{-1}$ in the dependence of membrane type, and used a mathematical model. Modrzejewska et al. [35] proposed the sorption kinetic model, which combine chemical reaction and intraparticle diffusion. The chemical reaction was characterized by the rate constant (1.69×10^{-5} s^{-1}), and by the diffusion coefficient (6.23×10^{-10} $m^2\,s^{-1}$). This value is only slightly higher than the diffusion coefficient of chitosan (4.0×10^{-10} $m^2\,s^{-1}$) determined on the basis of laser refraction experiment [36].

As described above, published studies are mainly focused on the adsorption processes [4,27–31]. Some of them deal with the diffusion through chitosan membranes and films [32–35]. Our approach is different. Since chitosan is considered as the material with high potential to immobilize toxic metal ions, it is desirable to investigate the interactions in detail. This study is focused on the reactivity mapping of chitosan in combination with the passing of diffusing metal ions through material containing chitosan as the active substance. In order to distinguish between the diffusivity through reactive and non-reactive medium, agarose hydrogel was chosen as basic non-reactive material. Agarose hydrogel proved to be the suitable medium for the investigation of diffusion of different substances. [37–39]. It can be enriched by active substance which can interact with diffusing particles and partially immobilize them. Therefore, the interactions of diffusing metal ions with active substance incorporated in the hydrogel can be studied in their motion. Thus, the main aim of this study is the detailed analysis of these two parallel processes.

2. Results and Discussion

2.1. Diffusion Experiments

The main goal of this study was to investigate the effect of chitosan on the mobility of copper(II) ions in agarose hydrogel. The diffusion experiments presented in this paper follow previous works which investigated the barrier effects of other biopolymer-humic acid on the transport of copper(II) ions by means of different methods [37,38,40–42]. We expected a similar effect of chitosan on the migration of metal ions as in the case of adding humic acids to hydrogel caused by interactions between copper(II) ions and biopolymer because of its chelating ability. Interactions of chitosan with metal ions can be realized due to its hydroxyl and amino groups, which participate in their adsorption processes [4,12,16,17,23,35]. The functional groups represent coordination and binding sites in chitosan structure. In general, other approaches to interactions between copper(II) ions and chitosan can be discussed. The first approach proposes the coordination site formed by four amino groups, another considers two amino and two hydroxy groups [16]. The chelation can be also realized by means of functional groups from the same chitosan chain, different chitosan chains, or by one amino group and one hydroxy group in coordination with two water molecules [12,43,44]. The ion exchange can participate in the interactions of chitosan with metal ions [26,45].

In Figure 1a, the concentration profiles of copper(II) ions in hydrogel enriched by chitosan (0.5 mg g^{-1}) is shown two different times. The increase in concentration (at a given distance from the interface) with prolongation of diffusion experiments was observed. A similar increase in the time development of concentration profiles was observed for pure agarose hydrogel. In Figure 1b, the comparison of concentration profiles for pure agarose hydrogel and hydrogel enriched by 1 mg g^{-1} of chitosan is shown. The concentration profiles were fitted by a mathematical model based on Fick's laws [46–48] and the following assumptions. In beginning, the cuvettes with hydrogel (free of copper(II)ions) were covered by the donor solution with initial concentration c_0. Hydrogels can be considered as semi-infinite mediums. It means that the hydrogel closed to bottom of cuvette remained free of copper(II) ions during whole diffusion experiment. The concentration changes of diffusing metal ions at the interface between hydrogel and donor solution are negligible (the maximal calculated decrease in the amount of copper(II) ions in donor solution is 0.5 %). It means that a concentration at the interface c_s can be considered as constant. The assumptions can be expressed as initial and boundary conditions in Table 1. In the case of non-reactive hydrogel, the time development of the concentration profile in hydrogel can be thus be expressed as Equation (1):

$$c = c_s \operatorname{erfc} \frac{x}{2\sqrt{D_h t}} \qquad (1)$$

where t is time, x is distance from interface, c is concentration of copper(II) ions, c_s is concentration at interface, and D_h is the diffusion coefficient of copper(II) ions in hydrogel [47,48]. The total diffusion flux m_t then depends on the square root of time according to Equation (2):

$$m_t = 2c_s \sqrt{\frac{D_h t}{\pi}} \qquad (2)$$

Table 1. Initial and boundary conditions of diffusion experiments.

Time t	Distance x	Concentration c
$t = 0$	$x > 0$	$c = 0$
$t > 0$	$x = 0$	$c = c_s$
$t > 0$	$x \to \infty$	$c = 0$

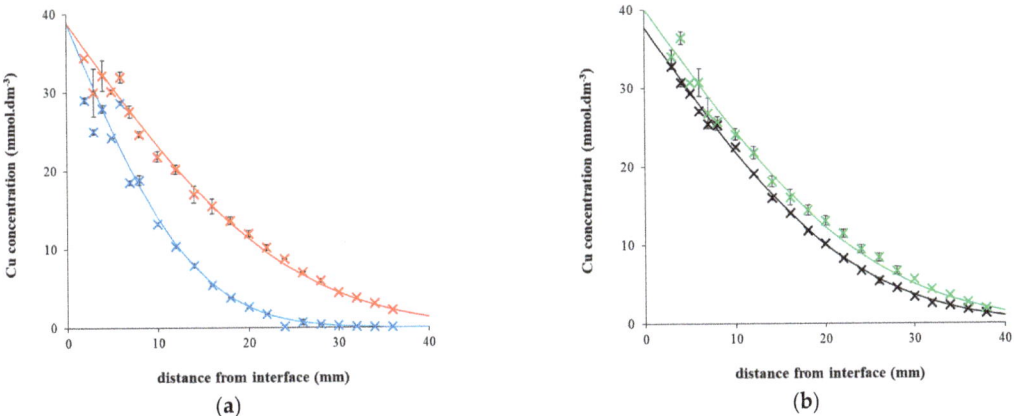

Figure 1. (a) Concentration profiles of copper(II) ions in agarose hydrogel enriched by chitosan (0.5 mg g^{-1}) after 24 h (blue) and 72 h (red). (b) Concentration profiles of copper(II) ions in pure agarose hydrogel (black) and hydrogel enriched by chitosan (1 mg g^{-1}; green) after 72 h. Experimental data are fitted by Equation (1).

Concentration profiles in Figure 1 are fitted by Equation (1). The diffusion coefficient used for the fitting was determined on the basis of Equation (2) from the slope of dependency shown in Figure 2.

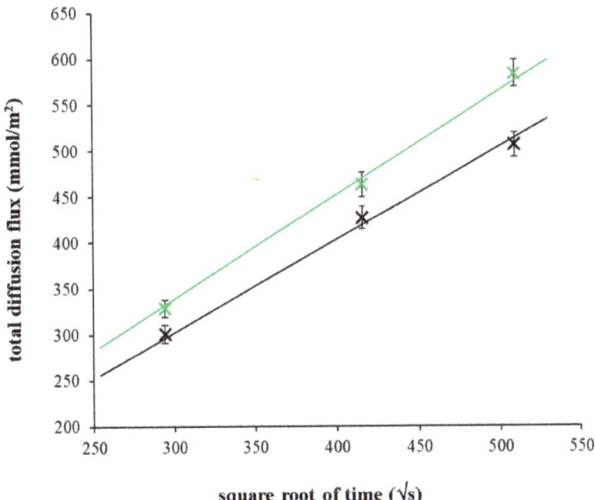

Figure 2. Experimental data fitting for the calculation of effective diffusion coefficients. The dependence of total diffusion flux of copper(II) ions in pure agarose hydrogel (black) and hydrogel enriched by chitosan (1 mg g^{-1}; green) on square root of time. Experimental data are fitted by Equation (2).

The diffusion coefficient D_h is valid for the diffusion in non-reactive hydrogel and its value is affected by the pore structure of hydrogel. If the particles diffuse in hydrogel, their motion is influenced by size of pore volume accessible for the diffusion. It can be expressed as the porosity φ; meaning that a portion of free volume joins the bulk volume of hydrogel. The diffusion in pores, which are not straight, also requires a longer diffusion path in comparison with the diffusion in homogeneous medium as, e.g., water [36–39].

The longer diffusion path can be characterized by tortuosity factor τ which is the ratio of squares of a real (effective) diffusion path in pore L_{ef} and a macroscopic distance L:

$$\tau = \left(\frac{L_{ef}}{L}\right)^2 \qquad (3)$$

Thus, the diffusion coefficient in non-reactive hydrogel can be calculated as the product of the structure factor of hydrogel (μ), and the diffusion coefficient of copper(II) ions in water (D):

$$D_h = \mu D = \frac{\varphi}{\tau} D. \qquad (4)$$

If the hydrogel is enriched by an active substance able chemically interact with diffusing particles, the determined value of diffusion coefficient also includes the influence of the interactions. If we assume a simple equilibrium between bound immobilized particles (concentration c_{im}) and free movable particles (concentration c_{free}), an apparent equilibrium constant K as the ratio between these two forms of particles can be included in the effective diffusion coefficient D_{ef} [38,39,46]:

$$D_{ef} = \frac{D_h}{\frac{c_{im}}{c_{free}} + 1} = \frac{D_h}{K+1} = \frac{\mu D}{K+1}. \qquad (5)$$

The symbol D_h in Equations (1) and (2) can be replaced by the effective diffusion D_{ef}, and the modified equations can be used in the case of the hydrogel containing active substance. The apparent equilibrium constant then can be calculated as:

$$K = \frac{D_h}{D_{ef}} - 1 \qquad (6)$$

The values of diffusion coefficients and concentration at interfaces for pure agarose hydrogels and hydrogels containing different amounts of chitosan are listed in Table 2.

Table 2. Values of concentration of copper(II) ions at interfaces and diffusion coefficients determined for studied hydrogels.

Chitosan Content (mg g^{-1})	Concentration at Interface (mmol dm^{-3})	Diffusion Coefficient (m^2 s^{-1})
0	37.73 ± 0.62	(6.25 ± 0.06) × 10^{-10}
0.2	38.48 ± 0.86	(6.03 ± 0.13) × 10^{-10}
0.5	38.86 ± 1.15	(6.98 ± 0.24) × 10^{-10}
1	40.15 ± 1.12	(7.41 ± 0.22) × 10^{-10}

It can be seen that both quantities depend on the content of chitosan in hydrogel. The concentration of Cu(II) ions in the interface gradually increases from pure agarose hydrogel up to the hydrogel with the highest chitosan content. The dependence of diffusion coefficient on the chitosan content is more complicated. According to Equation (5), the effective diffusion coefficient determined for hydrogel enriched by an active substance should be lower than the diffusion coefficient of the same diffusing particles in non-reactive hydrogel and in water. All calculated diffusion coefficients in Table 1 are lower than diffusion coefficient of Cu(II) ions in water [49]. This means that the pore structure of hydrogel decelerated the diffusion rate as a result of smaller pore volume accessible for the diffusion and longer diffusion paths in comparison with the diffusion in homogeneous medium as, e.g., water [36–39]. On the other hand, interactions between Cu(II) ions and chitosan in more enriched hydrogels had opposite effect than it was expected.

The ratios between values obtained for hydrogels containing chitosan and pure agarose hydrogel are shown in Figure 3. We can see that the ratio of diffusion coefficients is less than

1 only for the lowest chitosan content; other ratios are higher than 1. Effective diffusion coefficient of copper(II) ions in hydrogels enriched by chitosan were determined in the range of $(6–7.5) \times 10^{-10}$ m^2 s^{-1}. The values obtained in this work are similar to the diffusion coefficient determined for copper(II) ions in water (7.1×10^{-10} m^2 s^{-1}) [48,49]. Kyzas et al. [48] determined diffusion coefficients of copper(II) ions in cross-linked chitosan and grafted chitosan derivatives with poly(acrylamide) and poly(acrylic acid). They used the model of swollen spherical particles combining adsorption and diffusion processes. Obtained values were 3.64×10^{-10} m^2 s^{-1}, 2.52×10^{-10} m^2 s^{-1}, and 9.88×10^{-10} m^2 s^{-1}. Their results were in the same magnitude as our values. The authors also observed an increase in the diffusivity in chitosan grafted with poly(acrylic acid) (comparing with the diffusion in water). Zhang et al. [50] published diffusion coefficients in hydrogels used in DGT and DET techniques. The values were 6.20×10^{-10} m^2 s^{-1} for pure agarose hydrogel and 6.38×10^{-10} m^2 s^{-1} for water. The diffusion coefficient of copper(II) ions in hydrogel beads based on chitosan was determined as 6.23×10^{-10} m^2 s^{-1} [35]. Thus, our results agree with some previous works and determined diffusion coefficients have values characteristic for hydrogel samples. In contrast, our results do not agree with hypothesis expressed by Equations (5) and (6), which include a simple equilibrium between immobilized and free movable particle.

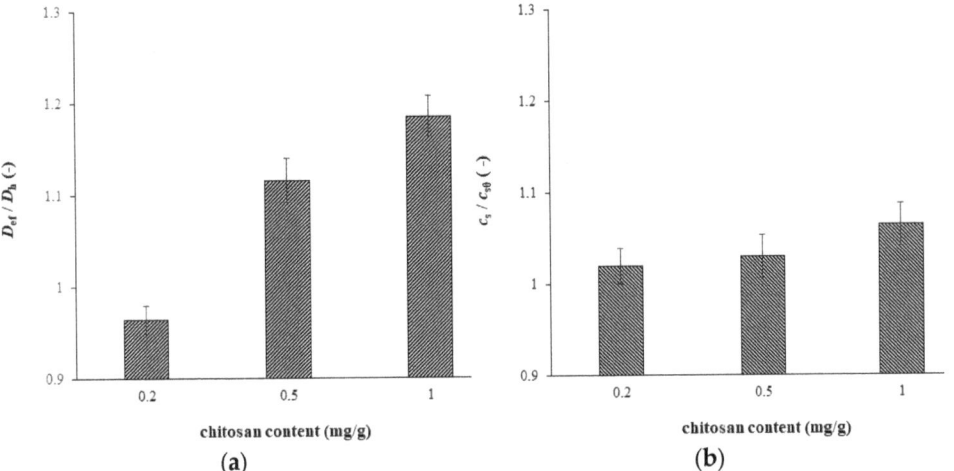

Figure 3. The ratios between effective diffusion coefficient D_{ef} obtained for hydrogels containing chitosan and D_h obtained for pure agarose hydrogel (**a**), and ratios between concentration of copper(II) ions at interface c_s obtained for hydrogels containing chitosan, and c_{s0} obtained for pure agarose hydrogel (**b**).

Thus, an explanation can be found in the more complex mechanism of interactions between copper(II) ions and chitosan. The affinity of chitosan to metal ions is well-known. Published studies are focused mainly on adsorption batch experiments, e.g., [17,19,25,27]. Some works combined adsorption models with intraparticle diffusion [4,35,51] or deals with the diffusion through chitosan membranes and films [8,34]. Our approach is different. We are focused on the interaction directly realized in the motion of copper(II) ions in hydrogel containing chitosan as an active substance. The mechanism of interactions is relatively complex, and influenced by many factors as heterogeneity of chitosan and ionic environment of aqueous solution.

As mentioned above, the interactions of chitosan with metal ions include adsorption, ion exchange, and chelation. The most important reaction and coordination sites are amino and hydroxyl groups [4,31,35,51,52]. The interaction between metal ions and chitosan is comprised of more reaction steps for the protonation or dissociation of functional groups, formation of chelate rings, and others. Modrzejewska [52] described several models of copper-chitosan complexes, and discussed its probability, binding possibility, and structure. On the basis of her own data and published results, she stated that the complexation of copper(II) ions is more effective for oligomers, and a polymerization degree is required. She proposed mechanisms of complexation for different functional groups of chitosan (NH_2 and OH). Guzman et al. [53] pointed to many differences in the abundant literature about metal binding on chitosan. They stated that an important underestimated parameter of the binding mechanism is the metal ion speciation (affected by pH). Similarly, this influence was discussed by Guibal et al. [27,28]. Published results showed that the interactions between chitosan and copper(II) ions cannot be described by a simple equilibrium between immobilized and free movable metal ions, and an apparent equilibrium constant K in Equations (5) and (6). The mechanism not only includes one reversible reaction, but it can be comprised of more consecutive reactions in the interactions can form more intermediates and final products (including, e.g., eliminating of ions). Nevertheless, the ratio between the diffusion coefficient in pure agarose hydrogel (D_h), and the hydrogel enriched by chitosan (D_{ef}) represents the important characteristics of interactions and their influence on the mobility of metal ions in hydrogel.

In addition, the interaction of copper(II) ions with chitosan contained in the hydrogel can result in insoluble or slightly soluble products. The products can be observed as colored layers shown in Figure 4. We can see that the Liesegang patterns [54–56] are stronger with higher amount of chitosan in hydrogel. In contrast, no colored layers were observed in pure agarose hydrogel. The Liesegang phenomena are probably the reason why the measuring error of concentration profiles measured in hydrogels enriched in chitosan is higher in the region closed to the interface (Figure 1). The formation of Liesegang structures in systems containing chitosan was studied in several works [57–60]. The Liesegang pattern occurs when a precipitation reaction is coupled to the mass transport of reagents in hydrogel [61]. The phenomenon is usually described by means of the spacing law assuming that the precipitate zones are members of a geometrical series and the time law expressing that the zones are directly proportional to square root of time [56,57,60]. Simultaneously, the chemical mechanism is described by simple model comprised of two consecutive reaction steps: the formation of soluble intermediate (I), and its transformation to precipitate (P): $A_{(aq)} + B_{(aq)} \rightarrow I_{(aq)} \rightarrow P_{(s)}$. Although this model is widely used, the mechanism is not able to express the mechanism of interaction between metal ions and chitosan. The first "imperfection" of the model is an absence of an equilibrium between precipitated and dissolved forms of metal-chitosan complexes. The periodic precipitation occurs when particles (ions) dissolved in water diffuse through hydrogel and can form a weakly soluble substance (complex, salt) with other substance containing in hydrogel. The hydrogel medium is important to prevent the sedimentation of formed precipitate [61]. In the case of ionic reaction, the soluble salt is formed in the initial step as the intermediate, and its transformation to precipitate proceeds if the product of concentrations of reacting ions (e.g., $[A^+] \times [B^-]$) exceeds the so-called solubility product K_{sp} [61,62]. The solubility product represents the simplest theoretical approach to the weakly soluble substances. It considers precipitate as salt, which can partially dissociate: $AB_{(s)} \leftrightarrow A^+_{(aq)} + B^-_{(aq)}$. The solubility product is then defined as the product of activities of both ions in its saturated solution. The activities can be (under certain circumstances) replaced by their concentrations [63]:

$$K_{sp} = a_{A^+} \cdot a_{B^-} \cong [A^+][B^-] \quad (7)$$

The value of K_{sp} is constant under given temperature and pressure. The solubility product is in fact a special type of equilibrium (dissociation) constant. Its value includes only activities (concentrations) of dissociated ions because the chemical potential of solid phase is constant and can be included in the equilibrium constant K_{sp}. This is formally expressed as its unit activity. In the case of dilute solution, the activities equal to concentrations can be supposed.

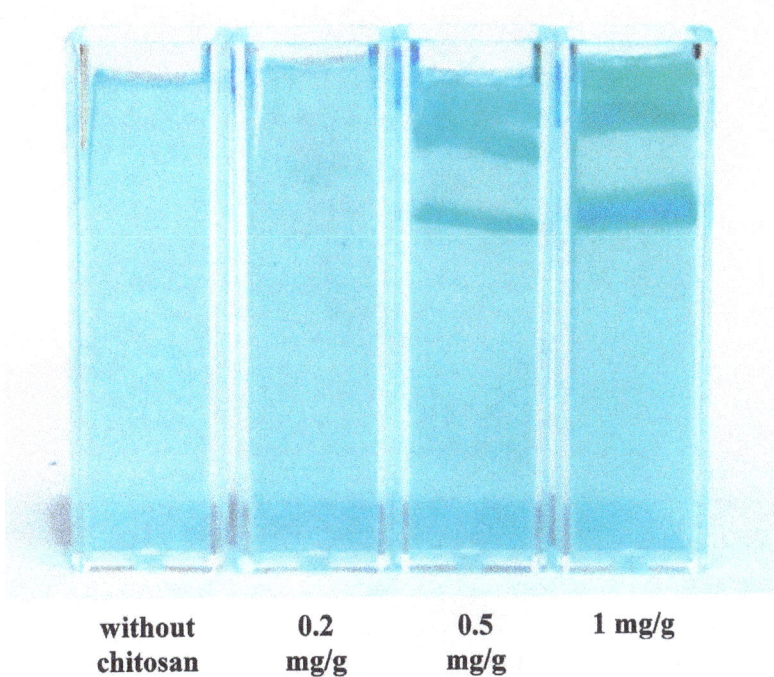

Figure 4. Formation of Liesegang patterns in agarose hydrogel enriched by chitosan.

The described simple dissociation model thus represents the reversible reaction characterized by the equilibrium constant K_{sp}. As mentioned above the model of chemical mechanism of Liesegang phenomena comprised of two consecutive reaction steps is lacking in an equilibrium between dissolved particles and precipitate. On the other hand, the equilibrium between solid salt and their dissolved ions is lacking in a soluble intermediate, and is too simple to characterize the Liesegang phenomenon, and the interactions between copper(II) ions and chitosan in studied hydrogels. Nevertheless, our findings showed that it is necessary to include reverse reactions and equilibrium between solid and dissolved particles into the reaction mechanism. In general, the above-described mechanism can be modified in the model of two consecutive reaction steps, which are both reversible reactions: $A_{(aq)} + B_{(aq)} \to I_{(aq)} \to P_{(s)}$. Both reaction steps can be characterized by its own equilibrium constant. This model can be re-written in the ionic form as: $A^{+}_{(aq)} + B^{-}_{(aq)} \leftrightarrow AB_{(aq)} \leftrightarrow AB_{(s)}$. A direct equilibrium between precipitate and dissociated ions is not involved here. It is an open-ended question if the solid precipitate can dissociate and form dissolved ions without the formation of dissolved intermediate. It was reported that some precipitation reactions are not governed by solubility product K_{sp}, but by the threshold of aggregation, which is related to the nucleation threshold concentration of $AB_{(aq)}$ intermediate (c^*) [61,62,64]. One example is the reaction system, where the precipitate can be dissolved in an excess of one reagent. In these systems, the mechanism also consists of two consecutive reaction steps: the precipitation $A_{(aq)} + B_{(aq)} \to P_{(s)}$, and the dissolution of precipitate accompanied by the formation of its complex with the dissolved reagent $P_{(s)} + A_{(aq)} \to P\text{-}A_{(aq)}$. This type of reaction proceeding diffusion through hydrogel can result in a seemingly moving precipitation zone [64,65].

The combination of precipitation reaction with mass transport of a reagent through hydrogel containing active substance can lead to more complex reaction mechanism involving reverse reactions and potential redissolution of precipitate. In the case of known reaction mechanism, we can modify Fick's law and add the reaction rate r into the equation for non-stationary diffusion.

$$\frac{\partial c}{\partial t} = D_h \frac{\partial^2 c}{\partial x^2} + r \tag{8}$$

Then, we can express the rate of decrease in the concentration of diffusing particles according to the reaction mechanism. In the simplest case (reaction A + B without reverse reaction), the reaction rate r can be expressed as the product of rate constant k, concentration of diffusing particles and concentration of active substance in hydrogel. The problem is that Equation (8) is not directly (analytically) soluble. An analytical solution can only be obtained for the simple equilibrium between free movable and immobilized particles (see above). As it was found, our results are not in agreement with this simple model. Some authors derived more complex mathematical models for the description of Liesegang phenomenon, and used numerical methods for their solution [56,61,64]. Iszák and Lagzi [56] suggested a new universal model for the formation of Liesegang patterns based on the time dependent quantities of the precipitate. Saad et al. [64] stated that Liesegang systems with re-dissolving reaction exhibit deterministic chaos in the dynamic oscillations of the numbers of bands. Nabika et al. [61] reviewed existing models of the Liesegang phenomenon including different pre- and post-nucleation scenarios. They discussed also the model including three consecutive processes (reaction, nucleation, and aggregation): $A_{(aq)} + B_{(aq)} \rightarrow I_{(aq)} \rightarrow I_{(s)} \rightarrow P_{(s)}$. The model calculates with three different "threshold" constants, rate constants, and diffusion coefficients.

As can be seen, the situation in the hydrogel systems in which the diffusion is accompanied by the chemical reaction and precipitation is complicated. In most cases, suggested mechanisms and models involved more consecutive steps without reverse reactions. In contrast, they assumed a supersaturation or exceeding of the solubility product and different "threshold" constants which can be considered as equilibrium constants. The aim of this study is not to solve the mechanism of Liesegang phenomenon; it is discussed in connection with the effective diffusion coefficient. In beginning, we assumed the simple equilibrium between free diffusing copper(II) ions and their immobilized form, but this equilibrium should lead to the decrease in diffusion coefficient, which was not confirmed. It is a question if the decrease in diffusion coefficient obtained for hydrogel containing the lowest amount of chitosan means a different mechanism of Liesegang phenomena. According to Equation (6), the apparent equilibrium constant K should be equal to 3.72×10^{-2}. It is relatively small value showing on very low degree of immobilization. Since the donor solutions with the same initial concentration were used in our experiments, the differences between diffusion coefficients determined for pure and enriched hydrogels should be caused by the different content of chitosan.

Some published models of Liesegang mechanism assume the re-dissolving of precipitate caused by an excess of diffusing ions. In contrast, the content of active substance (chitosan) incorporated in hydrogels changed in our experiments obtained a "break" in the ratio between D_{ef} and D_h (see Table 2 and Figure 3); thus, it probably cannot be caused by changing the mechanism from a simple immobilization equilibrium into more complex mechanism involving the re-dissolution. As mentioned above, we can define two deficiencies: an absence of reverse reactions (and related equilibriums) in proposed mechanisms, and too many parameters in published models, which can only be solved numerically. The other problem is that our results cannot be explained by the re-dissolving because the diffusivity of metal ions is dependent on the content of chitosan "fixed" in hydrogel. We have also to take into consideration that the decrease in diffusion coefficient is not necessarily connected with the decrease in the diffusion rate. The rate of diffusion is strongly dependent on the concentration gradient, and it increases with the immobilization of metal ions. The reason is that if the copper(II) ion is immobilized by chitosan, it disappears from the solution and cannot participate on the diffusion. The resulting diffusion rate is a combination of both influences. Although the effective diffusion coefficient D_{ef} cannot by simply expressed by means of the apparent equilibrium constant K, we believe that its value can be considered as a parameter characterizing degree of interactions between diffusing metal ions and chitosan as active substance incorporated in hydrogel. In the given conditions, a more precise definition of D_{ef} is not possible. Different published models and mechanisms require numerical methods for solution and modified Fick´s law, Equation (8), together with initial and boundary conditions, defined on the basis of the experimental arrangement and selected mechanism, cannot provide an analytical solution. Therefore, we are not able to directly express D_{ef} as a function of parameters characterizing proceedings chemical reactions (equilibrium and rate constants).

2.2. Rheological Experiments

The diffusion of metal ions in agarose hydrogel is influenced by two basic factors: their interactions with chitosan as active substance (if incorporated), and the structure of hydrogel. It is not easy to distinguish between the effect of hydrogel structure and the effect of their reactivity. It is well-known that diffusivity of spherical particles is dependent on their size (radius r) and viscosity of the diffusion medium η [46,47,63,66]:

$$D = \frac{k_b T}{6\pi \eta r} \quad (9)$$

where k_b is Boltzmann constant and T is temperature. Equation (9) can be used in the case of Newtonian liquid medium characterized by one value of dynamic viscosity η at given temperature T. Rheological behavior of hydrogels is usually more complex [3,37,66]. They can be considered as viscoelastic substances combining behavior of viscous liquids and elastic materials. Therefore, their rheological behavior cannot be characterized by one value of viscosity. Our rheological measurements provided information about the viscoelasticity of the studied hydrogels and changes caused by the addition of chitosan.

Figure 5 shows the dependencies of the storage and loss moduli on the frequency. The strain (0.1 %) was chosen from linear viscoelastic region, where the moduli are strain independent. The storage modulus G' proportional to the elastic properties increased with increasing oscillation frequency for both types of hydrogels. As can be seen, its values are much higher for pure agarose hydrogel. They decreased gradually with increasing content of chitosan (not shown) up to approximately 60% of their initial values. It means that the addition of chitosan resulted in more liquid character of hydrogel and the pure agarose hydrogel is the most resistant to mechanical stresses. The frequency dependencies of loss modulus G'' had different shapes. While its values measured for hydrogel containing chitosan are approximately constant up to 0.1 Hz, then strongly increases, the dependence obtained for pure agarose hydrogel had a flat maximum around 0.2 Hz.

Values obtained for pure agarose hydrogel are much lower in comparison with hydrogels containing chitosan which confirmed its high resistance to stresses. The frequency dependence of ratio between both moduli had, for pure agarose hydrogel, am identical shape as the loss modulus (Figure 6). A maximum was also observed for hydrogel enriched with chitosan. The "most liquid character" of chitosan was achieved at lower oscillation frequency 0.14 Hz. In contrast to pure agarose hydrogel, the ratio then decreased up to 0.63 Hz, and increases at higher frequencies. The highest value of G''/G' ratio was comparable with the maximum. The complex viscosity was lower for hydrogels containing chitosan and decreased gradually with increasing oscillation frequency for all studied hydrogels.

There was no cross-point of storage and loss moduli was observed, and the elastic character predominated at all frequencies for all studied hydrogels. Our results showed that the rheological behavior of hydrogels was changed by the addition of chitosan. The behavior of enriched hydrogels shifted gradually to more liquid character as the chitosan content increased. There was no break in rheological properties when its addition was observed. All changes were gradual. As can be seen, the complex viscosity decreases with frequency gradually, the dependence is linear in logarithm axis (Figure 7). This means that the changes in rheological behavior of hydrogels cannot directly result in the break in the diffusivity of copper(II) ions described above. On the other hand, the transport properties of hydrogels were affected by a combination of two influences–rheological character of hydrogels and their reactivity (if contained chitosan as active substance). The hydrogels containing chitosan had lower abilities to resist mechanical stresses, which can be connected with their higher permeabilities. This effect can prevail over the effect of interactions between chitosan and copper(II) ions. The shifting to more liquid character for hydrogels containing chitosan can resulted in their higher permeability connected with higher effective diffusion coefficients obtained for more enriched hydrogels. The effect of chemical interactions in enriched hydrogels was strong and visually observable. On the other hand, the contents of chitosan in hydrogels were low. It means that the amounts of active sites in hydrogels were limited and the portion of copper(II) ions which can by immobilized. According to the value of apparent equilibrium constant K determined in previous chapter, only less than 4% of copper(II) ions can be immobilized.

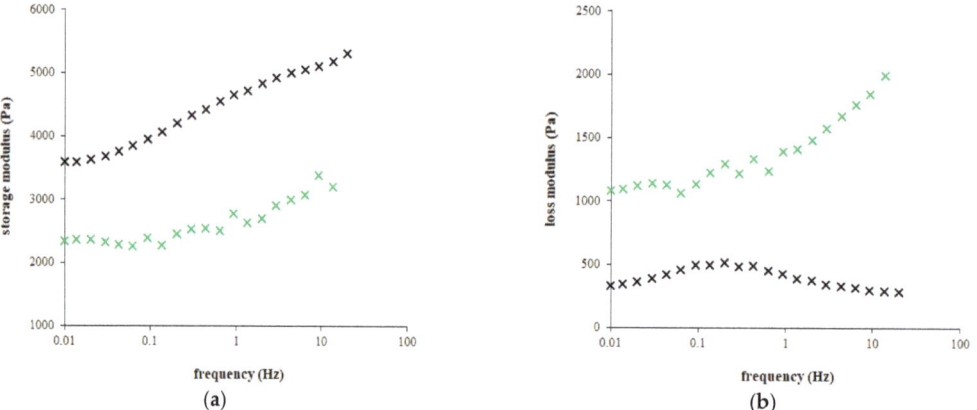

Figure 5. Storage modulus (**a**) and loss modulus (**b**) measured for pure agarose hydrogel (black) and hydrogel enriched by chitosan (1 mg g^{-1}; green). Measurement was realized in linear viscoelastic region (strain 0.1 %).

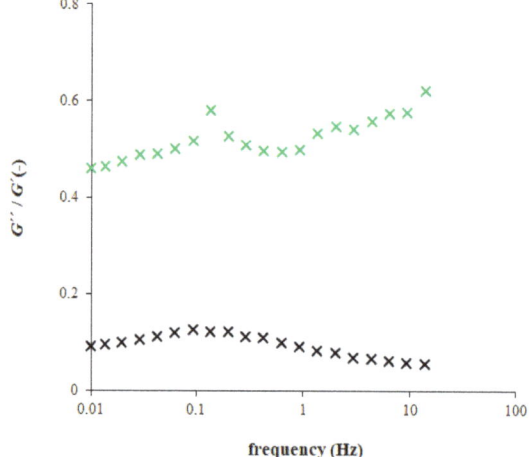

Figure 6. The ratio between loss (G'') and storage (G') moduli (pure agarose hydrogel–black; hydrogel enriched by chitosan (1 mg g^{-1})-green).

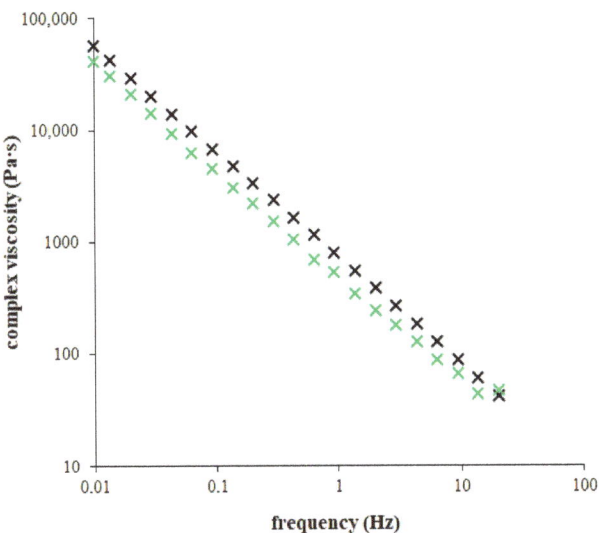

Figure 7. Complex viscosity measured for pure agarose hydrogel (black) and hydrogel enriched by chitosan (1 mg g^{-1}; green). Measurement was realized in linear viscoelastic region (strain 0.1 %).

3. Conclusions

Despite the contents of chitosan in the hydrogels were very low, its addition resulted in changes in their viscoelasticity and permeability. The changes were influenced by two contrary effects. Although the storage modulus was higher than the low one and the elastic character predominated for all studied hydrogels, the addition of chitosan caused the hydrogels to become more liquid, and, therefore, more permeable for diffusing particles. In contrast, the interactions of copper(II) ions in chitosan and formation of precipitate caused the decrease in free movable ions in hydrogel and changes in the effective diffusion coefficient. The value of D_{ef} included both the influence of hydrogel structure and the effect of immobilization of copper(II) ions. These contrary effects influenced final value of diffusivity in varying degrees according to the given content of chitosan. Its low content resulted in the decrease in the value of D_{ef}. We can believe that the effect of immobilization of copper(II) ions prevailed above the changes in theological character of hydrogel and its "more liquid character" can be considered less strong. Higher amounts of chitosan in hydrogels shifted their rheological behavior more to liquid, and the permeability of hydrogels were much higher than the effect of chemical interactions between copper(II) ions and chitosan. Although the simple model of the equilibrium between immobilized and free copper(II) ions cannot express the mechanism of the interactions, we can state that the portion of immobilized metal ions is relatively low. It means that the possibility to immobilize metal ions is limited by the content of chitosan. The diffusion of copper(II) ions in hydrogels enriched by chitosan can thus be accelerated by the more liquid character of hydrogel, and the increase in concentration gradient or the diffusion, which can be suppressed by the decrease in the diffusion coefficient as the effect of chemical reaction.

4. Materials and Methods

4.1. Chemicals

Chitosan (medium molecular weight), agarose (routine use class) and copper(II) sulfate (p.a.) were purchased from Sigma Aldrich (St. Luis, MO, USA). Acetic acid for the preparation of chitosan solution was purchased from Lachner (Neratovice, Czech Republic).

The exact molecular weights of chitosan and agarose were determined by means of size exclusion chromatography, coupled with multiangle static light scattering, differential refractive index, and UV/VIS detection (SEC chromatographic system from Agilent Technologies, detectors from Wyatt Technology). The exact molecular weights were 251 ± 4 kDa for chitosan and 146 ± 3 kDa for agarose.

Deacetylation degree of chitosan was determined by potentiometric titration described by Garcia et al. [67]. The degree was determined as $83.8 \pm 0.2\%$ mol.

4.2. Preparation of Hydrogels

The preparation of hydrogels was based on the thermo-reversible gelation of agarose solution described in previous works [36–39,66]. Agarose hydrogel gelatinized from the solution of agarose in water. Agarose content in hydrogel was 10 mg g^{-1}. The mixture was slowly heated with continuous stirring up to 80 °C, stirred at this temperature order to obtain a transparent solution, and finally sonicated (1 min) to remove gasses. Afterwards, it was slowly poured into the PMMA spectrophotometric cuvette (inner dimensions: $10 \times 10 \times 42$ mm). The cuvette orifice was immediately covered with a pre-heated plate of glass to prevent drying and shrinking of gel. The flat surface of the boundary of resulting hydrogels was provided by wiping an excess solution away. Gentle cooling of cuvettes at the laboratory temperature led to the gradual gelation of the mixture.

Agarose-chitosan hydrogels were prepared from agarose solution mixed with the solution of chitosan. An accurately weighed amount of chitosan was dissolved in 50 cm^3 of acetic acid (5% wt.). The solution was titrated by 1M NaOH up to pH equal to 7 and diluted by distilled water (the final volume was 100 cm^3). Agarose content in hydrogel was 10 mg g^{-1}, the contents of chitosan were 0.2, 0.5, and 1 mg g^{-1}, respectively.

4.3. Diffusion Experiments

Three cuvettes containing the same type of hydrogel were placed into 300 cm^3 of donor solution (0.1 M CuSO$_4$). The solution was stirred continuously by the magnetic stirrer and the copper(II) ions were left to diffuse from the solution into the hydrogels through the square orifices of the cuvettes. Diffusion experiments were triplicated, meaning that three different vessels for the same type of hydrogel (nine cuvettes in total) were used. The durations of the diffusion experiments were 24, 48, and 72 h, respectively. In these time intervals, the cuvettes were taken out of the solution, and the UV-VIS spectra were measured in dependence on distances from the interface between hydrogel and donor solution. The cuvettes were taken out of the solution and the spectra were measured on Varian Cary 50 UV–VIS spectrophotometer (Agilent Technologies, Palo Alto, CA, USA) equipped with the special accessory providing controlled fine vertical movement of the cuvette in the spectrophotometer [38,39]. Concentration of the copper(II) ions was determined at different positions in the hydrogels by means of calibration line. The spectra were calibrated for the hydrogels with the known concentration, homogeneously distributed in the whole volume of the hydrogel.

All experiments were performed at laboratory temperature (25 ± 1 °C). Data are presented as average values with standard deviation bars.

4.4. Rheological Experiments

The hydrogels were sliced to obtain cylindrical samples suitable for rheological measurements (1 mm in thickness). Each hydrogel sample was placed between two titanium plates (40 mm in diameter) of an AR-G2 rheometer (TA Instruments, Ltd., New Castle, DE, USA) equipped with Rheology Advantage Instrument Control AR software. (v5.5.24, TA Instruments, Ltd., New Castle, DE, USA). Silicon oil was used to prevent drying of the hydrogels. The hydrogels were left to relax for 15 min before measurements were made. Measurements were performed at 25 ± 1 °C. Rheological behavior of hydrogels was characterized with respect to their viscoelastic character. The storage modulus G' (proportional to the extent of the elastic component), loss modulus G'' (proportional to the extent of the viscous component), and complex viscosity η^* were determined. All experiments were triplicated, and average values are presented.

Funding: This research was funded by the National Program for Sustainability I (Ministry of Education, Youth and Sports), grant number REG LO1211, Materials Research Centre at FCH BUT-Sustainability and Development.

Institutional Review Board Statement: Not applicable.

Informed Consent Statement: Not applicable.

Conflicts of Interest: The author declares no conflict of interest.

References

1. Duarte, M.L.; Ferreira, M.C.; Marvao, M.R.; Rocha, J. An Optimised Method to Determine the Degree of Acetylation of Chitin and Chitosan by FTIR Spectroscopy. *Int. J. Biol. Macromol.* **2002**, *31*, 1–8. [CrossRef] [PubMed]
2. Paulino, A.T.; Simionato, J.-I.; Garcia, J.C.; Nozaki, J. Characterization of Chitosan and Chitin Produced from Silkworm Crysalides. *Carbohydr. Polym.* **2006**, *64*, 98–103. [CrossRef]
3. Martinez-Ruvalcaba, A.; Chornet, E.; Rodrigue, D. Viscoelastic Properties of Dispersed Chitosan/Xanthan Hydrogels. *Carbohydr. Polym.* **2007**, *67*, 586–595. [CrossRef]
4. Wan Ngah, W.S.; Fatinathan, S. Adsorption of Cu(II) Ions in Aqueous Solution Using Chitosan Beads, Chitosan-GLA Beads and Chitosan-Alginate Beads. *Chem. Eng. J.* **2008**, *143*, 62–72. [CrossRef]
5. Mourza, V.K.; Inamdar, N.N. Chitosan-Modifications and Applications: Opportunities Galore. *React. Funct. Polym.* **2008**, *68*, 1013–1051. [CrossRef]
6. Jayakumar, R.; Menon, D.; Manzoor, K.; Nair, S.V.; Tamura, H. Biomedical Applications of Chitin and Chitosan Based Nanomaterials. *Carbohydr. Polym.* **2010**, *82*, 227–232. [CrossRef]
7. Kamari, A.; Pulford, I.D.; Hargreaves, J.S.J. Chitosan as a Potential Amendment to Remediate Metal Contaminated Soil—A Characterisation Study. *Colloid. Surface. B* **2011**, *82*, 71–80. [CrossRef]
8. Bensaha, S.; Slimane, S.K. A Comparative Study on the Chitosan Membranes Prepared from Acetic Acid and Glycine Hydrochloride for Removal of Copper. *Russ. J. Appl. Chem.* **2016**, *89*, 1991–2000. [CrossRef]
9. Zhang, J.; Wang, Q.; Wang, A. Synthesis and Characterization of Chitosan-g-poly(acrylic acid)/Attapulgite Superabsorbent Composites. *Carbohydr. Polym.* **2007**, *68*, 367–374. [CrossRef]
10. Narayanan, A.; Kartik, R.; Sangeetha, E.; Dhamodharan, R. Super Water Absorbing Polymeric Gel from Chitosan, Citric Acid and Urea: Synthesis and Mechanism of Water Absorption. *Carbohydr. Polym.* **2018**, *191*, 152–160. [CrossRef]
11. Funakoshi, T.; Majima, T.; Iwasaki, N.; Yamane, S.; Masuko, T.; Minami, A.; Harada, K.; Tamura, H.; Tokura, S.; Nishimura, S.-I. Novel Chitosan-Based Hyaluronan Hybrid Polymer Fibers as a Scaffold in Ligament Tissue Engineering. *J. Biomed. Mater. Res. A* **2005**, *74*, 338–346. [CrossRef] [PubMed]
12. Yu, K.; Ho, J.; McCandlish, E.; Buckley, B.; Patel, R.; Li, Z.; Shapley, N.C. Copper Ion Adsorption by Chitosan Nanoparticles and Alginate Microparticles for Water Purification Applications. *Colloid. Surface. A* **2013**, *425*, 31–41. [CrossRef]
13. Yasmeen, S.; Lo, M.K.; Bajracharya, S.; Roldo, M. Injectable Scaffolds for Bone Regeneration. *Langmuir* **2014**, *30*, 12977–12985. [CrossRef]
14. Kyzas, G.Z.; Bikiaris, D.N.; Lambropoulou, D. Effect of Humic Acid on Pharmaceuticals Adsorption using Sulfonic Acid Grafted Chitosan. *J. Mol. Liq.* **2017**, *230*, 1–5. [CrossRef]
15. Jakubec, M.; Klimša, V.; Hanuš, J.; Biegaj, K.; Heng, J.Y.Y.; Štěpánek, F. Formation of Multi-Compartmental Drug Carriers by Hetero-Aggregation of Polyelectrolyte Microgels. *Colloid. Surface. A* **2017**, *522*, 250–259. [CrossRef]
16. Zhao, F.; Binyu, Y.; Zhengrong, Y.; Wang, T.; Wen, X.; Liu, Z.; Zhao, C. Preparation of Porous Chitosan Gel Beads for Copper(II) Ion Adsorption. *J. Hazard. Mater.* **2007**, *147*, 67–73. [CrossRef]
17. Babel, S.; Kurniawan, T.A. Low-Cost Adsorbents for Heavy Metals Uptake from Contaminated Water: A Review. *J. Hazard. Mater.* **2003**, *97*, 219–243. [CrossRef]
18. Li, N.; Bai, R. Copper Adsorption on Chitosan-Cellulose Hydrogel Beads: Behaviors and Mechanisms. *Sep. Purif. Technol.* **2005**, *42*, 237–247. [CrossRef]
19. Schmuhl, R.; Krieg, H.M.; Keizer, K. Adsorption of Cu(II) and Cr(VI) Ions by Chitosan: Kinetics and Equilibrium Studies. *Water SA* **2001**, *27*, 1–8. [CrossRef]
20. Elshaarawy, R.F.M.; El-Azim, H.A.; Hegazy, W.H.; Mustafa, F.H.A.; Talkhan, T.A. Poly(Ammonium/ Pyridinium)-Chitosan Schiff Base as a Smart Biosorbent for Scavenging of Cu^{2+} Ions from Aqueous Effluents. *Polym. Test.* **2020**, *83*, 106244. [CrossRef]
21. Kara, A.; Demirbel, E. Physicochemical Parameters of Cu^{2+} Ions Adsorption from Aqueous Solution by Magnetic-Poly(Divinylbenzene-N-vinylimidazole) Microbeads. *Sep. Sci. Technol.* **2012**, *47*, 709–722. [CrossRef]
22. Bassi, R.; Prasher, S.O.; Simpson, B.K. Removal of Selected Metal Ions from Aqueous Solutions Using Chitosan Flakes. *Sep. Sci. Technol.* **2000**, *35*, 547–560. [CrossRef]
23. Ahmad, M.; Manzoor, K.; Ikram, S. Versatile Nature of Hetero-Chitosan Based Derivatives as Biodegradable Adsorbent for Heavy Metal Ions; A Review. *Int. J. Biol. Macromol.* **2017**, *105*, 190–203. [CrossRef]

24. Yang, X.; Wan, Y.; Zheng, Y.; He, F.; Yu, Z.; Huang, J.; Wang, H.; Ok, Y.S.; Jiang, Y.; Gao, B. Surface Functional Groups of Carbon-Based Adsorbents and Their Roles in the Removal of Heavy Metals from Aqueous Solutions: A Critical Review. *Chem. Eng. J.* **2019**, *366*, 608–621. [CrossRef]
25. Zhang, Y.; Zhao, M.; Cheng, Q.; Wan, C.; Han, X.; Fan, Z.; Su, G.; Pan, D.; Li, Z. Research Progress of Adsorption and Removal of Heavy Metals by Chitosan and Its Derivatives: A Review. *Chemosphere* **2021**, *279*, 130927. [CrossRef]
26. Lee, S.T.; Mi, F.L.; Shen, Z.J.; Shyu, S.S. Equilibrium and Kinetic Studies of Copper(II) Ion Uptake by Chitosan-Tripolyphosphate Chelating Resin. *Polymer* **2021**, *42*, 1879–1892. [CrossRef]
27. Guibal, E. Interactions of Metal Ions with Chitosan-Based Sorbents: A Review. *Sep. Purif. Technol.* **2004**, *38*, 43–74. [CrossRef]
28. Guibal, E.; Jansson-Charrier, M.; Saucedo, I.; Le Cloirec, P. Enhancement of Metal Ion Sorption Performances of Chitosan: Effect of the Structure on the Diffusion Properties. *Langmuir* **1995**, *11*, 591–598. [CrossRef]
29. Jansson-Charrier, M.; Guibal, E.; Roussy, J.; Delanghe, B.; Le Cloirec, P. Vanadium (IV) Sorption by Chitosan: Kinetics and Equilibrium. *Wat. Res.* **1996**, *30*, 465–475. [CrossRef]
30. Karthikeyan, G.; Anbalagan, K.; Muthulakshmi Andal, N. Adsorption Dynamics and Equilibrium Studies of Zn (II) onto Chitosan. *J. Chem. Sci.* **2004**, *116*, 119–127. [CrossRef]
31. Milosavljevic, N.B.; Ristic, M.D.; Peric-Grujic, A.A.; Filipovic, J.M.; Strbac, S.B.; Rakocevic, Z.L.; Kalagasidis Krusic, M.T. Removal of Cu^{2+} Ions Using Hydrogels of Chitosan, Itaconic and Methacrylic Acid: FTIR, SEM/EDX, AFM, Kinetic and Equilibrium Study. *Colloid. Surface. A* **2011**, *388*, 59–69. [CrossRef]
32. de Vasconcelos, C.L.; Rocha, A.N.L.; Pereira, M.R.; Foncesa, J.L.C. Electrolyte Diffusion in a Chitosan Membrane. *Polym. Int.* **2001**, *50*, 309–312. [CrossRef]
33. Krajewska, B. Diffusion of Metal Ions through Gel Chitosan Membranes. *React. Funct. Polym.* **2001**, *47*, 37–47. [CrossRef]
34. Yoshida, C.M.P.; Bastos, C.E.N.; Franco, T.T. Modeling of Potassium Sorbate Diffusion through Chitosan Films. *LWT—Food Sci. Technol.* **2010**, *43*, 584–589. [CrossRef]
35. Modrzejewska, Z.; Rogacki, G.; Sujka, W.; Zarzycky, R. Sorption of Copper by Chitosan Hydrogel: Kinetics and Equilibrium. *Chem. Eng. Process.* **2016**, *109*, 104–113. [CrossRef]
36. Mankidy, B.D.; Coutinho, C.A.; Gupta, V.K. Probing the Interplay of Size, Shape, and Solution Environment in Macromolecular Diffusion Using a Simple Refraction Experiment. *J. Chem. Educ.* **2010**, *87*, 515–518. [CrossRef]
37. Klučáková, M.; Smilek, J.; Sedláček, P. How Humic Acids Affect the Rheological and Transport Properties of Hydrogels. *Molecules* **2019**, *24*, 1545. [CrossRef]
38. Klučáková, M. Agarose Hydrogels Enriched by Humic Acids as Complexation Agent. *Polymers* **2020**, *12*, 687. [CrossRef]
39. Sedláček, P.; Smilek, J.; Klučáková, M. How the Interactions with Humic Acids Affect the Mobility of Ionic Dyes in Hydrogels—2. Non-Stationary Diffusion Experiments. *React. Funct. Polym.* **2014**, *75*, 41–50. [CrossRef]
40. Klučáková, M.; Pekař, M. Study of Structure and Properties of Humic and Fulvic Acids. IV. Study of Interactions of Cu^{2+} Ions with Humic Gels and Final Comparison. *J. Polym. Mater.* **2003**, *20*, 155–162.
41. Klučáková, M.; Kalina, M.; Sedláček, P.; Grasset, L. Reactivity and Transport Mapping of Cu(II) Ions in Humic Hydrogels. *J. Soil. Sediment.* **2014**, *14*, 368–376. [CrossRef]
42. Klučáková, M.; Kalina, M.; Smilek, J.; Laštůvková, M. The Transport of Metal Ions in Hydrogels Containing Humic Acids as Active Complexation Agent. *Colloid. Surface. A* **2018**, *557*, 116–122. [CrossRef]
43. Chiessi, E.; Paradossi, G.; Venanzi, M.; Pispisa, B. Copper Complexes Immobilized to Chitosan. *J. Inorg. Biochem.* **1992**, *46*, 109–118. [CrossRef] [PubMed]
44. Monteiro Jr, O.A.C.; Airoldi, C. Some Thermodynamic Data on Copper–Chitin and Copper–Chitosan Biopolymer Interactions. *J. Colloid Interface Sci.* **1999**, *212*, 212–219. [CrossRef]
45. Onsoyen, E.; Skaugrud, O. Metal Recovery Using Chitosan. *J. Chem. Technol. Biotechnol.* **1990**, *49*, 395–404. [CrossRef]
46. Crank, J. *The Mathematics of Diffusion*, 1st ed.; Clarendon Press: Oxford, UK, 1956; pp. 26–41.
47. Cussler, E.L. *Diffusion: Mass Transfer in Fluid Systems*, 2nd ed.; Cambridge University Press: Cambridge, MA, USA, 1984; pp. 13–49.
48. Kyzas, G.Z.; Kostoglou, M.; Layaridis, N.K. Copper and Chromium(VI) Removal by Chitosan Derivatives—Equilibrium and Kinetic Studies. *Chem. Eng. J.* **2009**, *152*, 440–448. [CrossRef]
49. Lobo, V.M.M.; Quaresma, J.L. *Handbook of Electrolyte Solutions*; Physical Science Data Series 41; Elsevier: Amsterdam, The Netherlands, 1989.
50. Zhang, H.; Davison, W. Diffusional Characteristics of Hydrogels Used in DGT and DET Techniques. *Anal. Chim. Acta* **1999**, *398*, 329–340. [CrossRef]
51. Chu, H.H. Removal of Copper from Aqueous Solution by Chitosan in Prawn Shell: Adsorption Equilibrium and Kinetics. *J. Hazard. Mater.* **2002**, *90*, 77–95. [CrossRef]
52. Modrzejewska, Z. Sorption Mechanism of Copper in Chitosan Hydrogel. *React. Funct. Polym.* **2013**, *73*, 719–729. [CrossRef]
53. Guzman, J.; Saucedo, I.; Revilla, J.; Navarro, R.; Guibal, E. Copper Sorption by Chitosan in the Presence of Citrate Ions: Influence of Metal Speciation on Sorption Mechanism and Uptake Capacities. *Int. J. Biol. Macromol.* **2003**, *33*, 57–65. [CrossRef]
54. Lagzi, I. Controlling and Engineering Precipitation Patterns. *Langmuir* **2012**, *28*, 3350–3354. [CrossRef] [PubMed]
55. Meng, X.; Mi, Y.; Jia, D.; Guo, N.; An, Y.; Miao, Y. Polymorphs Co Hydroxides Formed between Hydrazine and Co^{2+} as Liesegang Bands in Semisolid Agar Gel. *J. Mol. Liq.* **2019**, *285*, 416–423. [CrossRef]

56. Izsák, F.; Lagzi, I. A New Universal Law for the Liesegang Pattern Formation. *J. Chem. Phys.* **2005**, *122*, 184707. [CrossRef] [PubMed]
57. Li, B.; Gao, Y.; Feng, Y.; Ma, B.; Zhu, R.; Zhou, Y. Formation of Concentric Multilayers in a Chitosan Hydrogel Inspired by Liesegang Ring Phenomen. *J. Biomater. Sci.* **2011**, *22*, 2295–2304. [CrossRef]
58. Gegel, N.; Babicheva, T.; Shipovskaya, A. Morphology of Chitosan-Based Hollow Cylindrical Materials with a Layered Structure. *BioNanoScience* **2018**, *8*, 661–667. [CrossRef]
59. Babicheva, T.S.; Konduktorova, A.A.; Shmakov, S.L.; Shipovskaya, A.B. Formation of Liesegang Structures under the Conditions of the Spatiotemporal Reaction of Polymer-Analogous Transformation (Salt Base) of Chitosan. *J. Phys. Chem. B* **2020**, *124*, 9255–9266. [CrossRef]
60. Kumar, P.; Sebok, D.; Kukovecs, A.; Horvath, D.; Toth, A. Hierarchical Self-Assembly of Metal-Ion-Modulated Chitosan Tubules. *Langmuir* **2021**, *37*, 12690–12696. [CrossRef] [PubMed]
61. Nabika, H.; Itatani, M.; Lagzi, I. Pattern Formation in Precipitation Reactions: The Liesegang Phenomenon. *Langmuir* **2020**, *36*, 481–497. [CrossRef]
62. Shimizu, Y.; Matsui, J.; Unoura, K.; Nabika, H. Liesegang Mechanism with a Gradual Phase Transition. *J. Phys. Chem. B* **2017**, *121*, 2495–2501. [CrossRef]
63. Monk, P. *Physical Chemistry. Understanding Our Chemical World*, 1st ed.; John Wiley & Sons Ltd.: Chichester, UK, 2004; pp. 177–229.
64. Saad, M.; Safieddine, A.; Sultan, R. Revisited Chaos in a Diffusion-Precipitation-Redissolution Liesegang System. *J. Phys. Chem. B* **2018**, *122*, 6043–6047. [CrossRef]
65. Nakouzi, E.; Steinbock, O. Self-Organization in Precipitation Reactions Far from the Equilibrium. *Sci. Adv.* **2016**, *2*, e1601144. [CrossRef] [PubMed]
66. Sedláček, P.; Smilek, J.; Klučáková, M. How Interactions with Polyelectrolytes Affect Mobility of Low Molecular Ions—Results from Diffusion Cells. *React. Funct. Polym.* **2013**, *73*, 1500–1509. [CrossRef]
67. Garcia, L.G.S.; de Melo Guedes, G.M.; da Silva, M.L.Q.; Castelo-Branco, D.S.C.M.; Sidrim, J.J.C.; de Aguiar Cordeiro, R.; Rocha, M.F.G.; Vieira, R.S.; Brilhante, R.S.N. Effect of the Molecular Weight of Chitosan on its Antifungal Activity against *Candida* spp. in Planktonic Cells and Biofilm. *Carbohydr. Polym.* **2018**, *195*, 662–669. [CrossRef] [PubMed]

Disclaimer/Publisher's Note: The statements, opinions and data contained in all publications are solely those of the individual author(s) and contributor(s) and not of MDPI and/or the editor(s). MDPI and/or the editor(s) disclaim responsibility for any injury to people or property resulting from any ideas, methods, instructions or products referred to in the content.

Article

Cobalt- and Copper-Based *Chemiresistors* for Low Concentration Methane Detection, a Comparison Study

Paul Chesler [1], Cristian Hornoiu [1,*], Mihai Anastasescu [1], Jose Maria Calderon-Moreno [1], Marin Gheorghe [2] and Mariuca Gartner [1]

[1] "Ilie Murgulescu" Institute of Physical Chemistry—Romanian Academy, Splaiul Independentei 202, 060021 Bucharest, Romania
[2] NANOM MEMS SRL, Strada George Cosbuc 9, Rasnov, 505400 Brașov, Romania
* Correspondence: chornoiu@icf.ro

Abstract: Methane is a colorless/odorless major greenhouse effect gas, which can explode when it accumulates at concentrations above 50,000 ppm. Its detection cannot be performed without specialized equipment, namely sensing devices. A series of MOX sensors (*chemiresistors* type), with CoO and CuO sensitive films were obtained using an eco-friendly and low-cost deposition technique (sol–gel). The sensing films were characterized using AFM and SEM as thin film. The transducers are based on an alumina wafer, with Au or Pt interdigital electrodes (IDE) printed onto the alumina surface. The sensor response was recorded upon sensor exposure to different methane concentrations (target gas) under lab conditions (dried target and carrier gas from gas cylinders), in a constant gas flow, with target gas concentrations in the 5–2000 ppm domain and a direct current (DC) applied to the IDE as sensor operating voltage. Humidity and cross-sensitivity (CO_2) measurements were performed, along with sensor stability measurements, to better characterize the obtained sensors. The obtained results emphasize *good 3-S* sensor parameters (sensitivity, partial selectivity and stability) and also short response time and complete sensor recovery, completed by a low working temperature (220 °C), which are key factors for further development of a new commercial chemiresistor for methane detection.

Keywords: sol–gel; thin films; alumina wafer; IDE; lab conditions; 5 ppm CH_4; low-cost/eco-friendly; cross-sensitivity/humidity test

Citation: Chesler, P.; Hornoiu, C.; Anastasescu, M.; Calderon-Moreno, J.M.; Gheorghe, M.; Gartner, M. Cobalt- and Copper-Based *Chemiresistors* for Low Concentration Methane Detection, a Comparison Study. *Gels* **2022**, *8*, 721. https://doi.org/10.3390/gels8110721

Academic Editors: Luca Burratti, Paolo Prosposito and Iole Venditti

Received: 20 October 2022
Accepted: 6 November 2022
Published: 8 November 2022

Publisher's Note: MDPI stays neutral with regard to jurisdictional claims in published maps and institutional affiliations.

Copyright: © 2022 by the authors. Licensee MDPI, Basel, Switzerland. This article is an open access article distributed under the terms and conditions of the Creative Commons Attribution (CC BY) license (https://creativecommons.org/licenses/by/4.0/).

1. Introduction

Methane (CH_4) is a gas with a major greenhouse effect, being predominantly present in the agricultural areas of the planet. It can accumulate gradually, up to explosive concentrations (50,000 ppm or 5% volume), its formation being particularly favored by warm climates and humidity. Human activities that emit methane include leaks from natural gas systems and the existence of landfills on the outskirts of human settlements. In industry, methane is emitted during LPG refining or from mining activities (coal). Methane is also emitted from natural sources, such as natural wet areas (swamps). Natural processes in the soil and chemical reactions in the atmosphere help to remove CH_4 from the atmosphere. The lifetime of methane in the atmosphere is much shorter than that of carbon dioxide (CO_2), but CH_4 is more efficient at capturing radiation than CO_2. Quantitatively speaking, the comparative impact of CH_4 is 25 times greater than that of CO_2 over a period of 100 years [1].

NIOSH (National Institute for Occupational Safety and Health's) established a maximum limit of 1000 ppm [2], for an exposure time of 8 h at the workplace, so its detection is very important for safety reasons. Detection of methane without special devices is impossible, methane being odorless and colorless.

The development of gas detectors has increased dramatically in the past decades, starting in 1953, when Brattain and Bardeen discovered that when a gas is adsorbed on the surface of a semiconductor, a change in the electrical conductance of this material occurs [3]. In 1968, Taguchi released the first commercially available gas sensor for the detection of hydrocarbons [4]. Since then, gas sensors, having the advantage of being reduced in size [5] and also cheap devices that can be mass-produced, were used to monitor environmental pollution, obtain global contamination maps [6], monitordomestic safety, ensure public security, monitor automotive safety, monitor air quality and more recently, make medical diagnoses, such as exhaled breath analysis [7]. Gas detectors have been fabricated in many different ways (electrochemical and optical approaches), and solid-state gas sensors contain various gas sensing materials (e.g., metal oxides or MOX). For MOX gas sensors, the most widely accepted sensing mechanism can be explained by the resistance change, which is caused by the surface reaction between the target gas and the sensitive material deposited on the surface of the sensor (in this particular case the sensors are named *chemiresistors*), upon sensor exposure to different gaseous atmospheres [7]. *Chemiresistors* based on semiconductor metal oxides with low-costs, easy production, a compact size and simple electronics are the most widely used in gas detection applications, however, MOX-resistive sensors typically operate at high working temperatures, which limits their application as sensitive materials and leads to sensing material instability, increased power consumption and response drifts [6]. The key for obtaining an economically viable sensor is mainly the low-cost of the final product, which implies abundant raw materials for the sensor components and low-cost preparation techniques (sol–gel, hydrothermal, etc.) for the sensing element, combined with sensor working temperatures as low as possible (ideally room temperature). Ideal materials for gas-sensing applications should be characterized by *high 3-S* parameters (sensitivity, selectivity and stability). Other key features are fast response/recovery time [7].

Although CuO and CoO in different combinations were previously used as sensitive materials [8] for different gases (detection for VOC's, NH_3, carbon oxides, H_2S were summarized in ref. [7]), methane detection (in the percent concentrations range) using these oxides was very rarely reported, and usually high-cost preparation techniques are used to obtain the sensing oxides from their precursors (microwave in ref. [8], thermal oxidation in ref. [9].

In ref. [10], methane detection using Cu-doped CoO was reported, and sol–gel was used in the preparation of the sensitive pellets (compressed powders with silver painted electrodes on each pellet side), but no humidity/cross-sensitivity measurements were taken. Moreover, powder pellets imply usage of large quantities of sensitive material, a non-viable element from an economical point of view. Working temperature of the sensor was also high (300 °C), another important disadvantage.

The aim of this paper was to obtain a cheap, stable, highly sensitive and energy-efficient methane chemiresistor, using thin films of cobalt oxide (CoO) and copper oxide (CuO) as sensitive materials, deposited via a low-cost/eco-friendly technique on an own-design alumina transducer, having Au or Pt IDE's imprinted on the surface, for the purpose of excellent electrical conductivity. The oxides were used in pristine state, and the sensing performance was then evaluated in each separate case to see which sensor performs better for methane detection. Humidity and cross-sensitivity tests (with CO_2) were also performed to better define sensor characteristics.

2. Results and Discussion

2.1. Sensor Characterization

The sensor samples were investigated from a morphological point of view by Atomic Force Microscopy (AFM) after sensing experiments. Figure 1a,b present AFM images of the CoO film deposited by the sol–gel method on alumina substrates configured with Pt interdigitated electrodes (see the Materials and Methods section). The CoO film is characterized by a very high root mean square (RMS) roughness, ~350 nm at the scale of (8×8) μm^2, respectively, a peak-to-valley parameter of ~2477 nm, being the highest values recorded in this series of samples. The surface of the sample is characterized by the presence of protruding massive formations of material ("mountain"-like) but also by random valleys ("pits")—see the 2D (Figure 1a) and 3D (Figure 1b) AFM images from Figure 1, presented in so-called enhanced contrast view. Scanning a smaller area region, as shown in Figure 1b, suggests that there is a tendency to texturize the film, in the form of "ridges". At the small scale of (2 × 2) µm², the roughness parameters are significantly reduced, the CoO film having an RMS roughness of ~54 nm, respectively, a peak-to-valley parameter of ~291 nm. Figure 1c,d show AFM images of the CuO film deposited by sol–gel on Pt/alumina interdigitated electrodes. The CuO film shows uniform hills–valleys alternation, as depicted by the profile line from Figure 1c-right, with a level difference of about 900 nm along the selected line. On the entire scanned area of (8 × 8) µm², the global corrugation parameters have the following values: 306 nm for the RMS roughness, respectively, ~2104 nm for the peak-to-valley parameter. Thus, it can be noted that both films prepared by sol–gel are very rough. Scanning a smaller area of (2 × 2) µm² highlights the granular morphology of the CuO film, with a majority population of particles with a diameter in the range of 100–300 nm. The high roughness of the CoO and CuO films evidenced by AFM is an important advantage from the gas sensing point of view, surface roughness being a promoter for the gas adsorption process.

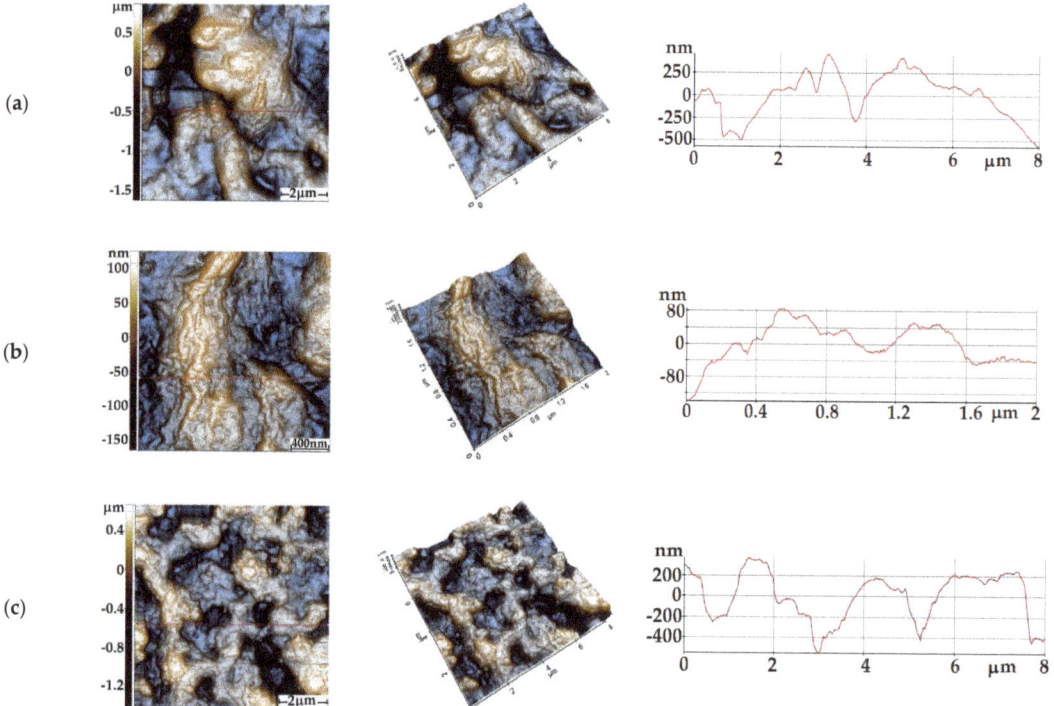

Figure 1. Cont.

Figure 1. AFM images of the CoO sensitive films at the scale of (8 × 8) µm² — (**a**) and (2 × 2) µm² — (**b**) and, respectively, of the CuO sensitive films at (8 × 8) µm² — (**c**) and (2 × 2) µm² — (**d**).

The sensors were then characterized by SEM (Figure 2a,b). The acquired high-RES images confirmed the facts evidenced by AFM investigations: both investigated sensing films having rough surfaces. The morphology of the sensing films deposited by sol-gel adapts very well to the morphology of polycrystalline IDE (Pt/Au) and also to the alumina substrates.

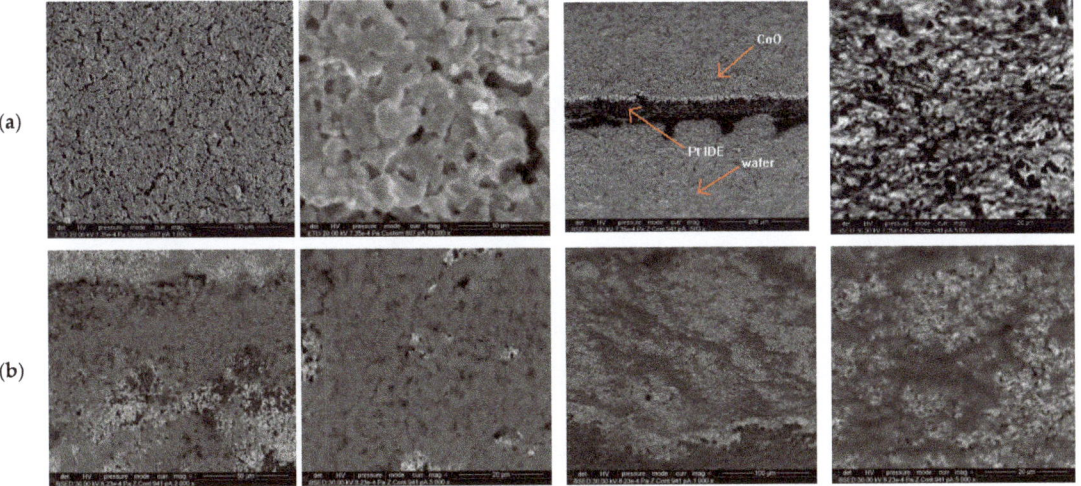

Figure 2. SEM images of: (**a**) CoO sensitive film; (**b**) CuO sensitive film, at different magnification factors; last two images of each group (**a**,**b**) are tilted.

The pores are more visible in SEM images, a network of surface channels being better evidenced in the case of CoO. The third image (tilted) in the Figure 2a group shows the low thickness of the CoO sensitive film (below 1 micron thickness value—the sensitive film may be characterized as *thin*; it appears as a gray coating with a white edge) deposited on a Pt interdigit, which appears to be much thicker than the deposited sensitive layer. As stated before, this increases the accessibility towards the surface adsorption centers for the target gas molecules, increasing the overall sensing capabilities of the oxide film.

2.2. Gas Sensing Experiments

All sensing tests were performed *in triplicate* to ensure signal reproducibility. The following sensors were prepared (listed in Table 1) and abbreviated accordingly (as resulted from the synthesis process):

Table 1. The investigated sensors and their composition.

Sensor Abbreviation	Sensitive Film	Transducer (IDE/Wafer)
S3	CuO	Au/Al_2O_3
S4	CuO	Pt/Al_2O_3
S5	CoO	Pt/Al_2O_3

It can be observed that the sensors having CuO sensitive film are available with two IDE types, gold (Au) or platinum (Pt), in order to investigate the influence that the IDE material may have over the sensing experiment. Figure 3a shows the response/recovery characteristics of the sensors presented in Table 1. The sensor with Pt IDE (**S4**) has a slightly higher working temperature—T_w (220 °C, comparing with 210 °C for the other two sensors-**S3** and **S5**). Sensor response is comparable when using Pt or Au as IDE (**S4**, **S3**) except in the methane high-concentration range, where cobalt-based sensor-**S5** seems to be performing slightly better than the copper-based sensors -**S3**, **S4**, thus supporting the literature findings regarding cobalt oxide [11]. In Figure 3b, a working temperature experiment is depicted for 2000 ppm of CH_4 injected in the sensing cell equipped with **S5**. It can be observed that with the increasing working temperature, the response/recovery of the sensor decreases; therefore 210 °C is considered T_w for this sensor.

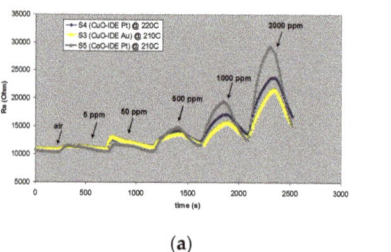
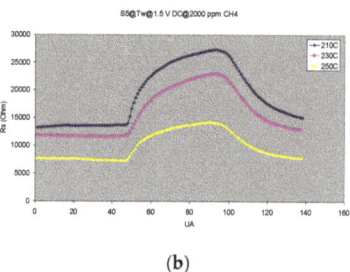

(a) (b)

Figure 3. (a) The response/recovery of the tested sensors for CH_4 concentrations in the range of 5–2000 ppm at T_w specific to the investigated sensors (210–220 °C); (b) response/recovery of the **S5** sensor for 2000 ppm CH_4 at various working temperatures (T_w).

Sensor response increases with increasing target gas concentrations (Figure 3a), the highest response being recorded for the concentration of 2000 ppm. All sensors also detect CH_4 in extremely low concentrations (5 ppm). These detection limits do not represent the lower/higher sensor detection limits, but they are imposed by the technical limitations of the experimental setup, the target gas from the cylinder being diluted by the carrier gas (from 5000 ppm in inert gas, as provided by the gas manufacturer, to a maximum concentration of 2000 ppm or a minimum of 5 ppm, according to the mass-flow controller calibration curves). The response of the sensors is fast (250 s), and the recovery is complete (250 s), making it possible to resume the experiments after the corresponding recovery cycle without sensor replacement.

Cross-Sensitivity Tests (Relative Humidity and CO_2 Measurements)

To better characterize the sensors, cross-sensitivity tests were performed. Thus, the sensors were exposed firstly to a humid atmosphere containing a standard 52% relative humidity (52% RH). Secondly, the sensors were exposed to CO_2, having concentrations in the 5000–20,000 ppm range, limits imposed by the MFC system in the experimental setup, as stated before. Both these analytes usually accompany methane in the atmospheric environment, being also the main products resulting from the burning of methane when used as gas fuel. The reaction that stands as the basis for the methane principle of detection occurs with an electrical resistance variation, which means there is a change in the charge carrier concentration detected by the measuring equipment (RLC bridge). This mechanism is proposed by Shaalan et al. [8].

$$CH_4 + 4O_{(ads)}^- \rightarrow CO_{2(gas)} + 2H_2O_{(gas)} + 4e^- \quad (1)$$

The increasing resistance upon p-type sensitive materials (CuO or CoO) exposure to CH_4 is explained in the mentioned reference as follows: "firstly, the gas reacts with adsorbed negative oxygen ions on the surface, leading to electron injection into p-type oxide. Secondly, this injected electron recombines with a hole in the oxide, reducing its positive free carriers, thus an increase in sensor electrical resistance occurs". The influence of humidity on the gas sensing properties of metal-oxide-based devices has been extensively investigated, particularly for tin oxide SnO_2 but is still not comprehensively understood to date [12]. The sensor response to humidity and CO_2 is shown in Figures 4 and 5.

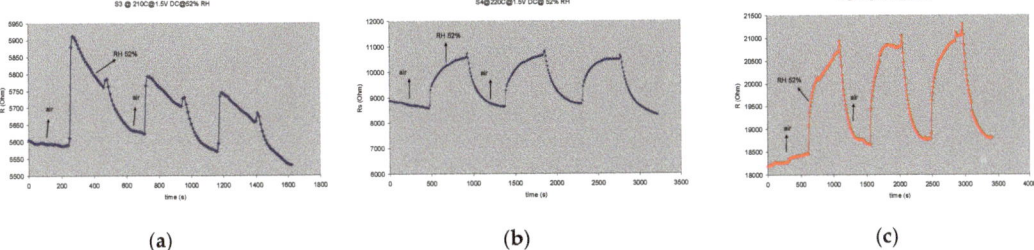

Figure 4. The response/recovery characteristics for 3 successive injections of 52% RH at corresponding T_w: (**a**) **S3** sensor, T_w = 210 °C; (**b**) **S4** sensor T_w = 220 °C; (**c**) **S5** sensor, T_w = 210 °C.

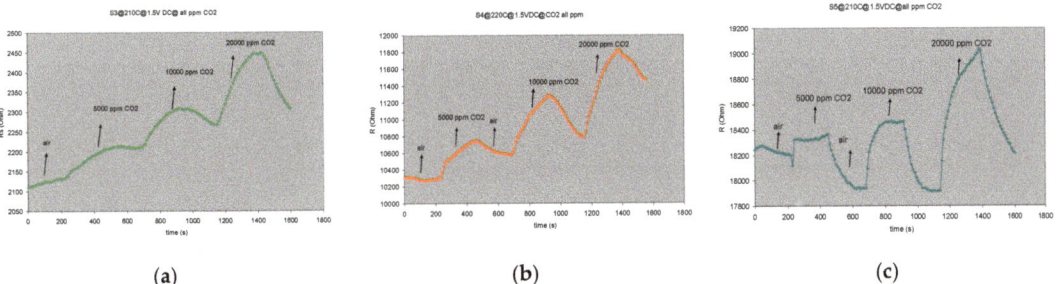

Figure 5. The response/recovery characteristics for 3 successive injections of CO_2 (5000–20,000 ppm) at corresponding T_w: (**a**) **S3** sensor, T_w = 210 °C; (**b**) **S4** sensor T_w = 220 °C; (**c**) **S5** sensor, T_w = 210 °C.

Figure 6 shows the cross-response of the investigated sensors for the main target gas and the interfering species (resulting from the oxidation reaction that takes place on the surface of the sensor), products which may affect sensor response. It can be seen that for all the investigated sensors responses to 2000 ppm, CH_4 is almost twice the response to the other interfering species (52% RH and 20,000 ppm CO_2). The best selectivity is recorded again for the **S5** cobalt-based sensor, where response for CH_4 is more than double compared with its response towards CO_2 and humidity. This particular sensor may be considered as *partially selective* for methane.

As seen in Figure 7, the sensor is stable over a 6-month period between tests, the response to methane in identical experimental conditions being virtually unchanged. Taking into account the results presented in Figures 4–7, we can state that the sensors meet the 3-*S* parameter requirements: sensitivity, selectivity (partial, for the **S5** sensor sample) and stability, completed by a relatively low working temperature and also short response and full sensor recovery characteristics.

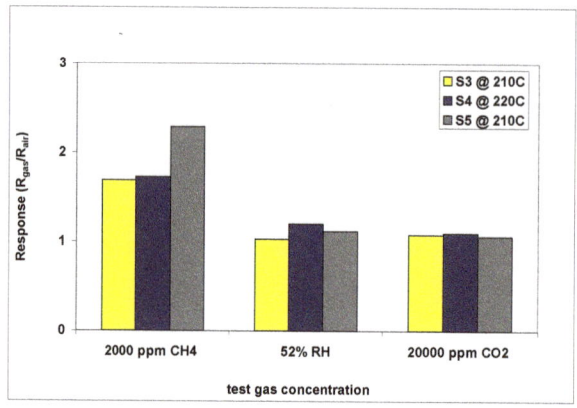

Figure 6. S3, S4, S5 cross-response for different tested target gas concentrations.

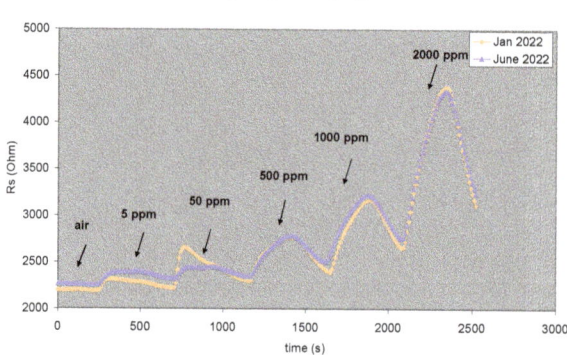

Figure 7. S3 sensor stability test (6-month period).

3. Conclusions

Sensors with CoO and CuO (MOX based *chemiresistors*) sensitive films have been prepared and characterized. The surface of the sensitive oxide has a high-roughness degree and a large network of channels and pores, as evidenced by AFM and SEM measurements. The films are thin (below 1 micron thick, as shown by SEM). All these characteristics promote gas sensing.

The prepared sensors are stable, partially selective and capable of detecting low concentrations of methane (5 ppm), with fast response (250 s) and a full recovery (250 s). Response for interfering species was recorded (CO_2 and humidity), but it was relatively low (about 50% from sensors response to methane). Better selectivity was recorded for the cobalt-based sensor towards high methane concentrations. A detection mechanism was formulated in agreement with literature findings. All the investigated sensors are energy-efficient, being characterized by a relatively low working temperature (max. 220 °C). Amongst the investigated sensors, the CoO-based *chemiresistors* are characterized by higher 3-S parameters (sensitivity, selectivity and stability) compared to the CuO based *chemiresistors*, which place them in a favorable position for further development of a new commercial methane detection MOX based chemiresistor.

4. Materials and Methods

Thick sensitive films were obtained using the sol–gel spinning method (1000 rotations/min). As precursors, the basic carbonates of the respective metals were used $Cu(CO_3)_2Cu(OH)_2$ for CuO and $Co(CO_3)Co(OH)_2$ for CoO.

The deposited film was stabilized by heating it at 400 °C for 10 min.

Own-design alumina transducers were used, with the following dimensions: $5 \times 10 \times 0.6$ mm (Figure 8). The transducers contained Pt or Au IDE's on one side and a Pt heater on the opposite side of the transducer. They were imprinted on the alumina surface using serigraphy-based technology.

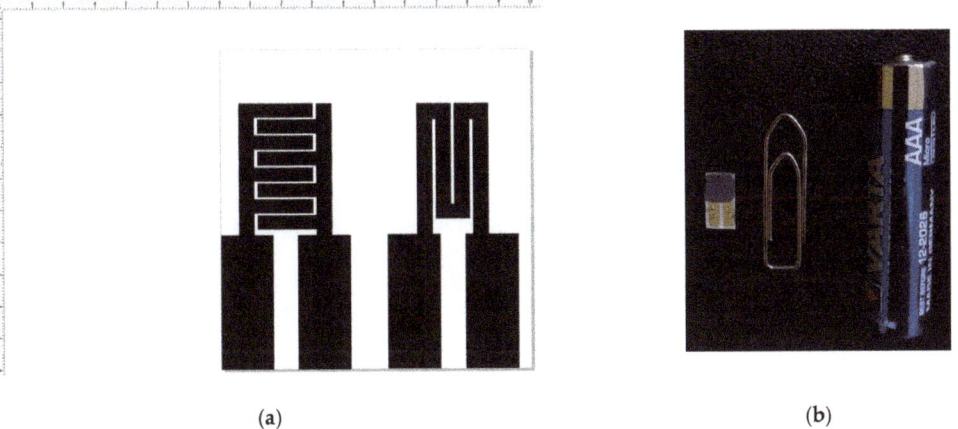

(a) (b)

Figure 8. Alumina transducer prototypes for the obtained methane sensors: (**a**) platinum or gold IDE and platinum heater circuit; (**b**) actual S3 sensor sample compared with various objects for sizing purposes.

All sensor measurements were performed under laboratory conditions using dry, high purity gases (5.0), purchased from specialized gas-suppliers (SIAD Romania). The gas concentrations in the cylinders were: 5000 ppm CH_4 and 50,000 ppm for CO_2 (both in inert gas), as stated by the certification labels. The operating voltage of the sensor was set at 1.5 V direct current (DC), the tested working temperatures (T_w) were in the range situated between room temperature and 220 °C (specific for each sensor used), and the sensing experiments were carried out in a continuous flow of gas (maximum 180 mL/min).

Two separate gas lines were used for the sensing experiments, one ensuring the flow of the carrier gas (dry air), the other ensuring the flow of the target gas. The target gas concentrations were achieved using a calibrated system of mass-flow controllers (MFC). The two separate gas flows were mixed inside a special glass vessel, shown in the scheme of the experimental installation (Figure 9), using an on-off valve system, thus diluting the target gas with the carrier gas.

The gas route continues to an own-design sensor cell, which contains the investigated sensor. In the sensor cell, a chemical reaction takes place on the surface of the sensor, which leads to a change in its electrical resistance, a variation recorded by the Hioki 3522-50 RLC bridge connected to the sensing cell. This bridge uses a "custom-made" acquisition software [13–15], based on the Labview platform, developed by our group. Thus, the analog signal taken from the chemiresistor was converted into a digital signal using a GPIB interface connected to the output of the RLC bridge. The resulting digital signal is then transformed into an *xy* graph using the data acquisitioning computer [16–18].

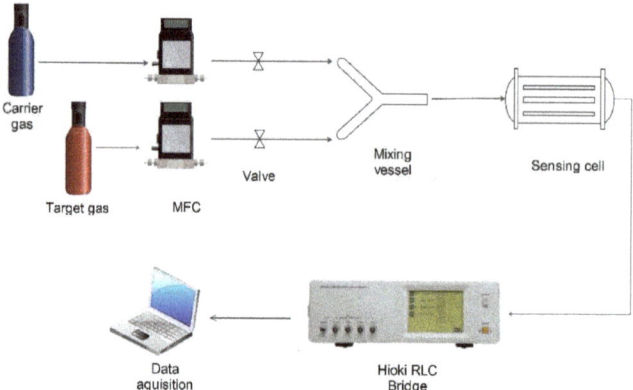

Figure 9. Schematic representation of the gas sensing experimental setup.

The sensors were characterized (after sensing experiments) by AFM and SEM to observe surface morphology for the sensing CuO and Co O films.

AFM measurements were carried out in non-contact mode [19] with XE-100 (Park Systems), using sharp tips (NCHR from Nanosensors), having less than 8 nm tip radius, ~125 μm length, ~30 μm mean width, thickness ~4 μm, ~42 N/m force constant and ~330 kHz resonance frequency. The XEI (v.1.8.0) image processing program developed by Park Systems was used for displaying the images and subsequent statistical data analysis.

The microstructure of the samples was investigated by SEM using a high-resolution microscope (FEI, Quanta 3D FEG). The analyses were performed in high vacuum mode at high accelerating voltages (30 kV), and the sensors were analyzed directly [19–22] (samples were immobilized on a double-sided carbon tape, without coating).

Author Contributions: Conceptualization, M.G. (Marin Gheorghe) and P.C.; methodology, P.C. and C.H.; software, P.C. and C.H.; validation, P.C.; formal analysis, P.C.; investigation, P.C., M.A. and J.M.C.-M.; resources, M.G. (Marin Gheorghe); data curation, P.C.; writing—original draft preparation, P.C.; writing—review and editing, P.C. and M.A.; visualization, P.C.; supervision, M.G. (Mariuca Gartner); project administration, M.G. (Mariuca Gartner); funding acquisition, M.G. (Mariuca Gartner). All authors have read and agreed to the published version of the manuscript.

Funding: This research was funded by Romanian National Authority for Scientific Research on Innovation, CCCDI-UEFISCDI, grant number "PN-III-P2-2.1-PED-2019-2073, Contract 308PED/2020".

Institutional Review Board Statement: Not applicable.

Informed Consent Statement: Not applicable.

Data Availability Statement: The datasets generated during and/or analyzed during the current study are available from the corresponding author on reasonable request.

Acknowledgments: The paper was carried out within the research program "Science of Surfaces and Thin Layers" of the "Ilie Murgulescu" Institute of Physical Chemistry of the Romanian Academy.

Conflicts of Interest: The authors declare no conflict of interest. The funders had no role in the design of the study; in the collection, analyses, or interpretation of data; in the writing of the manuscript; or in the decision to publish the results.

References

1. IPCC. *Climate Change 2007: The Physical Science Basis. Contribution of Working Group I to the Fourth Assessment Report of the Intergovernmental Panel on Climate Change*; Solomon, S., Qin, D., Manning, M., Chen, Z., Marquis, M., Averyt, K.B., Tignor, M., Miller, H.L., Eds.; Cambridge University Press: Cambridge, UK; New York, NY, USA, 2007; p. 996.
2. Available online: http://medbox.iiab.me/modules/en-cdc/www.cdc.gov/niosh/ipcsneng/neng0291.html (accessed on 1 October 2022).

3. Brattain, W.H.; Bardeen, J. Surface Properties of Germanium. *Bell Syst. Tech. J.* **1953**, *3*, 1–41. [CrossRef]
4. Taguchi, N. Gas Detecting Device. U.S. Patent 3,695,848 A, 3 October 1972.
5. Taya, S. Slab Waveguide with Air Core Layer and Anisotropic Left-Handed Material Claddings as a Sensor. *Opto-Electron. Rev.* **2014**, *22*, 252–257. [CrossRef]
6. Sayago, I.; Santos, J.P.; Sánchez-Vicente, C. The Effect of Rare Earths on the Response of Photo UV-Activate ZnO Gas Sensors. *Sensors* **2022**, *22*, 8150. [CrossRef]
7. Rydosz, A. The Use of Copper Oxide Thin Films in Gas-Sensing Applications. *Coatings* **2018**, *8*, 425. [CrossRef]
8. Shaalan, N.M.; Rashad, M.; Abdel-Rahim, M.A. CuO Nanoparticles Synthesized by Microwave-Assisted Method for Methane Sensing. *Opt. Quantum Electron.* **2016**, *48*, 531. [CrossRef]
9. Jayatissa, A.H.; Samarasekara, P.; Kun, G. Methane Gas Sensor Application of Cuprous Oxide Synthesized by Thermal Oxidation. *Phys. Status Solidi* **2009**, *206*, 332–337. [CrossRef]
10. Ahmadpour, A.; Mehrabadi, Z.S.; Esfandyari, J.R.; Koolivand-Salooki, M. Modeling of Cu Doped Cobalt Oxide Nanocrystal Gas Sensor for Methane Detection: ANFIS Approach. *J. Chem. Eng. Process Technol.* **2012**, *3*, 1–6.
11. Bratan, V.; Vasile, A.; Chesler, P.; Hornoiu, C. Insights into the Redox and Structural Properties of CoOx and MnOx: Fundamental Factors Affecting the Catalytic Performance in the Oxidation Process of VOCs. *Catalysts* **2022**, *12*, 1134. [CrossRef]
12. Steinhauer, S. Gas Sensors Based on Copper Oxide Nanomaterials: A Review. *Chemosensors* **2021**, *9*, 51. [CrossRef]
13. Bratan, V.; Chester, P.; Hornoiu, C.; Scurtu, M.; Postole, G.; Pietrzyk, P.; Gervasini, A.; Auroux, A.; Ionescu, N.I. In Situ Electrical Conductivity Study of Pt-Impregnated VOx/Gamma-Al2O3 Catalysts in Propene Deep Oxidation. *J. Mater. Sci.* **2020**, *55*, 10466–10481. [CrossRef]
14. Bratan, V.; Munteanu, C.; Chesler, P.; Negoescu, D.; Ionescu, N.I. electrical characterization and the catalytic properties of SnO_2/TiO_2 catalysts and their Pd-supported equivalents. *Rev. Roum. Chim.* **2014**, *59*, 335–341.
15. Brătan, V.; Chesler, P.; Vasile, A.; Todan, L.; Zaharescu, M.; Căldăraru, M. Surface Properties and Catalytic Oxidation on V_2O_5-CeO_2 Catalysts. *Rev. Roum. Chim.* **2011**, *56*, 1055–1065.
16. Firtat, B.; Moldovan, C.; Brasoveanu, C.; Muscalu, G.; Gartner, M.; Zaharescu, M.; Chesler, P.; Hornoiu, C.; Mihaiu, S.; Vladut, C.; et al. Miniaturised MOX Based Sensors for Pollutant and Explosive Gases Detection. *Sens. Actuators B Chem.* **2017**, *249*, 647–655. [CrossRef]
17. Chesler, P.; Hornoiu, C.; Bratan, V.; Munteanu, C.; Gartner, M.; Ionescu, N.I. Carbon monoxide sensing properties of TiO_2. *Rev. Roum. Chim.* **2015**, *60*, 227–232.
18. Chesler, P.; Hornoiu, C.; Bratan, V.; Munteanu, C.; Postole, G.; Ionescu, N.I.; Juzsakova, T.; Redey, A.; Gartner, M. CO Sensing Properties of SnO_2–CeO_2 Mixed Oxides. *React. Kinet. Mech. Catal.* **2016**, *117*, 551–563. [CrossRef]
19. Chelu, M.; Chesler, P.; Anastasescu, M.; Hornoiu, C.; Mitrea, D.; Atkinson, I.; Brasoveanu, C.; Moldovan, C.; Craciun, G.; Gheorghe, M.; et al. ZnO/NiO Heterostructure-Based Microsensors Used in Formaldehyde Detection at Room Temperature: Influence of the Sensor Operating Voltage. *J. Mater. Sci. Mater. Electron.* **2022**, *33*, 19998–20011. [CrossRef]
20. Chesler, P.; Hornoiu, C.; Mihaiu, S.; Munteanu, C.; Gartner, M. Tin-Zinc Oxide Composite Ceramics for Selective CO Sensing. *Ceram. Int.* **2016**, *42*, 16677–16684. [CrossRef]
21. Chesler, P.; Hornoiu, C.; Mihaiu, S.; Vladut, C.; Moreno, J.M.C.; Anastasescu, M.; Moldovan, C.; Firtat, B.; Brasoveanu, C.; Muscalu, G.; et al. Nanostructured SnO_2-ZnO Composite Gas Sensors for Selective Detection of Carbon Monoxide. *Beilstein J. Nanotechnol.* **2016**, *7*, 2045–2056. [CrossRef] [PubMed]
22. Duta, M.; Predoana, L.; Calderon-Moreno, J.M.; Preda, S.; Anastasescu, M.; Marin, A.; Dascalu, I.; Chesler, P.; Hornoiu, C.; Zaharescu, M.; et al. Nb-Doped TiO_2 Sol–Gel Films for CO Sensing Applications. *Mater. Sci. Semicond. Process.* **2016**, *42*, 397–404. [CrossRef]

Article

A Peptide-Based Hydrogel for Adsorption of Dyes and Pharmaceuticals in Water Remediation

Anna Fortunato and Miriam Mba *

Dipartimento di Scienze Chimiche, Università degli Studi di Padova, Via Marzolo 1, 35131 Padova, Italy
* Correspondence: miriam.mba@unipd.it

Abstract: The removal of dyes and pharmaceuticals from water has become a major issue in recent years due to the shortage of freshwater resources. The adsorption of these pollutants through nontoxic, easy-to-make, and environmentally friendly adsorbents has become a popular topic. In this work, a tetrapeptide–pyrene conjugate was rationally designed to form hydrogels under controlled acidic conditions. The hydrogels were thoroughly characterized, and their performance in the adsorption of various dyes and pharmaceuticals from water was investigated. The supramolecular hydrogel efficiently adsorbed methylene blue (MB) and diclofenac (DCF) from water. The effect of concentration in the adsorption efficiency was studied, and results indicated that while the adsorption of MB is governed by the availability of adsorption sites, in the case of DCF, concentration is the driving force of the process. In the case of MB, the nature of the dye–hydrogel interactions and the mechanism of the adsorption process were investigated through UV–Vis absorption spectroscopy. The studies proved how this dye is first adsorbed as a monomer, probably through electrostatic interactions; successively, at increasing concentrations as the electrostatic adsorption sites are depleted, dimerization on the hydrogel surface occurs.

Keywords: peptide; hydrogel; supramolecular gel; water remediation; dye adsorption; pharmaceuticals adsorption; diclofenac; methylene blue

1. Introduction

Pharmaceutically active compounds (PhACs) are emerging contaminants that can be found in industrial and domestic wastewater, surface water, ground water, and, more problematically, in drinking water [1]. They include nonsteroidal anti-inflammatory drugs (NSAIDs) (ibuprofen, diclofenac, naproxen), antibiotics (amoxicillin, ciprofloxacin, erythromycin, nalidixic acid), anticonvulsants (carbamazepine, primidone), antidepressants (diazepam, meprobamate), and hormones (estrogen), among others. Tonnes of pharmaceutically active compounds are used every year to treat human and animal diseases in hospitals, households, aquaculture, or feedstock facilities. However, they are not fully adsorbed, and the unmetabolized drugs are excreted, entering wastewater or surface and groundwater. Inappropriate disposal of unwanted drugs is another source of contamination [2]. These wastewaters end up in wastewater treatment plants (WWTPs), which are not able to efficiently remove this class of pollutants before final recirculation as drinking water. For example, a concentration of 67 µg/L of carbamazepine was found in the effluents of a WWTP in Madrid, which treats water in an area where there is a university campus, several pharmaceutical industries, and a hospital [3]. Furthermore, leaching or inappropriate waste management in the pharmaceutical industry also contributes to the introduction of PhACs into the environment. For example, a concentration of 17.48 mg/L of ciprofloxacin was detected in wastewater effluents from a pharmaceutical industry in Croatia [4]. Low concentrations of PhACs in drinking water do not constitute an immediate public health risk; however, the effects due to long-term exposure are not known. On the other hand, the adverse effects of PhACs in the environment have already been reported in the literature, including the development of antibiotic resistance [5]. Thus, the development of processes for the removal of PhACs has become of major importance. New processes are needed to remove low concentrations of PhACs (ng to µg per liter) arriving to WWTPs, but there is also a need for in situ facilities that are capable of removing higher concentrations of these pollutants in industry effluents (µg-mg per liter) [6].

In addition to PhACs, dyes are a critical source of water pollution that make water unfit for drinking. They are widely used in different industries, particularly the textile industry. Due to their toxicity, the presence of organic dyes in water causes health disorders and disrupts aquatic life [7]. Among these dyes, methylene blue (MB) is released in quantity by industry into water sources, posing a health threat due to its toxicity, non-biodegradability, and carcinogenic property [8].

Various advanced technologies have been developed for the removal of dyes and PhACs from water, including advanced oxidation techniques, membrane filtration, adsorption, and membranes with sorption capacities [9–11]. Adsorption stands out due to its low energy consumption, easy operation, and high efficiency [12,13]. In this context, hydrogels have emerged as a suitable alternative. Hydrogels are ideal adsorbents because of their high hydrophilicity, large surface area, and the presence of multiple functional groups on the porous structure [14–16].

Short peptide self-assembly is a rapidly extending research field. Short peptides combine inherent biocompatibility and biodegradability with ease of synthesis and scalability. In addition, peptide self-assembly is well-understood, and both properties and self-assembly can be tuned through control of the peptide sequence [17–22]. Short peptides self-assemble in water, through a combination of different noncovalent interactions, to give hydrogels [23–25]. These supramolecular soft materials have mainly found applications in biomedical fields [26–29], but also find applications in the development of conductive materials [30,31], catalysis [32], crystal growth [33], and water remediation. The efficiency of peptides and peptide-based hydrogels for oil spill recovery [34–36], removal of dyes [37–39], and metal cations [38,40] from water is well-documented in the literature. However, their application to the removal of PhACs from water is still at its beginning. Only recently, Giuri et al. reported the use of Boc-protected tripeptide hydrogels for the removal of diclofenac from water [41].

We recently reported the pyrene–pentapeptide conjugate **1** (Figure 1), and demonstrated that metal cations trigger its self-assembly in water to form metallo-hydrogels that efficiently adsorbed cationic dyes from water [42]. Although hydrogel formation was also achieved by simple pH variation in the absence of cationic metals, these pH-triggered hydrogels suffered from low mechanical stability and fell apart upon dye adsorption. To increase the environmental compatibility of the adsorbent, it is desirable to develop a metal-free peptide hydrogel. In this paper, we report a novel pyrene–tetrapeptide conjugate **2** able to form self-supporting hydrogels at acidic pH. The hydrogels efficiently removed cationic and zwitterionic dyes from water, and promising results were obtained for the adsorption of PhACs (Figure 1).

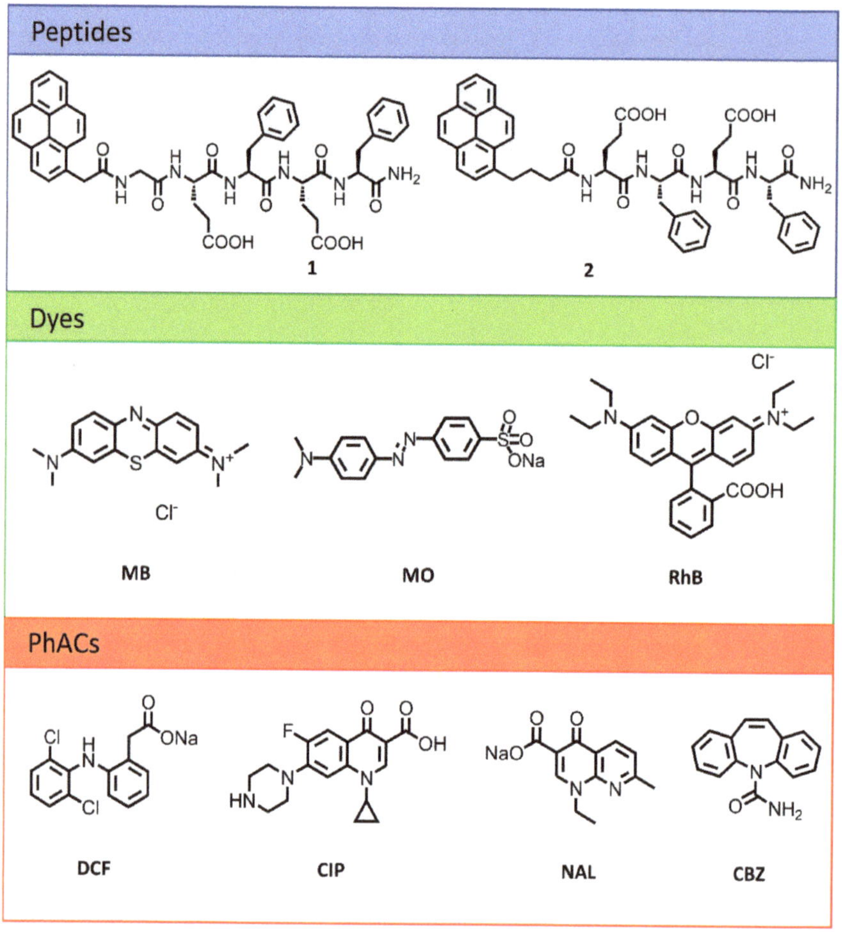

Figure 1. Structure of previously reported peptide **1**, and structures of peptide **2**, dyes, and PhACs employed in the present work.

2. Results and Discussion

2.1. Peptide Design and Synthesis

The structure of peptide **2** has alternating polar charged glutamic acids and hydrophobic phenylalanine residues. This pattern is known to favor the formation of β-sheets [43]. Here, the C-terminal has been amidated, while the N-terminal has been capped with a pyrene moiety, so the solubility of **2** in water is governed by the charge state of the Glu side chains. The large aromatic surface of pyrene can establish additional π–π interactions that facilitate self-assembly and interaction with aromatic pollutants. In addition, the fluorescence of this chromophore may provide information about self-assembly and allows for easy detection of peptide leaching [44]. The main differences between molecules **1** and **2** are the absence of the Gly residue in **2** and the introduction of a larger methylene spacer between the chromophore and the peptide. These structural modifications favor hydrogel formation at a lower gelator concentration of **2** compared to **1**.

Peptide **2** was synthesized using standard Fmoc-based solid-phase peptide synthesis protocols on rink-amide resin, as described in the Materials and Methods section. The peptide was obtained in 85% yield, and its identity was assessed by nuclear magnetic resonance spectroscopy (NMR), electrospray mass spectrometry (ESI-MS), and Fourier transform infrared spectroscopy (FT-IR) (see Scheme S1 and Figures S2–S4 of the Supporting Information).

2.2. Gelation Ability

The presence of two hydrophilic pH-sensitive amino acid residues (Glu) provides good solubility in basic water and introduces pH responsiveness.

As expected, **2** is not soluble in Milli-Q water alone (even after heating), but a transparent solution was obtained after the addition of 1.8 equivalents of NaOH 1M solution in combination with sonication. Acidification of the solution using both HCl and glucono-δ-lactone (GdL) led to the formation of self-supporting hydrogels, as assessed by the vial inversion test (Figure 2A and Figures S5 and S6 of the Supporting Information). HCl gels were formed in 10 min, while GdL gels required up to 18 h. Visually, the HCl gel was non-homogeneous, whereas the GdL gel was semitransparent and uniform. This phenomenon is well-known in the literature. When a strong acid such as HCl is used to trigger gelation, gel formation takes place faster than the diffusion of the acid in the solution, and gel spots can be found where the acid was added. Instead, GdL dissolves and diffuses into the solution before hydrolysis, leading to homogeneous gels. The minimum gelation concentration was found to be 0.1% for the GdL gel and 0.3% for the HCl gel. Interestingly, a gel–sol transition was not observed when heating any of the gels (Tgel); instead, at temperatures above 70 °C, the gels shrank, expelling part of the entrapped water (see Figure S5 of the Supporting Information).

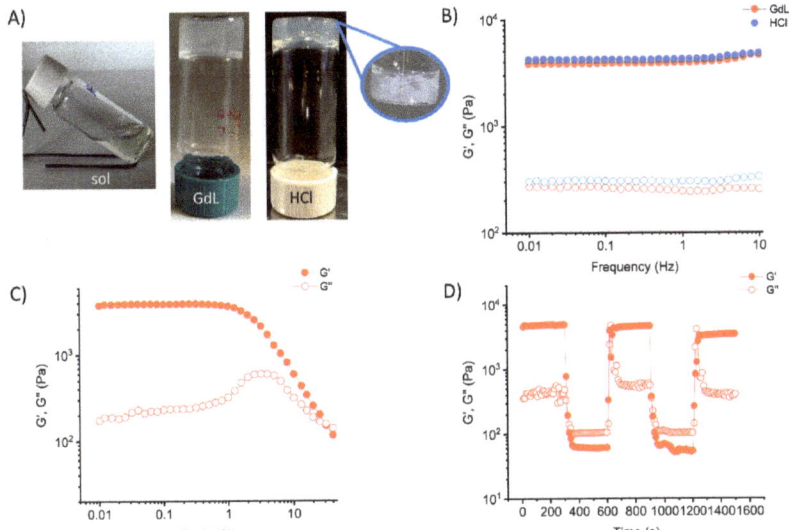

Figure 2. (**A**) Photographs of vials containing from left to right: the peptide at 0.5% in solution, the GdL gel, and the HCl gel; (**B**) frequency sweep of the HCl gel (blue) and the GdL gel (red) at 0.5% concentration, where G' values are indicated with filled circles and G" is indicated with open circles; (**C**) strain sweep of the GdL gel at 0.5% concentration; (**D**) dynamic strain amplitude cyclic test of the GdL gel at 0.5% concentration.

2.3. Gel Characterization

The following studies were performed at a concentration of 0.5%, as the gels could be manipulated without damage.

The gel nature of the samples was confirmed by oscillatory rheology. Frequency sweep experiments (Figure 2B) showed that the elastic modulus (G') of both gels was almost independent of the frequency in the studied range, and approximately one order of magnitude higher than the viscous modulus (G″), as expected for supramolecular gels [45]. Notably, the behavior of the two samples in amplitude sweep experiments differed significantly. The GdL gel displayed a linear viscoelastic behavior up to 1% of strain. Along with increasing applied strain, both G′ and G″ deviated progressively from linearity until the crossover point was reached at 30% of the strain, where the gel–sol transition took place (Figure 2C). In contrast, we were unable to obtain amplitude sweep tests for the HCl sample, since hydrogel fracture occured even at low strain values. Again, the differences between the two samples may be related to the different gelation kinetics. The formation of macroscopic gel domains in the case of HCl gels also resulted in the formation of regions that were weaker and easily broken under strain. We also evaluated the self-recovery ability of the GdL gel using dynamic strain amplitude experiments (Figure 2D). The gel was subjected to several cycles of high strain values (150% strain) to force the gel–sol transition, which were alternated with low strain values (0.1% strain) to allow recovery. As shown in Figure 2D the GdL gel maintained its stability and showed mechanical recovery, though the storage modulus decreased by 8% after three cycles.

Because of the superior mechanical properties and reproducibility of gelation of GdL gels, we decided to continue our studies using exclusively these samples (Data for the HCl gels are given in Figures S7–S9 of the Supporting Information). From the morphological point of view, the GdL gel was investigated via transmission electron microscopy (TEM). Micrographs of the xerogels showed the formation of dense networks of tape-like fibrils (Figure 3). A closer look reveals narrower flat structures sprouting from the tape-like fibrils. Because this peptide is expected to give β-sheet bilayer structures, with the hydrophobic residues buried inside, a possible packing mode is the formation of β-sheet bilayers that are associated through edge-to-edge interactions, mediated by pyrene moieties [46].

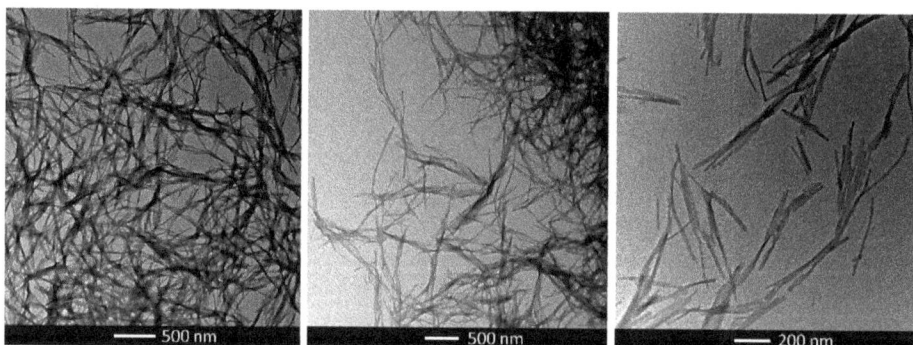

Figure 3. TEM micrographs of GdL xerogels at 0.5% concentration.

The π–π interactions were studied using UV–Vis absorption and emission spectroscopies (Figure 4A and Figure S10 of the supporting Inormation). First, we investigated the behavior of a 50 µM solution of peptide at basic pH. The solution displays the typical absorption profile of pyrene: the S0→S2 transition gives an intense band between 360 nm and 290 nm with the 0-0 peak at 343 nm, whereas the peaks at 327 nm, 313 nm, and 300 nm are the vibronic replicas [44]. At a higher concentration (0.5%), the UV–Vis of the basic solution is characterized by a broadening and redshift of the absorption band. The 0-0 peak moves to 352 nm, and the relative intensity between 0-0 and its vibronic replicas changes. Concerning the emission spectra, the diluted solution shows a band ranging from 350 nm to 500 nm, which corresponds to the nonaggregated pyrene monomer. On the other hand, the 0.5% solution exhibits a broad and unstructured profile with a peak located at 420 nm and a long tail up to 650 nm. For the gelled sample, the broadening of the absorption band further increases, while in the emission spectra, a new broad band ranging from 350 to 700 nm appears. On the basis of these results, we can support the hypothesis that peptide monomers exist only in diluted basic solutions, while at higher concentrations, the hydrophobic pyrene cores tend to interact to each other through π–π stacking, an interaction that is reinforced upon gelation.

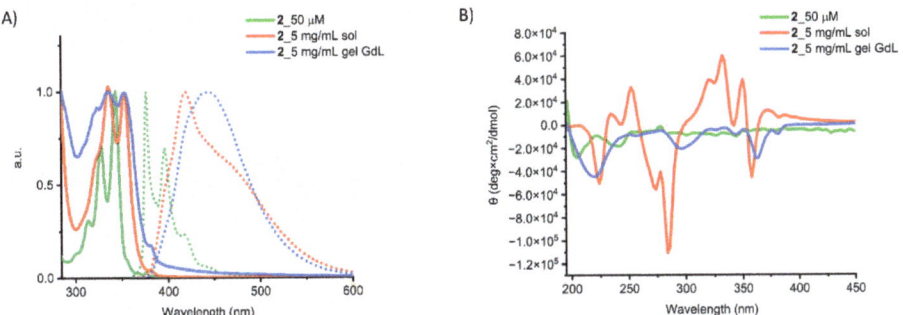

Figure 4. (**A**) UV–Vis absorption (solid line), emission spectra (dotted line), and (**B**) CD spectra of peptide **2** in solution and the GdL gel.

The secondary structure was studied by CD spectroscopy (Figure 4B). In the monomer state, the CD spectrum shows the typical profile of a random coil with disordered structure, characterized by a minimum at 204 nm. When the concentration of the solution was increased to 0.5%, a minimum at 225 nm was observed, indicating the formation of β-sheet structures [47,48]. However, the spectrum is dominated by strong Cotton effects observed in the absorption region of the pyrene chromophore, which are one order of magnitude more intense than the band at 225 nm. Thus, in agreement with absorption and emission data, CD suggests that in concentrated solution, strong π–π interactions between pyrene moieties become established, resulting in a supramolecular chiral arrangement of the chromophores [49]. Upon gelation, the negative band moves to shorter wavelengths, suggesting an increase of β-sheet content. Cotton effects are still observed in the pyrene absorption region, but with an intensity comparable to that of the β-sheet band.

2.4. Adsorption of Dyes

First, GdL gels were tested for the adsorption of methylene blue (MB). Three different experiment setups were investigated: (a) the gel was placed inside a syringe, and the MB solution was allowed to flow through the gel (Figure 5); (b) a preformed gel was immersed in a beaker containing the MB solution; (c) the gel was formed in a vial and the MB solution was casted on top of it. These preliminary studies were carried out under the same conditions (1 mL of GdL gel, 5 mL of MB solution at 50 mg/mL). A simple comparison by the naked eye of the solutions of pollutants after 24 h confirmed that, between the three setups, the first gave the best performances. Therefore, it was adopted for the subsequent studies (see Supporting Information for more details).

Next, we extended our studies to methyl orange (MO) and rhodamine B (RhB) and determined the dye removal efficiency (RE) by UV–Vis absorption spectroscopy using Equation (1) as follows:

$$\text{RE (\%)} = \frac{(C_0 - C_f)}{C_0} \times 100 \tag{1}$$

where C_0 is the initial concentration of pollutant and C_f is the concentration of pollutant in the eluted solution. A very good efficiency was obtained for the adsorption of MB (RE of 90%), while for both MO and RhB, the gel showed a poorer performance. The RE was 50% for RhB and only 16% for the negatively charged MO (Table 1, entries 1–3). No leaching of the gel components was detected in any case and the gel maintained its mechanical stability after dye adsorption, as assessed by inversion of the syringe (Figure 5C and Figure S11 of the Supporting Information). The better adsorption of MB can be justified by electrostatic interactions between the positively charged MB and deprotonated sites present on the GdL gel, which act simultaneously to hydrogen bonding, while the acidic groups present on

RhB and MO may prevent efficient adsorption because of electrostatic repulsions between negative charges.

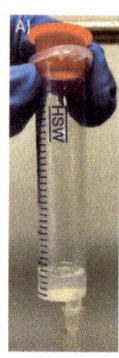
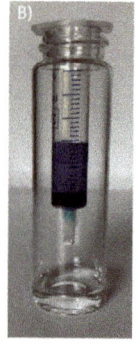

Figure 5. Adsorption experiment setup: (**A**) the gel is placed inside a syringe; (**B**) the solution of MB is allowed to flow through the gel; (**C**) upon elution, the gel maintains its mechanical stability and does not flow when the syringe is turned upside down.

Table 1. Removal efficiency of dyes. 1 mL of GdL-gel at 0.5% (adsorbent dosage 5 mg) was used.

Entry	Dye	C_0 (mg/L)	V (mL)	m_{dye} (µg)	RE [1] (%)
1	MO	50	5	250	16
2	RhB	50	5	250	50
3	MB	50	5	250	90
4	MB	25	5	125	100
5	MB	100	5	500	91
6	MB	50	2.5	125	99.5
7	MB	50	10	500	90

[1] Average values over 3 tests.

In the case of MO, UV–Vis analysis revealed not only a decreased concentration of the pollutant but also a red-shift of the absorption maxima. Analysis of the pH showed that the solution of the pollutant is more acidic upon contact with the hydrogel (pH 4.5). We attributed the shift of the absorption band to the acidic conditions, and, in agreement, the UV–Vis spectrum of MO registered at pH 4.5 was superimposable to that registered upon elution through the gel (see Supporting Information for details). At this pH, MO (pKa = 3.47) is not fully protonated, but both the neutral and anionic forms coexist.

The effect of the initial concentration of the dye on adsorption was investigated for MB. Data are summarized in Table 1, entries 3–5. To investigate the effect of concentration, additional solutions at 25 mg/L and 100 mg/L were prepared, and 5 mL were eluted through 1 mL of gel. Data reveal that RE decreases when increasing the initial concentration of MB from 25 to 50 mg/L, and then increases slightly when going from 50 to 100 mg/L.

The decrease of RE as initial concentration of MB increases can be related to the saturation of active adsorption sites on the hydrogel. The increase of RE with initial concentration deserves more attention. It is noted in the literature that depending on the concentration, MB exits in aqueous solution as monomer, dimer, trimer, or tetramer [50]. Moreover, it may adsorb to the adsorbent surface in any of its forms [51–53]. To gain more insight, we registered the UV–Vis spectra of the adsorbed dye and compared them to the UV–Vis spectra of the initial solutions of MB. Figure 6 shows the UV–Vis spectra of MB in aqueous solution at 25 and 50 mg/L (the one at 100 mg/L gave a saturated signal) and adsorbed MB on the hydrogel after elution of 5 mL of initial solutions of different concentrations.

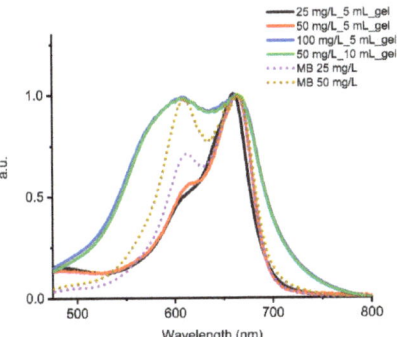

Figure 6. UV–Vis absorption spectra of the initial MB solutions (dotted line) and MB adsorbed on the hydrogel (solid line) upon elution of 5 mL of a 25 mg/L (black), 50 mg/L (red), and 100 mg/L (blue) solution of MB, and upon elution of 10 mL of a 50 mg/L (green) solution of MB.

In the initial solutions (Figure 6, dotted lines), MB shows two bands at 664 nm and 610 nm that have been attributed to the monomer and dimer forms, respectively [50]. As initial concentration decreases, the intensity of the monomer band increases at the expense of the dimer band, indicating, as expected, a decrease in dimer content as MB concentration decreases. In the UV–Vis absorption spectrum of the hydrogel upon elution of a 25 mg/L solution, the dimer band is attenuated significantly, and the spectrum resembles those reported for the monomeric MB in aqueous solution or adsorbed on a surface [54]. Thus, MB initially adsorbs as a monomer even if the initial solution has a high content of dimer. We hypothesized that this first adsorption process is governed by strong MB-adsorbent electrostatic interactions that predominate over the MB–MB interaction responsible for the dimer formation. Upon elution of a 50 mg/L solution, the hydrogel shows that the dimer band intensity increases slightly. When dye concentration is further increased to 100 mg/L, the spectrum changes significantly. The absorption band is broader; monomer and dimer bands have similar intensities, and an intense blue-shifted tail appears. Thus, at higher concentrations, MB first saturates electrostatic adsorption sites, and only the monomer form is found on the surface of the hydrogel. Afterwards, probably with the high concentration as driving force, extra MB molecules are adsorbed, now through π–π interactions and formation of the dimeric form of MB on the surface of the adsorbent is favored. The blue-shifted tail may indicate the formation of higher order aggregates of MB, but we cannot rule out MB–pyrene interactions.

We also investigated how the variation of the volume of initial solution affects the adsorption capability of the hydrogel (Table 1, entries 3, 6, 7). To do so, 2.5, 5, and 10 mL of a MB solution at 50 mg/L were eluted through 1 mL of GdL-gel. We found the same trend when initial MB concentrations were varied at a fixed volume. Thus, the adsorption process seems to be mainly affected by the mass ratio of adsorbent/MB, and it takes place as described in the previous paragraph. Indeed, the UV–Vis spectrum of the hydrogel upon elution of 10 mL of a 50 mg/L solution of MB is superimposable to that obtained after elution of 5 mL of a 100 mg/L solution of the dye (Figure 6).

2.5. Adsorption of PhACs

We then tested the GdL gel for the adsorption of a series of representative PhACs. We chose the NSAIDs diclofenac sodium salt (DCF), the antibiotics ciprofloxacin (CIP) and nalidixic acid sodium salt (NAL) and the anticonvulsant carbamazepine (CBZ). In all cases, 5 mL of a 50 mg/mL pollutant solution was eluted through 1 mL of GdL gel, and the pollutant concentration in the eluted solution was determined by UV–Vis spectroscopy. Under these conditions, we found that the best performances were obtained with DCF (RE 75%), followed by CIP (RE 40%), while CBZ and NAL showed the worst results (RE 21% and 24%, respectively) (Table 2, entries 1–4). That CBZ, with the largest π-surface, is adsorbed worse than DCF suggests that π–π interactions are not the main forces involved in the adsorption process. On the other hand, electrostatic interactions may be somewhat favorable for the case of CIP, containing a basic amine, but this do not justify the better performance in the adsorption of DCF. Both DCF and CIP structures contain groups capable of stablish hydrogen bonds, but also contain halogen atoms, so halogen bonding may also play an important role in the adsorption process. To study the possible role of halogen bonding, we tested the adsorption of the dehalogenated DCF analogue. However, we found that this compound is not stable under the working conditions, and the results were not conclusive (see Supporting Information for details).

Table 2. Removal efficiency of PhACs. 1 mL of GdL gel at 0.5% (adsorbent dosage 5 mg) was used.

Entry	Dye	C_0 (mg/L)	V (mL)	m_{PhAC} (µg)	RE [1] (%)
1	NAL	50	5	250	21
2	CBZ	50	5	250	24
3	CIP	50	5	250	40
4	DCF	50	5	250	75
5	DCF	25	5	125	59
6	DCF	100	5	500	89
7	DCF	50	2.5	125	83
8	DCF	50	10	500	53

[1] Average values over 3 tests.

The effect of the initial concentration on adsorption was investigated for DCF, which gave the best RE among the PhACs tested. Data are summarized in Table 2, entries 4–6. The data reveal that RE increases as the initial concentration increases. On the other hand, when the volume of the solution was changed at a fixed concentration of 50 mg/L, the RE increased when the volume decreased (Table 2, entries 4, 7, 8).

The increase in RE with initial pollutant concentration has been explained in the literature as an improved diffusion and mass transfer process that is favored at higher concentrations, accompanied by an augmented number of collisions between pollutant molecules and the adsorbent, which favors adsorption [55,56]. When the volume of the solution is increased at a fixed initial concentration, we hypothesized that the concentration of the solution on top of the gel does not remain constant throughout the experiment and that diffusion of DCF into the hydrogel may occur faster than the elution process, such that the concentration of DCF in the remaining solution decreases overtime. As a consequence, the driving force of the concentration loses its positive effect.

3. Conclusions

We have described a new peptide–pyrene hybrid gelator that forms stable hydrogels at acidic pH. Hydrogels obtained by acidification with GdL were successfully employed for the removal of a number of dyes and PhACs from water. In particular, the best results were achieved for the dye MB and the anti-inflammatory drug DCF. The effect of initial concentration and volume was studied for the two aforementioned molecules. Data revealed that adsorption of MB is mainly affected by the mass ratio of adsorbent/hydrogel. First, the monomeric form of the cationic dye adsorbed through electrostatic interactions until saturation of the charged adsorption sites. Afterwards, higher concentrations of dye were adsorbed through dimerization/polymerization of the dye on the hydrogel surface. On the other hand, DCF adsorption was favored at higher concentrations, and no saturation behavior was detected in the range of concentrations investigated. In the case of DCF, the concentration appears to be the driving force of the adsorption process: it improves the mass transfer and increases the chance of collision between DCF molecules and the active sites of the hydrogel.

4. Materials and Methods

4.1. General Methods

ESI-MS spectra were recorded using an ESI-TOF MarinerTM BiospectrometryTM Workstation of Applied Biosystems by flow injection analysis (mobile phase methanol with 0.1% formic).

^{1}H and ^{13}C were recorded in deuterated dimethyl sulfoxide at 25 °C at a frequency of 300 MHz. The residual solvent peak was used as internal reference. Chemical shifts (δ) are expressed in parts per million (ppm). The multiplicity of a signal is indicated as s (singlet), d (doublet), t (triplet), dd (doublet of doublets), dt (doublet of triplets), td (Triplet of doublets), q (quartet) and m (multiplet). The acronym "br" indicates a broadened signal.

FT-IR spectra were measured on a FT-IR Perkin-Elmer 1720X spectrophotometer, in KBr disk, using a resolution of 2 cm^{-1}, a total of 100 scans were averaged.

UV–Vis absorption spectra were measured on a Varian Cary 50 spectrophotometer at 25 °C. For solution samples, a reduced volume quartz cell with 0.1 cm, 0.4 cm, or 1 cm optical path was used. Gels were analysed using a rectangular cell with detachable windows and 0.2 mm of optical. Gels were prepared in a glass vial; a small amount was transferred to the sample chamber and the cell was carefully closed to avoid the formation of bubbles

Emission spectra were measured on a Varian CaryEclipse spectrophotometer at 25 °C. A quartz cell with optical path of 10 × 2 mm and volume 500 µL or 10 × 4 mm and volume 1400 µL was used for the solutions. Gels were analysed using a 500 µL quartz cell with an optical path of 10 × 2 mm. Gels were prepared directly inside the cuvette

CD spectra were measured on a Jasco J-1500 instrument at 25 °C and were baseline corrected, a total of 64 measurements were averaged. The spectra are expressed in terms of total molar ellipticity (deg·cm^2·dmol^{-1}). Solutions were analysed using a reduced volume quartz cell with 0.1 cm, 0.4 cm or 1 cm optical path. Gels were analysed using a rectangular cell with detachable windows and optical path of 0.2 mm. Gels were prepared in a glass vial; a small amount was transferred to the sample chamber, and the cell was carefully closed.

TEM images were recorded on a Jeol 300 PX instrument using glow discharged carbon coated grids. Gels were diluted prior to analysis; 10 µL of the sample were then deposited directly on a grid, and the excess of sample was removed with #50 hardened Whatman filter paper; no staining was used. The images were analyzed with the ImageJ program.

Oscillatory rheology was performed on a Kinexus Lab+ rheometer with parallel plate geometry. 1 mL of hydrogel was prepared in a glass mold with a diameter of 2 cm and placed onto the lower plate. A thermal cover was used, and temperature was set at 25 °C using a Peltier temperature controller. Frequency sweeps were recorded between 10–0.001 Hz at a constant strain. Strain sweeps were carried out between 0.01–110% at a constant frequency of 1 Hz. Step strain experiments were performed applying cycles of deformation and recovery steps. The first step (rest conditions) was performed at a fixed frequency of 0.1 Hz and at a strain of $\gamma = 0.1\%$ (within the LVE region) for a period of 5 min. The deformation step was performed at a fixed frequency of 0.1 Hz, applying a constant strain of $\gamma = 150\%$, (above the LVE region) for a period of 5 min. Deformation and recovery steps were repeated three times.

4.2. Synthesis of 2

All reagents and solvents were purchased from Irish Biotech or Merck and were used as received. Compound **2** was synthesized by solid phase 9-fluorenylmethoxycarbonyl (Fmoc) chemistry on Rink amide resin using stablished procedures [42].

ESI-MS: [M + Na]$^+$ calculated for C$_{48}$H$_{49}$N$_5$O$_9$ 862.34, found: 862.5.

^{1}H-NMR (DMSO-d_6, 300 MHz): δ 12.01 (bs, COOH), 8.36 (d, 1H, J = 9.3 Hz), 8.31–7.88 (m, 11H), 7.33 (d, 1H, J = 8.1 Hz), 7.29–6.94 (m, 12H, overlapping signal Ar of Phe and NH$_2$), 4.63–4.33 (m, 2H, H$_\alpha$, Phe), 4.33–4.10 (m, 4H, H$_\alpha$ Glu), 3.06–2.89 (m, 2H, H$_\beta$, Phe), 2.87–2.65 (m, 2H, H$_\beta$, Phe), 2.35–2.07 (m, 6H, overlapping signal H$_\gamma$ Glu and methylene linker), 2.04–1.92 (m, 2H, methylene linker), 1.89–1.5 (m, 4H, H$_\beta$, Glu) ppm.

^{13}C-NMR (DMSO-d_6, 75 MHz): δ 174.43, 174.40, 173.08, 172.76, 171.83, 171.46, 171.12, 138.18, 138.13, 137.04, 131.35, 130.90, 129.76, 129.62, 129.56, 128.63, 128.47, 128.38, 128.04, 127.92, 127.67, 126.96, 126.67, 126.59, 125.39, 125.23, 124.69, 124.62, 124.01, 54.12, 52.62, 52.43, 37.97, 37.53, 35.22, 32.66, 30.61, 30.51, 27.91, 27.77, 27.58, 18.31, 12.53.

FT-IR (KBr): $\tilde{\nu}$ (cm^{-1}) = 3395, 3286, 3086, 3061, 3032, 2940, 1734, 1719, 1650, 1617, 1544, 1498, 1453, 1443, 1415, 1340, 1322, 1315, 1273, 1244, 1236, 1207, 1201, 1182, 1170, 842, 742, 719, 699.

4.3. Gel Preparation

HCl-triggered gelation: A known amount of **2** was introduced in a 4 mL glass vial. Then, 900 μL of MilliQ water was added, followed by NaOH 1M (1.8 eq). A clear solution was obtained with the aid of sonication. Then, MilliQ water and HCl solution (0.5M, 2 eq.) were added to obtain a self-supporting gel with the desired final concentration. Gel formation was assessed by the vial inversion test.

GdL-triggered gelation: A known amount of **2** was introduced in a 4 mL glass vial. Then, 900 μL of milliQ water was added. NaOH 1M (1.8 eq.) was added, and a clear solution was obtained with the aid of sonication. The desired final concentration was reached through the addition of milliQ water. Then, a known amount of GdL (3 eq.) was added, and the mixture was vortexed and left at room temperature overnight. Gel formation was assessed by the vial inversion test.

4.4. Gel Preparation in Syringes

A volume of 1 mL gel at 0.5% concentration was prepared as follows: 5 mg of **2** was introduced to a 4 mL vial. Then, 900 μL of MilliQ water was added. NaOH 1M (1.8 equiv. 11 μL) was added, and a clear solution was obtained with the aid of sonication. MilliQ water was then added to reach a volume of 1 mL. Then, GdL (3 eq., 3.5 mg) was added, and the mixture was vortexed and immediately transferred into a 5 mL syringe that was sealed at the bottom. The mixture was allowed to rest at room temperature overnight to allow gel formation. Gel formation was assessed by the syringe inversion test.

4.5. Adsorption Experiments

Dye and PhAC stock solutions at 200 mg/L were prepared in MilliQ water. All other solutions were prepared by dilution of stock solutions. For all pollutants, UV–Vis absorption calibration curves were obtained in the range of concentrations between 5 mg/L and 50 mg/L. By plotting the maximum absorbance vs. concentration, the molar extinction coefficient for each dye and PhAC was determined (see Supporting Information for details). Adsorption experiments were performed in triplicate; only the average RE value is given. Here, 1 mL of gel at 0.5% concentration was prepared in a syringe. A known volume of dye or PhAC solution was loaded on top of the gel, and the solution was allowed to flow through the gel by gravity. The eluted solution was recovered in a vial. When the pollutant solution had passed through the gel, the eluted solution was analyzed by UV–Vis spectroscopy. The concentration of the pollutant was determined using the Lambert Beer equation $A = \varepsilon C l$, where A is the absorbance, ε the molar extinction coefficient (mol L^{-1} cm^{-1}), C the concentration of pollutant (mol L^{-1}), and l the path length (cm).

Supplementary Materials: The following are available online at https://www.mdpi.com/article/10.3390/gels8100672/s1, Scheme S1: synthesis of **2**, Figure S1: ESI-MS of compound **2**, Figure S2: ^{1}H NMR (DMSO-d6, 300 MHz) of **2**, Figure S3: ^{13}C-NMR (DMSO-d_6, 300 MHz) of **2**, Figure S4: FT-IR spectrum of **2** (KBr disk), Figure S5: Photographs of 0.1% GdL gel before and after heating Figure S6: Photos of HCl-triggered gel, Figure S7: Normalized absorption (solid-line) and emission (dotted-line) spectra of HCl-triggered gel (0.5%), Figure S8: CD spectrum of HCl-triggered gel (0.5%), Figure S9: TEM images of HCl-triggered gel, Figure S10: Non-normalized emission spectra of **2** diluted solution (green), 0.5% solution (red) and GdL gel 0.5% (blue), Figure S11: (Top) inversion of the syringe after the process to demonstrate the stability of the gel after the adsorption of pollutants; (bottom) UV-vis spectrum of eluted MB showing the absence of **2** in the eluted sample, Table S1:

Calibration curve slope and relative wavelength, Table S2. Removal efficiency and adsorption capacities for the adsorption of dyes and PhACs. Figure S12: UV-vis normalized absorption spectra of dyes and PhACs used, Figure S13: Schematic representation of gel formation, Scheme S2: synthesis of dehalogenated DCF, Figure S14: ESI-MS of dehalogenated DCF, Figure S15: ^{1}H NMR (MeOD, 300 MHz) of dehalogenated DCF.

Author Contributions: Conceptualization, A.F. and M.M.; investigation, A.F. and M.M.; writing—original draft preparation, A.F. and M.M.; writing—review and editing, M.M.; visualization, A.F.; supervision, M.M.; project administration, M.M. All authors have read and agreed to the published version of the manuscript.

Funding: This research was funded by University of Padova (P-DiSC#05BIRD2021-UNIPD).

Institutional Review Board Statement: Not applicable.

Informed Consent Statement: Not applicable.

Data Availability Statement: The data presented in this study are available in the article.

Conflicts of Interest: The authors declare no conflict of interest.

References

1. Patel, M.; Kumar, R.; Kishor, K.; Mlsna, T.; Pittman, C.U.; Mohan, D. Pharmaceuticals of Emerging Concern in Aquatic Systems: Chemistry, Occurrence, Effects, and Removal Methods. *Chem. Rev.* **2019**, *119*, 3510–3673. [CrossRef] [PubMed]
2. Glassmeyer, S.T.; Hinchey, E.K.; Boehme, S.E.; Daughton, C.G.; Ruhoy, I.S.; Conerly, O.; Daniels, R.L.; Lauer, L.; McCarthy, M.; Nettesheim, T.G.; et al. Disposal practices for unwanted residential medications in the United States. *Environ. Int.* **2009**, *35*, 566–572. [CrossRef]
3. Valcárcel, Y.; González Alonso, S.; Rodríguez-Gil, J.L.; Gil, A.; Catalá, M. Detection of pharmaceutically active compounds in the rivers and tap water of the Madrid Region (Spain) and potential ecotoxicological risk. *Chemosphere* **2011**, *84*, 1336–1348. [CrossRef]
4. Dolar, D.; Ignjatić Zokić, T.; Košutić, K.; Ašperger, D.; Mutavdžić Pavlović, D. RO/NF membrane treatment of veterinary pharmaceutical wastewater: Comparison of results obtained on a laboratory and a pilot scale. *Environ. Sci. Pollut. Res.* **2012**, *19*, 1033–1042. [CrossRef] [PubMed]
5. Majumder, A.; Gupta, B.; Gupta, A.K. Pharmaceutically active compounds in aqueous environment: A status, toxicity and insights of remediation. *Environ. Res.* **2019**, *176*, 108542. [CrossRef]
6. Kaya, S.I.; Gumus, E.; Cetinkaya, A.; Zor, E.; Ozkan, S.A. Trends in on-site removal, treatment, and sensitive assay of common pharmaceuticals in surface waters. *TrAC Trends Anal. Chem.* **2022**, *149*, 116556. [CrossRef]
7. Kant, R. Textile dyeing industry an environmental hazard. *Nat. Sci.* **2011**, *4*, 22–26. [CrossRef]
8. Khan, I.; Saeed, K.; Zekker, I.; Zhang, B.; Hendi, A.H.; Ahmad, A.; Ahmad, S.; Zada, N.; Ahmad, H.; Shah, L.A. Review on methylene blue: Its properties, uses, toxicity and photodegradation. *Water* **2022**, *14*, 242. [CrossRef]
9. Gadipelly, C.; Pérez-González, A.; Yadav, G.D.; Ortiz, I.; Ibáñez, R.; Rathod, V.K.; Marathe, K.V. Pharmaceutical Industry Wastewater: Review of the Technologies for Water Treatment and Reuse. *Ind. Eng. Chem. Res.* **2014**, *53*, 11571–11592. [CrossRef]
10. Katheresan, V.; Kansedo, J.; Lau, S.Y. Efficiency of various recent wastewater dye removal methods: A review. *J. Environ. Chem. Eng.* **2018**, *6*, 4676–4697. [CrossRef]
11. Polak, D.; Zielińska, I.; Szwast, M.; Kogut, I.; Małolepszy, A. Modification of Ceramic Membranes with Carbon Compounds for Pharmaceutical Substances Removal from Water in a Filtration—Adsorption System. *Membranes* **2021**, *11*, 481. [CrossRef] [PubMed]
12. de Andrade, J.R.; Oliveira, M.F.; da Silva, M.G.C.; Vieira, M.G.A. Adsorption of Pharmaceuticals from Water and Wastewater Using Nonconventional Low-Cost Materials: A Review. *Ind. Eng. Chem. Res.* **2018**, *57*, 3103–3127. [CrossRef]
13. Dutta, S.; Gupta, B.; Srivastava, S.K.; Gupta, A.K. Recent advances on the removal of dyes from wastewater using various adsorbents: A critical review. *Mater. Adv.* **2021**, *2*, 4497–4531. [CrossRef]
14. Seida, Y.; Tokuyama, H. Hydrogel Adsorbents for the Removal of Hazardous Pollutants-Requirements and Available Functions as Adsorbent. *Gels* **2022**, *8*, 220. [CrossRef]
15. Godiya, C.B.; Ruotolo, L.A.M.; Cai, W.Q. Functional biobased hydrogels for the removal of aqueous hazardous pollutants: Current status, challenges, and future perspectives. *J. Mater. Chem. A* **2020**, *8*, 21585–21612. [CrossRef]
16. Okesola, B.O.; Smith, D.K. Applying low-molecular weight supramolecular gelators in an environmental setting—self-assembled gels as smart materials for pollutant removal. *Chem. Soc. Rev.* **2016**, *45*, 4226–4251. [CrossRef]
17. Apostolopoulos, V.; Bojarska, J.; Chai, T.-T.; Elnagdy, S.; Kaczmarek, K.; Matsoukas, J.; New, R.; Parang, K.; Lopez, O.P.; Parhiz, H. A global review on short peptides: Frontiers and perspectives. *Molecules* **2021**, *26*, 430. [CrossRef]
18. Das, R.; Gayakvad, B.; Shinde, S.D.; Rath, J.; Jain, A.; Sahu, B. Ultrashort Peptides-A Glimpse into the Structural Modifications and Their Applications as Biomaterials. *ACS Appl. Bio Mater.* **2020**, *3*, 5474–5499. [CrossRef]

19. Hu, X.; Liao, M.; Gong, H.; Zhang, L.; Cox, H.; Waigh, T.A.; Lu, J.R. Recent advances in short peptide self-assembly: From rational design to novel applications. *Curr. Opin. Colloid Interface Sci.* **2020**, *45*, 1–13. [CrossRef]
20. Ekiz, M.S.; Cinar, G.; Khalily, M.A.; Guler, M.O. Self-assembled peptide nanostructures for functional materials. *Nanotechnology* **2016**, *27*, 402002. [CrossRef]
21. Levin, A.; Hakala, T.A.; Schnaider, L.; Bernardes, G.J.L.; Gazit, E.; Knowles, T.P.J. Biomimetic peptide self-assembly for functional materials. *Nat. Rev. Chem.* **2020**, *4*, 615–634. [CrossRef]
22. Edwards-Gayle, C.J.C.; Hamley, I.W. Self-assembly of bioactive peptides, peptide conjugates, and peptide mimetic materials. *Org. Biomol. Chem.* **2017**, *15*, 5867–5876. [CrossRef] [PubMed]
23. Fichman, G.; Gazit, E. Self-assembly of short peptides to form hydrogels: Design of building blocks, physical properties and technological applications. *Acta Biomater.* **2014**, *10*, 1671–1682. [CrossRef] [PubMed]
24. Mondal, S.; Das, S.; Nandi, A.K. A review on recent advances in polymer and peptide hydrogels. *Soft Matter* **2020**, *16*, 1404–1454. [CrossRef] [PubMed]
25. Kurbasic, M.; Garcia, A.M.; Viada, S.; Marchesan, S. Tripeptide Self-Assembly into Bioactive Hydrogels: Effects of Terminus Modification on Biocatalysis. *Molecules* **2021**, *26*, 173. [CrossRef]
26. Lee, J.; Kim, C. Peptide Materials for Smart Therapeutic Applications. *Macromol. Res.* **2021**, *29*, 2–14. [CrossRef]
27. Das, S.; Das, D. Rational design of peptide-based smart hydrogels for therapeutic applications. *Front. Chem.* **2021**, *9*, 770102. [CrossRef]
28. La Manna, S.; Di Natale, C.; Onesto, V.; Marasco, D. Self-Assembling Peptides: From Design to Biomedical Applications. *Int. J. Mol. Sci.* **2021**, *22*, 12662. [CrossRef]
29. Li, J.; Xing, R.; Bai, S.; Yan, X. Recent advances of self-assembling peptide-based hydrogels for biomedical applications. *Soft Matter* **2019**, *15*, 1704–1715. [CrossRef]
30. Ardoña, H.A.M.; Tovar, J.D. Peptide π-electron conjugates: Organic electronics for biology? *Bioconjugate Chem.* **2015**, *26*, 2290–2302. [CrossRef]
31. Panda, S.S.; Katz, H.E.; Tovar, J.D. Solid-state electrical applications of protein and peptide based nanomaterials. *Chem. Soc. Rev.* **2018**, *47*, 3640–3658. [CrossRef]
32. Tena-Solsona, M.; Nanda, J.; Díaz-Oltra, S.; Chotera, A.; Ashkenasy, G.; Escuder, B. Emergent Catalytic Behavior of Self-Assembled Low Molecular Weight Peptide-Based Aggregates and Hydrogels. *Chem. Eur. J.* **2016**, *22*, 6687–6694. [CrossRef]
33. Conejero-Muriel, M.; Contreras-Montoya, R.; Díaz-Mochón, J.J.; de Cienfuegos, L.Á.; Gavira, J.A. Protein crystallization in short-peptide supramolecular hydrogels: A versatile strategy towards biotechnological composite materials. *CrystEngComm* **2015**, *17*, 8072–8078. [CrossRef]
34. Bachl, J.; Oehm, S.; Mayr, J.; Cativiela, C.; Marrero-Tellado, J.J.; Díaz Díaz, D. Supramolecular Phase-Selective Gelation by Peptides Bearing Side-Chain Azobenzenes: Effect of Ultrasound and Potential for Dye Removal and Oil Spill Remediation. *Int. J. Mol. Sci.* **2015**, *16*, 11766–11784. [CrossRef]
35. Chetia, M.; Debnath, S.; Chowdhury, S.; Chatterjee, S. Self-assembly and multifunctionality of peptide organogels: Oil spill recovery, dye absorption and synthesis of conducting biomaterials. *RSC Adv.* **2020**, *10*, 5220–5233. [CrossRef]
36. Basak, S.; Nanda, J.; Banerjee, A. A new aromatic amino acid based organogel for oil spill recovery. *J. Mater. Chem.* **2012**, *22*, 11658–11664. [CrossRef]
37. Debnath, S.; Shome, A.; Dutta, S.; Das, P.K. Dipeptide-based low-molecular-weight efficient organogelators and their application in water purification. *Chem. Eur. J.* **2008**, *14*, 6870–6881. [CrossRef]
38. Mondal, B.; Bairagi, D.; Nandi, N.; Hansda, B.; Das, K.S.; Edwards-Gayle, C.J.C.; Castelletto, V.; Hamley, I.W.; Banerjee, A. Peptide-Based Gel in Environmental Remediation: Removal of Toxic Organic Dyes and Hazardous Pb^{2+} and Cd^{2+} Ions from Wastewater and Oil Spill Recovery. *Langmuir* **2020**, *36*, 12942–12953. [CrossRef]
39. Wood, D.M.; Greenland, B.W.; Acton, A.L.; Rodríguez-Llansola, F.; Murray, C.A.; Cardin, C.J.; Miravet, J.F.; Escuder, B.; Hamley, I.W.; Hayes, W. pH-Tunable Hydrogelators for Water Purification: Structural Optimisation and Evaluation. *Chem. Eur. J.* **2012**, *18*, 2692–2699. [CrossRef]
40. Roy, K.; Chetia, M.; Sarkar, A.K.; Chatterjee, S. Co-assembly of charge complementary peptides and their applications as organic dye/heavy metal ion (Pb^{2+}, Hg^{2+}) absorbents and arsenic(iii/v) detectors. *RSC Adv.* **2020**, *10*, 42062–42075. [CrossRef]
41. Giuri, D.; D'Agostino, S.; Ravarino, P.; Faccio, D.; Falini, G.; Tomasini, C. Water Remediation from Pollutant Agents by the Use of an Environmentally Friendly Supramolecular Hydrogel. *Chemnanomat* **2022**, *8*, e202200093. [CrossRef]
42. Fortunato, A.; Mba, M. Metal Cation Triggered Peptide Hydrogels and Their Application in Food Freshness Monitoring and Dye Adsorption. *Gels* **2021**, *7*, 85. [CrossRef]
43. Bowerman, C.J.; Nilsson, B.L. Review self-assembly of amphipathic β-sheet peptides: Insights and applications. *Pept. Sci.* **2012**, *98*, 169–184. [CrossRef]
44. Winnik, F.M. Photophysics of preassociated pyrenes in aqueous polymer solutions and in other organized media. *Chem. Rev.* **1993**, *93*, 587–614. [CrossRef]
45. Dawn, A.; Kumari, H. Low Molecular Weight Supramolecular Gels Under Shear: Rheology as the Tool for Elucidating Structure–Function Correlation. *Chem. Eur. J.* **2018**, *24*, 762–776. [CrossRef]
46. Lee, N.R.; Bowerman, C.J.; Nilsson, B.L. Effects of Varied Sequence Pattern on the Self-Assembly of Amphipathic Peptides. *Biomacromolecules* **2013**, *14*, 3267–3277. [CrossRef]

47. Greenfield, N.J.; Fasman, G.D. Computed circular dichroism spectra for the evaluation of protein conformation. *Biochemistry* **1969**, *8*, 4108–4116. [CrossRef]
48. Kelly, S.M.; Jess, T.J.; Price, N.C. How to study proteins by circular dichroism. *Biochim. Biophys. Acta (BBA)-Proteins Proteom.* **2005**, *1751*, 119–139. [CrossRef]
49. Garifullin, R.; Guler, M.O. Supramolecular chirality in self-assembled peptide amphiphile nanostructures. *Chem. Commun.* **2015**, *51*, 12470–12473. [CrossRef]
50. Fernández-Pérez, A.; Marbán, G. Visible Light Spectroscopic Analysis of Methylene Blue in Water; What Comes after Dimer? *ACS Omega* **2020**, *5*, 29801–29815. [CrossRef]
51. Avena, M.J.; Valenti, L.E.; Pfaffen, V.; De Pauli, C.P. Methylene Blue Dimerization Does Not Interfere in Surface-Area Measurements of Kaolinite and Soils. *Clays Clay Miner.* **2001**, *49*, 168–173. [CrossRef]
52. Liu, B.; Wen, L.; Nakata, K.; Zhao, X.; Liu, S.; Ochiai, T.; Murakami, T.; Fujishima, A. Polymeric Adsorption of Methylene Blue in TiO_2 Colloids—Highly Sensitive Thermochromism and Selective Photocatalysis. *Chem. Eur. J.* **2012**, *18*, 12705–12711. [CrossRef]
53. Dolia, V.; James, A.L.; Chakrabarty, S.; Jasuja, K. Dissimilar adsorption of higher-order aggregates compared with monomers and dimers of methylene blue on graphene oxide: An optical spectroscopic perspective. *Carbon Trends* **2021**, *4*, 100066. [CrossRef]
54. Bergmann, K.; O'Konski, C.T. A Spectroscopic Study of Methylene Blue Monomer, Dimer, and Complexes with Montmorillonite. *J. Phys. Chem.* **1963**, *67*, 2169–2177. [CrossRef]
55. Rápó, E.; Tonk, S. Factors Affecting Synthetic Dye Adsorption; Desorption Studies: A Review of Results from the Last Five Years (2017–2021). *Molecules* **2021**, *26*, 5419. [CrossRef]
56. Akhtar, J.; Amin, N.A.S.; Shahzad, K. A review on removal of pharmaceuticals from water by adsorption. *Desalin. Water Treat.* **2016**, *57*, 12842–12860. [CrossRef]

Article

Thermoresponsive Cationic Polymers: PFAS Binding Performance under Variable pH, Temperature and Comonomer Composition

E. Molly Frazar, Anicah Smith, Thomas Dziubla and J. Zach Hilt *

Department of Chemical and Materials Engineering, University of Kentucky, Lexington, KY 40506, USA
* Correspondence: hilt@engr.uky.edu

Abstract: The versatility and unique qualities of thermoresponsive polymeric systems have led to the application of these materials in a multitude of fields. One such field that can significantly benefit from the use of innovative, smart materials is environmental remediation. Of particular significance, multifunctional poly(N-isopropylacrylamide) (PNIPAAm) systems based on PNIPAAm copolymerized with various cationic comonomers have the opportunity to target and attract negatively charged pollutants such as perfluorooctanoic acid (PFOA). The thermoresponsive cationic PNIPAAm systems developed in this work were functionalized with cationic monomers N-[3-(dimethylamino)propyl]acrylamide (DMAPA) and (3-acrylamidopropyl)trimethylammonium chloride (DMAPAQ). The polymers were examined for swelling capacity behavior and PFOA binding potential when exposed to aqueous environments with varying pH and temperature. Comonomer loading percentages had the most significant effect on polymer swelling behavior and temperature responsiveness as compared to aqueous pH. PFOA removal efficiency was greatly improved with the addition of DMAPA and DMAPAQ monomers. Aqueous pH and buffer selection were important factors when examining binding potential of the polymers, as buffered aqueous environments altered polymer PFOA removal quite drastically. The role of temperature on binding potential was not as expected and had no discernible effect on the ability of DMAPAQ polymers to remove PFOA. Overall, the cationic systems show interesting swelling behavior and significant PFOA removal results that can be explored further for potential environmental remediation applications.

Keywords: thermoresponsive; PFAS; water remediation; cationic hydrogel

Citation: Frazar, E.M.; Smith, A.; Dziubla, T.; Hilt, J.Z. Thermoresponsive Cationic Polymers: PFAS Binding Performance under Variable pH, Temperature and Comonomer Composition. *Gels* **2022**, *8*, 668. https://doi.org/10.3390/gels8100668

Academic Editors: Luca Burratti, Paolo Prosposito and Iole Venditti

Received: 20 September 2022
Accepted: 14 October 2022
Published: 18 October 2022

Publisher's Note: MDPI stays neutral with regard to jurisdictional claims in published maps and institutional affiliations.

Copyright: © 2022 by the authors. Licensee MDPI, Basel, Switzerland. This article is an open access article distributed under the terms and conditions of the Creative Commons Attribution (CC BY) license (https://creativecommons.org/licenses/by/4.0/).

1. Introduction

Stimuli-responsive polymers have long been an attractive option for a wide range of applications due to the transformations exhibited upon exposure to external stimuli [1]. Specifically, temperature-responsive polymers have been reported as especially useful in the fields of biomedicine [2,3], drug delivery [3–5], microfluidics [3,6], environmental remediation [7,8] and separations [9]. The unique qualities of these types of systems are many-fold but the most remarkable is a reversible phase change that occurs at a critical solution temperature in aqueous solvent. This behavior can be attributed to a disruption of intra- and intermolecular interactions that cause the polymer to expand or collapse [10]. Thermoresponsive hydrogels have the ability to swell in aqueous environments without dissolving, due to a volume phase transition around their characteristic critical temperature.

The most extensively studied thermoresponsive hydrogels are those based on poly(N-isopropylacrylamide) (PNIPAAm). PNIPAAm polymers undergo phase transitions from hydrophilic to hydrophobic at a lower critical transition temperature (LCST) around 32 °C [11,12]. As illustrated in Figure 1, PNIPAAm polymer chains are hydrated and begin to expand when external temperatures drop below the LCST, resulting in a swollen polymer state. Conversely, PNIPAAm polymer chains become hydrophobic if external temperatures rise above the LCST, causing the polymer network to collapse [12]. This phenomenon has been attributed to hydrogen bond formation/destruction between water molecules and the amide groups present in PNIPAAm [12,13]. Research reporting on the application of such stimuli-responsive hydrogels for environmental remediation purposes is not novel and several well-organized reviews discuss synthesis and application specifics [14–16]. In short, because thermoresponsive hydrogels, such as PNIPAAm, exhibit hydrophilic behavior at a certain temperature range, many aqueous contaminants are allowed to easily diffuse into the hydrogel-based sorbent. This type of binding model provides an alternative to that demonstrated by activated carbon or other traditional sorbents [16]. In addition, thermoresponsive stimulus changes can aid in the removal of environmental contaminants through expansion to drive sorption and contraction to drive desorption [17]. Copolymerization with various comonomers can yield polymers containing functionalities that modify the network properties. The addition of comonomers to PNIPAAm systems can cause shifts in the polymer LCST, subsequently affecting swelling behavior [18]. These materials, however, can be designed to exhibit characteristic properties (e.g., electrostatic interactions with contaminants of interest, such as ubiquitous per- and polyfluoroalkyl substances, PFAS), and the swelling/shrinking properties of these systems have created a platform for the development of useful sorbents that can potentially target and remove contaminants when needed [19].

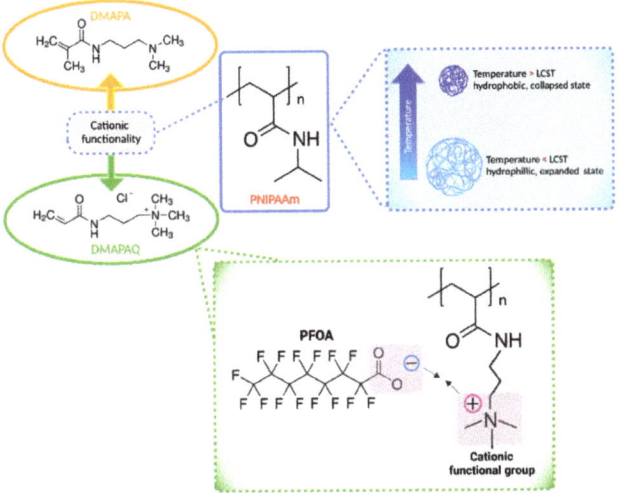

Figure 1. PNIPAAm-based polymers undergo phase changes from hydrophilic to hydrophobic at a lower critical transition temperature (LCST~32 °C) and can be modified with various comonomers to yield functionalities that attract environmental pollutants such as PFOA.

One category of environmental contaminants that could be targeted through application of stimuli-responsive hydrogels is poly- and perfluoroalkyl substances (PFAS). PFAS are a class of an ever-growing number of compounds (now believed to encompass approximately 12,000 compounds) and have been frequently used for their stain and water resistant properties [20]. They have been dubbed as "forever chemicals" due to their extremely persistent nature. In fact, no environmental half-life has been established thus far. Numerous research studies have linked human PFAS exposure to shocking health consequences such as several types of cancer, thyroid, kidney and liver disease, cholesterol dysregulation, developmental and reproductive issues, and immune suppression [21–28]. As such, there has never been a more suitable time for innovative and renewable materials research to address the current state of environmental pollution. Facile functionalization of thermoresponsive hydrogels can produce materials with extremely desirable physical characteristics for exactly this type of application [19].

The thermoresponsive cationic polymers examined in this work have been synthesized by free radical polymerization of a temperature-responsive platform of PNIPAAm crosslinked with N-N'-methylenebis(acrylamide) (NMBA) and various cationic comonomers: (1) dimethylamino propyl acrylamide (DMAPA), and its quaternized sister form, (2) dimethylamino propyl acrylamide, methyl chloride quaternary (DMAPAQ). Herein, we explore the effect of polymer composition, environmental pH, and temperature on hydrogel swelling behavior and fluorinated contaminant binding affinity. Perfluorooctanoic acid (PFOA) was chosen as a model PFAS contaminant and binding studies were conducted by treating PFOA-spiked water samples with the synthesized polymers. Cationic monomers are used to functionalize the polymers to equip them with positively charged moieties that interact and bond with deprotonated PFOA moieties through electrostatic interactions. It was hypothesized that increased addition of cationic comonomer content hinders thermoresponsive behavior and swelling capacity of the PNIPAAm hydrogels while conversely enhancing polymer PFOA affinity. Binding studies conducted at temperatures below polymer LCST are expected to result in higher removal efficiencies than those conducted above LCST temperature.

2. Results and Discussion

Various crosslinked NIPAAm-based copolymers with varying cationic comonomer type and concentrations were successfully synthesized by free-radical polymerization along with a crosslinked PNIPAAm hydrogel that did not contain a cationic comonomer. Thermoresponsive cationic polymers were synthesized by copolymerization of NIPAAm with DMAPA or DMAPAQ, and successful incorporation of the cationic comonomers was confirmed by FTIR analysis (Figures S1 and S2).

2.1. Swelling Behavior

2.1.1. Effect of Aqueous pH

As illustrated in Figure 2, aqueous pH had little discernible effect on PNIPAAm equilibrium mass swelling ratio when examined in both buffered and titrated solutions at 20 °C (held in isothermal water bath). Additionally, the difference in swelling ratio for PNIPAAm in both aqueous systems is not statistically significant and generally average out to be $Q_{eq,buff}$ = 5.4 ± 0.3 over the three pH systems (Table 1). Similarly, titrated aqueous solutions had little effect on the swelling ratio of PNIPAAm at various pH (average $Q_{eq,tit}$ = 5.0 ± 0.3), with the only slight difference occurring between the pH = 4 and 10 systems (Figure 2b). This could be attributed to the effect of increased hydrogen ion interaction with the isopropyl and amino functional groups present in PNIPAAm, allowing for more hydrated polymer chains. It can be noted, however, when comparing swelling kinetics between gels placed in a buffered vs. titrated solution, equilibrium is achieved much more rapidly for the titrated solution gels (Figure 2) likely due to increased electrolyte presence in the buffered solutions that can interact with the polymer chains and slow the hydration process.

At the low comonomer loading percentage, pH has little effect on DMAPA(1) and DMAPAQ(1) hydrogels (Figure 3a,d and Figure 4a,d) at 20 °C, just as is seen with the PNIPAAm gels. For DMAPA(5) and DMAPA(10) loading, however, swelling ratio decreases with increasing pH. This is likely due to the amide functional group present in DMAPA gels, which would potentially become deprotonated at high pH values and thus tend to result in more significant dehydration of the polymer chains. Additionally, note the behavior of DMAPA(5) and DMAPA(10) in the temperature-dependent swelling study, where pH 10 resulted in hydrogel behavior similar to that of PNIPAAm, even as swelling ratios for the pH 4 and pH 7 systems reached much higher values, indicating limited collapse and a significant loss in thermoresponsive behavior (Figures S3 and S4). This trend, however, is not as significantly noticed for the DMAPAQ polymers, although DMAPAQ(5) gels do show similar inclination (Figure S5).

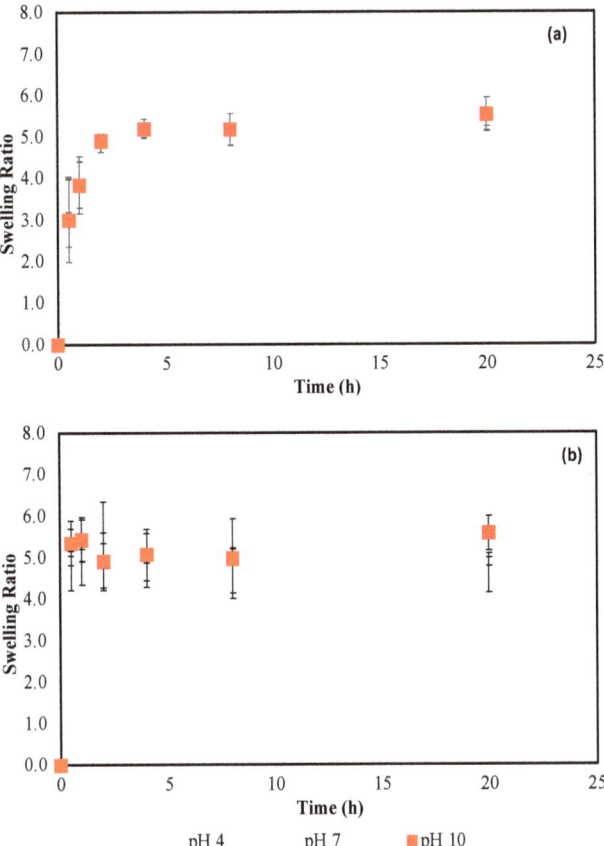

Figure 2. Kinetic swelling behavior of crosslinked PNIPAAm (95 mol%) in various aqueous pH: (**a**) buffered DI-H$_2$O at 20 °C and I = 0.15 M, (**b**) titrated DI-H$_2$O at 20 °C; n = 3, error bars represent ± STD.

Table 1. Comonomer loading percentages for polymer synthesis via free radical polymerization with corresponding equilibrium swelling ratios at 20 °C for various aqueous environments and pH values. Crosslinker loading was consistent for all systems at 5 mol% NMBA; n = 3, numbers in parenthesis represent +/− STD.

Polymer ID	Cationic Comonomer	Comonomer Loading (mol%)	NIPAAm Loading (mol%)	$Q_{eq,buff}$ pH 4	$Q_{eq,buff}$ pH 7	$Q_{eq,buff}$ pH 10	$Q_{eq,tit}$ pH 4	$Q_{eq,tit}$ pH 7	$Q_{eq,tit}$ pH 10
PNIPAAm	–	–	95	5.4(0.2)	5.4(0.2)	5.5(0.4)	4.6(0.4)	4.9(0.1)	5.6(0.4)
DMAPA(1)		1	94	4.5(0.4)	4.5(0.7)	4.6(0.0)	5.0(1.7)	6.1(1.1)	5.5(0.4)
DMAPA(5)	DMAPA	5	90	6.4(0.2)	6.1(0.1)	5.5(0.1)	8.9(1.4)	9.9(0.6)	9.4(0.6)
DMAPA(10)		10	85	6.9(0.1)	6.9(0.2)	5.8(0.0)	7.0(0.6)	6.3(0.3)	5.7(0.6)
DMAPAQ(1)		1	94	6.3(0.2)	6.1(0.2)	5.9(0.2)	8.2(1.0)	8.7(1.0)	8.3(0.2)
DMAPAQ(5)	DMAPAQ	5	90	5.5(0.1)	5.8(0.1)	5.2(0.2)	9.3(0.1)	9.7(0.6)	9.5(0.6)
DMAPAQ(10)		10	85	6.1(0.1)	6.2(0.2)	5.5(0.1)	6.1(0.1)	6.8(0.8)	6.9(1.3)

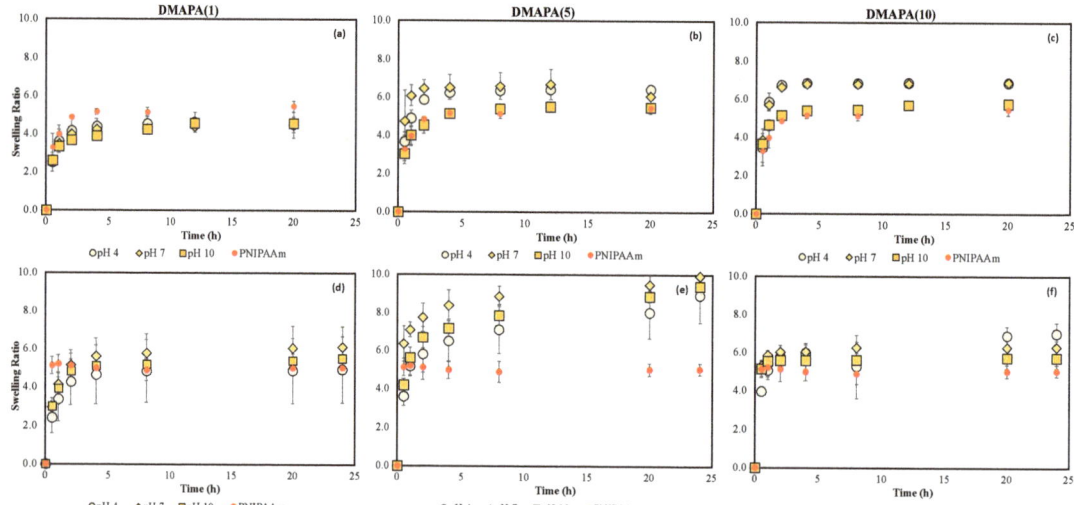

Figure 3. Kinetic swelling behavior of DMAPA hydrogels in the presence of varying pH aqueous solutions at 20 °C: buffered swelling behavior of (**a**) DMAPA(1), (**b**) DMAPA(5), (**c**) DMAPA(10), and titrated DI-H$_2$O swelling behavior of (**d**) DMAPA(1), (**e**) DMAPA(5), (**f**) DMAPA(10). Red circles indicate PNIPAAm swelling averages; n = 3, error bars represent ± STD.

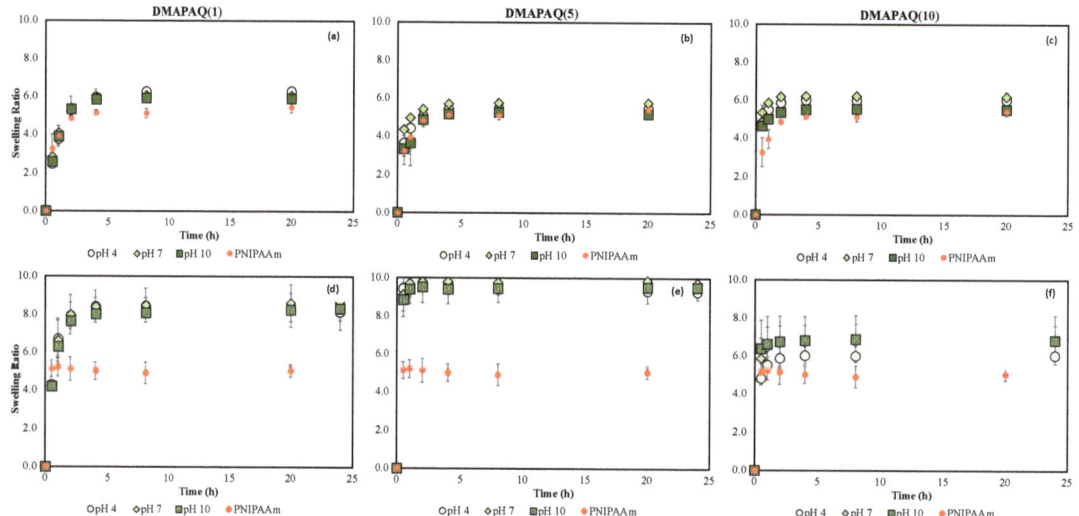

Figure 4. Kinetic swelling behavior of DMAPAQ hydrogels in the presence of varying pH aqueous solutions at 20 °C: buffered swelling behavior of (**a**) DMAPAQ(1), (**b**) DMAPAQ(5), (**c**) DMAPAQ(10), and titrated DI-H$_2$O swelling behavior of (**d**) DMAPAQ(1), (**e**) DMAPAQ(5), (**f**) DMAPAQ(10). Red circles indicate PNIPAAm swelling averages; n = 3, error bars represent ± STD.

DMAPAQ hydrogels show similar results to those of the DMAPA polymers, although they do not exhibit quite the drastic pH dependence (Figure 4. Swelling ratios of higher loading gels decrease when pH is raised to 10 for all temperatures (Figure S5). Unlike the DMAPA gels, however, thermoresponsive behavior only seems to be maintained for DMA-PAQ(5) at pH 10 and not for DMAPAQ(10) at pH 10 (Figures S4 and S5). This would suggest that the quaternary ammonium functional group present in DMAPAQ allows the polymer chain to stay hydrated even as hydrogen ion concentration decreases, bringing insight into how important amino functional groups are for polymer hydration/dehydration.

Comparisons between swelling studies conducted in buffered versus titrated solutions are reported to be significant, as shown in Figures 2–4. Swelling dependence on solution pH was insignificant for PNIPAAm, DMAPA, and DMAPAQ gels in titrated solutions, and only showed some discernible difference in buffered solution for DMAPA(5) and DMAPA(10) gels.

2.1.2. Effect of Comonomer Composition

To examine the impact of cationic comonomer addition, crosslinked PNIPAAm (95 mol%) was used as a control for all swelling studies, and as an effective visual comparison, the averaged PNIPAAm swelling ratios are included on all other copolymer swelling graphs as red circle markers. DMAPA(1) exhibits diminished swelling behavior in buffered solution, $Q_{eq,buff,pH7} = 4.5 \pm 0.7$ (Figure 3a), as opposed to PNIPAAm $Q_{eq,buff,avg} = 5.4 \pm 0.3$, whereas DMAPAQ(1) is seen to have slightly increased swelling behavior at $Q_{eq,buff,pH7} = 6.1 \pm 0.2$ (Figure 4a). Interestingly, out of all the examined systems, the addition of DMAPA(5) and DMAPAQ(5) appeared to have the greatest swelling when placed in the nonbuffered solution where polymer swelling capacity was significantly increased. It is unexpected that the 10 mol% loading would not further increase the swelling of these systems, since comonomer addition is expected to increase polymer hydrophilicity. It is our speculation that interactions between the comonomer and other functionalities present in the PNIPAAm and/or NMBA chains are hindering polymer swelling capacity. Using pH swings in congruence with temperature shifts could offer some additional functionality to the DMAPA and DMAPAQ monomers. It was observed, however, that higher aqueous pH values resulted in thermoresponsive swelling similar to that of PNIPAAm. For instance, for DMAPA(10) hydrogels in a buffered aqueous environment of pH = 7, a swelling ratio of $Q_{eq,buff,pH7} = 6.9 \pm 0.2$ was observed, but the gel was only able to contract back down to $Q_{eq,buff,pH7} = 4.2 \pm 0.4$ at 60 °C if left in the same pH environment. However, if the aqueous environment was altered to pH = 10 after the gel had reached equilibrium, it collapsed to $Q_{eq,buff,pH10} = 1.9 \pm 0.3$ at 60 °C (Figure 3c and Figure S4).

Comparison of the DMAPAQ gels placed in buffered versus titrated solutions reveals an interesting variance. The swelling ratios for DMAPAQ(1) and DMAPAQ(5) polymers were significantly stifled in buffered solutions and remained similar to that of pure PNIPAAm, likely due to interactions with electrolytes present in the buffer solution that were not present in the titrated solutions (Figure 4a,b,d,e). DMAPAQ(1)-titrated swelling ratios are about 20% higher than those of the buffered solution, while DMAPAQ(5)-titrated swelling ratios are around 60% higher than their buffered counterparts. Interestingly, this trend is not as prominent for the higher-loading DMAPAQ(10) gels (Figure 4c,f).

Ultimately, the results reported here show a greater swelling deviation from PNIPAAm for DMAPA and DMAPAQ when examined in titrated solution rather than buffered solution, particularly for DMAPAQ gels.

2.2. PFOA Binding Affinity

To investigate the role of amine-functionalized monomers in thermoresponsive polymer sorbents on legacy PFAS uptake, PFOA removal was evaluated in batch experiments using a higher concentration ([PFOA]$_0$ = 200 µg L^{-1}) than those found in most contaminated water sources, along with a relatively high polymer concentration (2500 mg L^{-1}) (Figure 5). Hydrogel affinity toward PFOA was demonstrated to be significantly affected by pH when placed in buffered solutions (Figure 5a,b). All polymeric systems in buffered solutions showed high removal efficiencies (>90%) for PFOA at pH = 4. This is most likely due to the nature of PFOA itself, which usually assumes a deprotonated state. Increasing pH presumably results in decreased electrostatic interactions between the negatively charged contaminant and positively charged polymer. Furthermore, although it was expected that higher adsorption would occur at temperatures below the PNIPAAm LCST, there appears to be no significant difference between binding studies conducted at 20 °C versus those conducted at 50 °C in buffered solution. DMAPA(5) was the only system that hinted at a thermoresponsive binding trend by achieving removal efficiencies of 73.4% at pH 7 when examined at 20 °C, which is below polymer LCST. Removal efficiency was decreased to 68.0% at pH 7 when binding temperature was above LCST. The addition of the cationic comonomers to PNIPAAm gels did not have a significant impact on PFOA binding affinity when the gels were examined in buffered solution, despite evidence that binding mostly occurs through ionic interactions between anionic PFOA heads and cationic polymer moieties. In fact, PNIPAAm outperformed DMAPAQ systems at buffered aqueous pH 7 and 10. DMAPA gels had slightly higher removal efficiencies than the other systems at pH 10, and this was one of the only instances where the 50 °C binding study showed higher removal than the 20 °C binding study. A higher percentage of removal efficiency achieved by PNIPAAm could most likely be attributed to ionic attraction of the hydrophilic carboxylic head of PFOA to the hydrophilic secondary amine functionality in PNIPAAm.

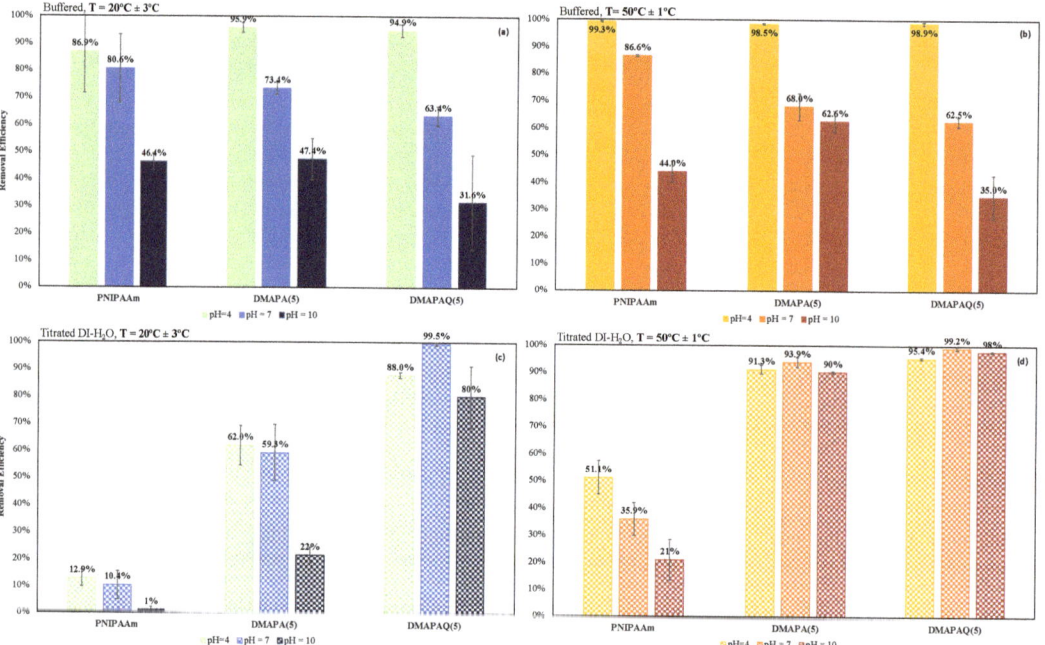

Figure 5. PFOA removal efficiency of PNIPAAm, DMAPA(5), and DMAPAQ(5) hydrogels at various aqueous pH values after 20 h: (**a**) buffered aqueous solution at T = 20 °C, (**b**) buffered aqueous solution at T = 50 °C, (**c**) titrated aqueous solution at T = 20 °C, and (**d**) titrated aqueous solution at T = 50 °C; n = 3, error bars represent ± STD.

Removal efficiencies for the DMAPA(5) and DMAPAQ(5) systems significantly improved when examined in titrated aqueous conditions as compared to those reported in buffered solutions. For example, DMAPA(5) examined in aqueous pH 7 at 50 °C had PFOA removal of 93.9% in titrated solution as compared to 68.0% in buffered solution. Similarly, DMAPAQ(5) saw an increase from 62.5% PFOA removal to 99.2% removal in the same conditions (Figure 5). As expected, the quaternary amine containing DMAPAQ performs consistently well across all pH values. PNIPAAm, DMAPA, and DMAPAQ gels achieved lower removal efficiencies when binding studies were conducted at 20 °C, contradictory to what was originally hypothesized. For gels in titrated aqueous solutions of pH 7 at both 20 and 50 °C, DMAPAQ(5) achieved near 100% removal (within error), which would bring PFOA measurements within acceptable range of EPA lifetime health advisory limits [29]. Absolute capture is an important measure when considering environmental contaminants such as PFOA. Determining an equilibrium state and binding equilibrium constant for the hydrogel sorbents are crucial factors for reporting reliable binding measurements. Additional studies with these materials are needed to establish appropriate contact time and concentration ranges.

3. Conclusions

In this work, a variety of thermoresponsive cationic hydrogels were successfully synthesized via free radical polymerization. The effect of pH on hydrogel swelling behavior was found to be insignificant for PNIPAAm and hydrogels containing loading percentages of 1 and 5 mol% cationic comonomer. Inclusion of cationic comonomers, however, did alter hydrogel swelling capacity, mostly due to losses in thermoresponsive behavior as the comonomer amount was increased. For all three cationic comonomer systems, high loading percentages led to significantly higher swelling ratios that also corresponded to a decrease in the inability to collapse as temperature was increased above the LCST. The only exceptions to this observed behavior were seen at aqueous pH = 10 where DMAPA(10) and DMAPAQ(5) behaved similarly to that of PNIPAAm (Figures S4 and S5). Sorption of PFOA was inversely related to buffered aqueous pH, while cationic monomer type had little noticeable consequence in the buffered solutions. A stark contrast is observed, however, when binding studies are conducted in titrated aqueous environments, indicating that buffer selection can deeply hinder contaminant removal efficiency because of competitive electrostatic interactions. These insights gained from hydrogel performance under variable pH, buffer, temperature, and comonomer composition provide us with a deeper understanding of which polymer functionalities are most beneficial when designing materials for PFAS remediation in aqueous environments.

4. Materials and Methods

4.1. Materials

N-isopropylacrylamide (NIPAAm, Sigma-Aldrich, St. Louis, MO, USA, 97%), N-[3-(dimethylamino)propyl]acrylamide (DMAPA, TCI, $\geq$98.0%), (3-acrylamidopropyl)trimethylammonium chloride (DMAPAQ, Sigma-Aldrich, St. Louis, MO, USA, 75 wt% in H_2O), crosslinker N-N'-methylenebis(acrylamide) (NMBA, BeanTown Chemical, 99%), initiator ammonium persulfate (APS, Sigma-Aldrich, St. Louis, MO, USA, $\geq$98%), catalyst N,N,N',N'-tetramethylethylenediamine (TEMED, VWR, $\geq$98%), perfluorooctanoic acid (PFOA, Sigma-Aldrich, St. Louis, MO, USA, 95%) were used as received. Ultrapure water (resistivity 18.2 MΩ) was used in all synthesis reactions and subsequent experiments.

4.2. Hydrogel Synthesis

Preparation of PNIPAAm hydrogel systems was conducted via free radical polymerization reactions (Table 1). The calculated amounts of NIPAAm and cationic comonomer were dissolved in ultrapure water (2 mL) with feed ratios of 95/0, 94/1, 90/5, or 85/10 mol% and crosslinker NMBA kept constant at 5 mol%. A 0.5 mg/mL initiator stock solution was prepared by dissolving 50 mg of APS in 1 mL ultrapure water and was added to the reactant solution at 0.1 wt% combined weight of NIPAAm, comonomer, and NMBA. The catalyst TEMED was added at 2 wt% combined weight of NIPAAm, comonomer, and NMBA. Vortex mixing was conducted for 10 s before the reactant solution was transferred to glass templates to create hydrogel sheets with dimensions of 1 mm by 12 mm by 24 mm. The polymerization reaction was allowed to proceed for 24 h at ambient conditions. The synthesized hydrogels were immersed in ultrapure water for an additional 24 h at ambient temperature to ensure removal of any unreacted monomers, initiator, or catalyst, during which the wash water was replaced with fresh water at least 3 times. After washing, the polymers were cut into small rectangular pieces and oven dried at 75 °C for 24 h. Dried polymers were either stored as is or ground with mortar and pestle into fine powder to be used for subsequent swelling and binding studies.

4.3. Hydrogel Characterization

4.3.1. FTIR Analysis

Attenuated total reflectance Fourier transform infrared (ATR-FTIR) was used to confirm successful incorporation of the cationic comonomers into the synthesized hydrogels with a Varian Inc. 7000e spectrometer. Dried samples were placed on a diamond ATR crystal and spectrums were obtained between 700 and 4000 cm^{-1}.

4.3.2. Kinetic Swelling Study

Swelling kinetics of each hydrogel were determined via gravimetric analysis in both buffered and titrated DI-H_2O aqueous solutions. Buffered solutions were prepared at pH = 4 and 7 using a phosphate citrate buffer and at pH = 10 using an ammonia buffer. Titrated DI-H_2O solutions were prepared through slow addition of either 1 M NaOH or 2 M HCl to DI-H_2O to achieve aqueous pH of 4, 7, or 10. An approximately 10.0 mg piece of dry gel was immersed in 5 mL of aqueous solution at 20 °C in an isothermal water bath and measurements were taken at regular intervals of 0.5, 1, 2, 4, 8, 12 and 20 h. To do so, each sample was removed from the solution, gently patted with a Kimwipe to remove excess surface water, and quickly weighed on an analytical balance. The equilibrium mass swelling ratio (Q_{eq}) of the hydrogel in buffered aqueous solution was calculated according to Equation (1):

$$Q_{eq} = \frac{m_s}{m_d} \quad (1)$$

where m_s (mg) and m_d (mg) are the sample weight of the swollen and dried hydrogel samples, respectively.

4.3.3. Temperature-Dependent Swelling Study

Temperature responsiveness of each hydrogel was examined by allowing an approximately 10.0 mg piece of dry gel to equilibrate in 5 mL of aqueous solution (pH = 4, 7, or 10 of buffered solution as described above) for 24 h at various solution temperatures. Swelling ratios were measured at temperatures of 10, 20, 25, 30, 35, 40, and 50 °C. Mass measurements were collected by the same method described in the kinetic swelling study section above. The mass swelling ratio was calculated using Equation (1).

4.3.4. PFOA Binding Affinity

Environmental remediation proof-of-concept experiments were conducted by examining the PFOA binding potential of the synthesized polymers through straightforward equilibrium binding studies. Approximately 2.5 mg/mL of dried granulated sorbent (only cationic polymers DMAPA(5) and DMAPAQ(5) were examined along with PNIPAAm and no sorbent as a negative control) was added to aqueous solutions (pH = 4, 7, or 10 of buffered or titrated solutions as described above) spiked with 200 ppb PFOA in glass vials. The system was then agitated on an orbital shaker for 20 h at either 20 or 50 °C. Subsequently, all samples (including controls) were filtered via 0.2 μm syringe tip filter before analysis via liquid chromatography mass spectrometry (LC–MS/MS). Ultraperformance liquid chromatography (UPLC) coupled with electrospray ionization tandem mass spectrometry was used for analysis of PFOA concentration. Instrumentation included a bench-top binary prominence Shimadzu chromatograph (Model: LC-20 AD) equipped with a SIL 20 AC HT autosampler interfaced with an AB SCIEX Flash Quant mass spectrometer (MS/MS) (Model: 4000 Q TRAP). Limit of detection (LOD) for target analytes were 0.25 ng/L at S/N 1/4 4. Seven calibration points with linear dynamic range (LDR) were within 2.5–320 ng/mL and had R^2 values of 0.99968. For all PFAS binding experiments, the pollutant removal efficiency by the hydrogel sorbents was calculated as:

$$Removal\ Percentage\ (\%) = \frac{C_0 - C_t}{C_0} \times 100 \qquad (2)$$

where C_0 (µg L^{-1}) is the initial concentration of PFAS and C_t (µg L^{-1}) is the concentration of PFAS at time (t). The initial concentration C_0 was obtained from the average concentration of negative control samples to account for loss of pollutant from experimental conditions.

Supplementary Materials: The following supporting information can be downloaded at: https://www.mdpi.com/article/10.3390/gels8100668/s1, Figure S1: FTIR spectra for DMAPA hydrogels; Figure S2: FTIR spectra for DMAPAQ hydrogels; Figure S3: Equilibrium temperature-responsive swelling behavior of crosslinked PNIPAAm (95 mol%) in various buffered aqueous pH solutions at t = 24 h; n = 3, error bars represent +/− STD; Figure S4: Equilibrium temperature-responsive swelling behavior of crosslinked DMAPA hydrogels in various pH buffered aqueous solutions at t = 24 h: (a) DMAPA(1), (b) DMAPA(5), and (c) DMAPA(10). Red circles indicate PNIPAAm swelling averages; N = 3, error bars represent +/− STD; Figure S5: Equilibrium temperature-responsive swelling behavior of crosslinked DMAPAQ hydrogels in various pH buffered aqueous solutions at t = 24 h: (a) DMAPAQ(1), (b) DMAPAQ(5), and (c) DMAPAQ(10). Red circles indicate PNIPAAm swelling averages; n = 3, error bars represent +/− STD.

Author Contributions: Conceptualization, E.M.F., T.D. and J.Z.H.; methodology, E.M.F. and A.S.; formal analysis, E.M.F.; investigation, E.M.F. and A.S.; writing—original draft preparation, E.M.F.; writing—review and editing, T.D. and J.Z.H.; visualization, E.M.F.; supervision, T.D. and J.Z.H.; project administration, J.Z.H.; funding acquisition, T.D. and J.Z.H. All authors have read and agreed to the published version of the manuscript.

Funding: This research was funded by the National Institute of Environmental Health Sciences (NIEHS) of the National Institute of Health under award number P42ES007380. The content is solely the responsibility of the authors and does not necessarily represent the official views of the National Institutes of Health.

Institutional Review Board Statement: Not applicable.

Informed Consent Statement: Not applicable.

Data Availability Statement: The data presented in this study are available on request from the corresponding author.

Conflicts of Interest: The authors declare no conflict of interest.

References

1. Wei, M.; Gao, Y.; Li, X.; Serpe, M.J. Stimuli-Responsive Polymers and Their Applications. *Polym. Chem.* **2017**, *8*, 127–143. [CrossRef]
2. Blum, A.P.; Kammeyer, J.K.; Rush, A.M.; Callmann, C.E.; Hahn, M.E.; Gianneschi, N.C. Stimuli-Responsive Nanomaterials for Biomedical Applications. *J. Am. Chem. Soc.* **2015**, *137*, 2140–2154. [CrossRef] [PubMed]
3. Frazar, E.M.; Shah, R.A.; Dziubla, T.D.; Hilt, J.Z. Multifunctional Temperature-Responsive Polymers as Advanced Biomaterials and Beyond. *J. Appl. Polym. Sci.* **2020**, *137*, 48770. [CrossRef] [PubMed]
4. Bawa, P.; Pillay, V.; Choonara, Y.E.; du Toit, L.C. Stimuli-Responsive Polymers and Their Applications in Drug Delivery. *Biomed. Mater.* **2009**, *4*, 22001. [CrossRef]
5. Schmaljohann, D. Thermo- and PH-Responsive Polymers in Drug Delivery. *Adv. Drug Deliv. Rev.* **2006**, *58*, 1655–1670. [CrossRef]
6. Liu, Y.; Zhang, K.; Ma, J.; Vancso, G.J. Thermoresponsive Semi-IPN Hydrogel Microfibers from Continuous Fluidic Processing with High Elasticity and Fast Actuation. *ACS Appl. Mater. Interfaces* **2017**, *9*, 901–908. [CrossRef]
7. Ganesh, V.A.; Baji, A.; Ramakrishna, S. Smart Functional Polymers—A New Route towards Creating a Sustainable Environment. *RSC Adv.* **2014**, *4*, 53352–53364. [CrossRef]
8. Gray, H.N.; Bergbreiter, D.E. Applications of Polymeric Smart Materials to Environmental Problems. *Environ. Health Perspect.* **1997**, *105*, 55–63. [CrossRef]
9. Ju, X.J.; Zhang, S.B.; Zhou, M.Y.; Xie, R.; Yang, L.; Chu, L.Y. Novel Heavy-Metal Adsorption Material: Ion-Recognition P(NIPAM-Co-BCAm) Hydrogels for Removal of Lead(II) Ions. *J. Hazard. Mater.* **2009**, *167*, 114–118. [CrossRef]

10. Phillips, D.J.; Gibson, M.I. Towards Being Genuinely Smart: "isothermally-Responsive" Polymers as Versatile, Programmable Scaffolds for Biologically-Adaptable Materials. *Polym. Chem.* **2015**, *6*, 1033–1043. [CrossRef]
11. Tang, S.; Bhandari, R.; Delaney, S.P.; Munson, E.J.; Dziubla, T.D.; Hilt, J.Z. Synthesis and Characterization of Thermally Responsive N-Isopropylacrylamide Hydrogels Copolymerized with Novel Hydrophobic Polyphenolic Crosslinkers. *Mater. Today Commun.* **2017**, *10*, 46–53. [CrossRef] [PubMed]
12. Schild, H.G. Poly (N-Isopropylacrylamide): Experiment, Theory and Application. *Prog. Polym. Sci.* **1992**, *17*, 163–249. [CrossRef]
13. Zhang, Q.; Weber, C.; Schubert, U.S.; Hoogenboom, R. Thermoresponsive Polymers with Lower Critical Solution Temperature: From Fundamental Aspects and Measuring Techniques to Recommended Turbidimetry Conditions. *Mater. Horizons* **2017**, *4*, 109–116. [CrossRef]
14. Shah, R.A.; Frazar, E.M.; Hilt, J.Z. Recent Developments in Stimuli Responsive Nanomaterials and Their Bionanotechnology Applications. *Curr. Opin. Chem. Eng.* **2020**, *30*, 103–111. [CrossRef]
15. Du, H.; Shi, S.; Liu, W.; Teng, H.; Piao, M. Processing and Modification of Hydrogel and Its Application in Emerging Contaminant Adsorption and in Catalyst Immobilization: A Review. *Environ. Sci. Pollut. Res.* **2020**, *27*, 12967–12994. [CrossRef]
16. Gutierrez, A.M.; Frazar, E.M.; Maria, M.V.; Paul, P.; Hilt, J.Z. Hydrogels and Hydrogel Nanocomposites: Enhancing Healthcare Through Human and Environmental Treatment. *Adv. Healthc. Mater.* **2022**, *11*, e2101820. [CrossRef]
17. Tang, L.; Wang, L.; Yang, X.; Feng, Y.; Li, Y.; Feng, W. Poly(N-Isopropylacrylamide)-Based Smart Hydrogels: Design, Properties and Applications. *Prog. Mater. Sci.* **2021**, *115*, 100702. [CrossRef]
18. Yoshida, R.; Sakai, S.K.; Okano, T.; Sakurai, Y. Modulating the Phase Transition Temperature and Thermosensitivity in N-Isopropylacrylamide Copolymer Gels. *J. Biomater. Sci. Polym. Ed.* **1994**, *6*, 585–598. [CrossRef] [PubMed]
19. Karbarz, M.; Mackiewicz, M.; Kaniewska, K.; Marcisz, K.; Stojek, Z. Recent Developments in Design and Functionalization of Micro- and Nanostructural Environmentally-Sensitive Hydrogels Based on N-Isopropylacrylamide. *Appl. Mater. Today* **2017**, *9*, 516–532. [CrossRef]
20. ITRC. Per- and Polyfluoroalkyl Substances (PFAS) Fact Sheets. 2017, pp. 1–2. Available online: https://pfas-1.itrcweb.org/ (accessed on 21 September 2020).
21. Bonefeld-Jorgensen, E.C.; Long, M.; Bossi, R.; Ayotte, P.; Asmund, G.; Krüger, T.; Ghisari, M.; Mulvad, G.; Kern, P.; Nzulumiki, P.; et al. Perfluorinated Compounds Are Related to Breast Cancer Risk in Greenlandic Inuit: A Case Control Study. *Environ. Health A Glob. Access Sci. Source* **2011**, *10*, 88. [CrossRef]
22. Barry, V.; Winquist, A.; Steenland, K. Perfluorooctanoic Acid (PFOA) Exposures and Incident Cancers among Adults Living near a Chemical Plant. *Environ. Health Perspect.* **2013**, *121*, 1313–1318. [CrossRef] [PubMed]
23. Melzer, D.; Rice, N.; Depledge, M.H.; Henley, W.E.; Galloway, T.S. Association between Serum Perfluorooctanoic Acid (PFOA) and Thyroid Disease in the U.S. National Health and Nutrition Examination Survey. *Environ. Health Perspect.* **2010**, *118*, 686–692. [CrossRef] [PubMed]
24. Shankar, A.; Xiao, J.; Ducatman, A. Perfluoroalkyl Chemicals and Chronic Kidney Disease in US Adults. *Am. J. Epidemiol.* **2011**, *174*, 893–900. [CrossRef] [PubMed]
25. Bassler, J.; Ducatman, A.; Elliott, M.; Wen, S.; Wahlang, B.; Barnett, J.; Cave, M.C. Environmental Perfluoroalkyl Acid Exposures Are Associated with Liver Disease Characterized by Apoptosis and Altered Serum Adipocytokines. *Environ. Pollut.* **2019**, *247*, 1055–1063. [CrossRef]
26. Graber, J.M.; Alexander, C.; Laumbach, R.J.; Black, K.; Strickland, P.O.; Georgopoulos, P.G.; Marshall, E.G.; Shendell, D.G.; Alderson, D.; Mi, Z.; et al. Per and Polyfluoroalkyl Substances (PFAS) Blood Levels after Contamination of a Community Water Supply and Comparison with 2013–2014 NHANES. *J. Expo. Sci. Environ. Epidemiol.* **2019**, *29*, 172–182. [CrossRef]
27. Oulhote, Y.; Steuerwald, U.; Debes, F.; Weihe, P.; Grandjean, P. Behavioral Difficulties in 7-Year Old Children in Relation to Developmental Exposure to Perfluorinated Alkyl Substances. *Environ. Int.* **2016**, *97*, 237–245. [CrossRef]
28. Corsini, E.; Luebke, R.W.; Germolec, D.R.; DeWitt, J.C. Perfluorinated Compounds: Emerging POPs with Potential Immunotoxicity. *Toxicol. Lett.* **2014**, *230*, 263–270. [CrossRef] [PubMed]
29. EPA. *Drinking Water Health Advisories for PFOA and PFOS*; US EPA: Washington, DC, USA, 2022.

Article

Design and Development of Fluorinated and Biocide-Free Sol–Gel Based Hybrid Functional Coatings for Anti-Biofouling/Foul-Release Activity

Silvia Sfameni [1,2], Giulia Rando [2,3], Maurilio Galletta [3], Ileana Ielo [3], Marco Brucale [4], Filomena De Leo [3], Paola Cardiano [3], Simone Cappello [5], Annamaria Visco [1,6], Valentina Trovato [7], Clara Urzì [3,*] and Maria Rosaria Plutino [2,*]

1. Department of Engineering, University of Messina, Contrada di Dio, S. Agata, 98166 Messina, Italy
2. Institute for the Study of Nanostructured Materials, ISMN—CNR, Palermo, c/o Department of ChiBioFarAm, University of Messina, Viale F. Stagno d'Alcontres 31, Vill. S. Agata, 98166 Messina, Italy
3. Department of ChiBioFarAm, University of Messina, Viale F. Stagno d'Alcontres 31, Vill. S. Agata, 98166 Messina, Italy
4. Institute for the Study of Nanostructured Materials, ISMN—CNR, Bologna, CNR Bologna Research Area, Via Piero Gobetti 101, 40129 Bologna, Italy
5. Institute for Biological Resource and Marine Biotechnology (IRBIM)—CNR of Messina, Spianata S. Raineri 86, 98122 Messina, Italy
6. Institute for Polymers, Composites and Biomaterials, CNR—IPCB, Via Paolo Gaifami 18, 95126 Catania, Italy
7. Department of Engineering and Applied Sciences, University of Bergamo, Viale Marconi 5, 24044 Dalmine, Italy
* Correspondence: urzicl@unime.it (C.U.); mariarosaria.plutino@cnr.it (M.R.P.)

Abstract: Biofouling has destructive effects on shipping and leisure vessels, thus producing severe problems for marine and naval sectors due to corrosion with consequent elevated fuel consumption and higher maintenance costs. The development of anti-fouling or fouling release coatings creates deterrent surfaces that prevent the initial settlement of microorganisms. In this regard, new silica-based materials were prepared using two alkoxysilane cross-linkers containing epoxy and amine groups (i.e., 3-Glycidyloxypropyltrimethoxysilane and 3-aminopropyltriethoxysilane, respectively), in combination with two functional fluoro-silane (i.e., 3,3,3-trifluoropropyl-trimethoxysilane and glycidyl-2,2,3,3,4,4,5,5,6,6,7,7,8,8,9,9-hexadecafluorononylether) featuring well-known hydro repellent and anti-corrosion properties. As a matter of fact, the co-condensation of alkoxysilane featuring epoxide and amine ends, also mixed with two opportune long chain and short chain perfluorosilane precursors, allows getting stable amphiphilic, non-toxic, fouling release coatings. The sol–gel mixtures on coated glass slides were fully characterized by FT-IR spectroscopy, while the morphology was studied by scanning electron microscopy (SEM), and atomic force microscopy (AFM). The fouling release properties were evaluated through tests on treated glass slides in different microbial suspensions in seawater-based mediums and in seawater natural microcosms. The developed fluorinated coatings show suitable antimicrobial activities and low adhesive properties; no biocidal effects were observed for the microorganisms (bacteria).

Keywords: sol–gel technique; anti-fouling properties; fouling release activity; marine bacteria; non-biocide release; amphiphilic coating

1. Introduction

Biofouling, defined as the undesired accumulation given by any association of microorganisms, algae, plants, and marine animals on submerged surfaces, causes over time bio-deterioration of exposed areas of ships, boats, ports, underwater cultural heritage, with progressive loss of economic value, as well as higher costs for their maintenance, and fuel consumption in the case of vessels [1]. In this regard, it has been calculated that the development of new anti-fouling technologies would reduce the fuel consumption for navigation (between 38 and 72%), saving 60 billion dollars and avoiding the emission of about 390 million tons of greenhouse gases each year [2].

In an underwater environment, biological colonization, at the base of biofouling phenomena, is a relatively fast process in which microbial species [3] and the adhesion time vary significantly depending on the geographic area, especially related to environmental conditions (i.e., salinity, pH, temperature, nutrient levels, solar irradiation, etc.) [4,5].

Several research studies have already been carried out on products that prevent marine fouling, namely anti-fouling (AF, hereafter) coatings [6–8]. An ideal AF coating should have the following properties: durability, resistance to external/mechanical agents, easiness to apply, low cost, and non-toxicity for non-target species and marine environment [9].

The AF coatings can be classified into two categories: biocide-release coatings and non-biocide-release coatings, see Figure 1 [10–12]. The biocide-release coatings are based on the dispersion of biocides from different types of polymeric hosting matrices and they are progressively released over time in seawater. Currently, these coatings are the most used; however, in this regard, the problem of toxicity and high maintenance costs still remain a critical issue [13–15]. On the other hand, non-biocide release AF coatings represent the non-toxic and environmentally friendly alternative to anti-fouling coatings containing biocides (also called fouling release activity, FR hereafter) [16,17].

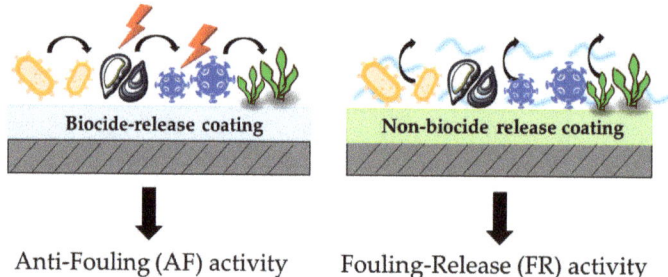

Figure 1. Schematization of anti-fouling and fouling release activity of sol–gel functional coatings.

Two strategies are exploited in the non-biocidal approach: (i) the "separation of stabilized biofoulants", which attempts to reduce as much as possible the force with which the micro-organisms adhere on a surface, facilitating their removal due to the weight of the deposits or by the flow of water generated during navigation; (ii) "prevention of the adhesion of biofoulants", which aims to avoid the formation of a stable fouling film, thus preventing the adhesion of organic molecules that will trigger the bio-settlement process. Many other aspects, related to the characteristics of the materials that can be used to optimize AF or FR strategies, play an important role in controlling the non-desired biofouling process [18–21]. The chemical, physical, mechanical, and structural properties, the mass (elastic modulus, coating thickness, etc.), and topography (i.e., the physical constraints) of the covered surface are all equally important parameters, as they will determine the character of the AF coating itself and the life span of the applied material [22,23]. The coatings designed to solve these needs are usually silicones- and fluoro-polymers based on the strategy of separation of biofoulants [24–26]. Reticulated polyurethane/polysiloxane systems [27–31] and combined coatings based on fluorine and silicon demonstrate high abilities in favoring the separation of stabilized biofoulants on surfaces immersed in the marine environment [32–36].

The enormous adaptability of sol–gel methodology, which is highly controlled, provides some advantages over previous existing procedures such as low process temperatures, good homogeneity products such as low thickness coatings can be obtained, and production of mixed oxides thanks to the stoichiometric control of the composition of the starting solution, better control of the porosity of the material produced by varying the heat treatment, and a high degree of purity, but it presents also some limitations as costs of starting materials, possible formation of fractures during the cross-linking phase and long process times [37–39].

Methods and applications have expanded in tandem with the growing interest in sol–gel technique [40–42].

In this regard, sol–gel-based coatings have already been widely used to enhance the surface properties of different substrates, with which they may bind covalently and steadily, and let the final coated surface show good chemical inertness, resistance to thermal and mechanical stress, and still no cytotoxicity towards human health and environment [43–45].

As a matter of fact, currently, fluorinated long-chain derivatives are still widely employed, thanks to very low surface energy, as functional hydro repellent additives or cross-linkers to improve water repellency, together with chemical- and photo-stability properties of sol–gel-based coating, as well as coated surface.

Hybrid organic–inorganic fluorinated materials were already prepared and used as hydrophobic coatings for conserving lithic substrates [46–48].

The aim of this work is to design and develop fouling release biocide-free coatings, bearing different long-chain fluorinated substituents in combination with other common sol–gel-based cross-linkers [49,50].

Despite the presence of nonpolar fluorinated substituents, very often (bio)fouling is still able to adhere on (super)hydrophobic coatings, leading to water penetration and subsequent coating breaking and run over [51].

Moreover, as mentioned before, hydrophobic coatings thanks to their low surface energy, present a low adhesion force towards polar marine foulants [52,53], but they are not so effective with non-polar foulants characterized by an adhesion highly related to the coating surface energy and wettability, such as some barnacle cyprids and algal zoospores [54] or other non-polar extracellular substances and cell walls components [55].

In this regard, different research studies are still devoted [56–58] to the design and synthesis of amphiphilic FR coatings, bearing both polar and nonpolar groups (i.e., fluoropolymers), either also as hyperbranched or mixed cross-linked networks. With this approach, the overall amount of functional hydrophobic copolymer may be drastically reduced (<15%), since it has been shown that at the water/coating interface, the hydrophobic brushes align with each other and aggregates in a self-assembled monolayer (SAM) form [59]. Furthermore, the presence of a dynamic surface with local variations in surface chemistry, topography, and mechanical properties of such amphiphilic coatings, leads to lower interfacial interactions and therefore a less fouling settlement [60].

Four types of silica precursors (see Figure 2) were used in order to combine the chemical-physical properties of the alkoxysilane cross-linkers subunits containing epoxy and amine groups (i.e., 3-Glycidyloxypropyltrimethoxysilane, GPTMS, and 3-aminopropyltriethoxysilane, APTES; respectively), with those of co-monomers containing functional fluorinated organic compounds (i.e., 3,3,3-trifluoropropyl-trimethoxysilane and glycidyl-2,2,3,3,4,4,5,5,6,6,7,7,8,8,9,9-hexadecafluorononylether).

Four developed (G_A, G_A_F3, G_A_F16, G_A_F3_F16) coatings are obtained by reaction of two bifunctional starting sol–gel precursors, namely APTES and GPTMS (in a concentration ratio of 1:2; hereafter also indicated as A and G, respectively, Figure 2); both of them are bearing a trialkoxysilyl group in one side, and an amine or an epoxy group, on the other, respectively. This bifunctionality of the two reacting ends of A and G favors the development of a stable sol–gel-based 3D matrix, thus guaranteeing complete and stable coverage of the treated coated surface.

Moreover, the amphiphilic coatings show to be very efficient as biocide-free FR coatings, thanks also to synergic actions coming from the polar fouling resistant a polyethylene oxide (PEO)- and polyether amine(PEA)-based, cross-linked matrix and the fouling release F3 and F16 copolymers. Finally, we thought it worthwhile to develop an asymmetric nanostructured hyperbranched polymeric coating using both F3 and F16 co-monomers, i.e., bearing both long and short perfluorinated chains.

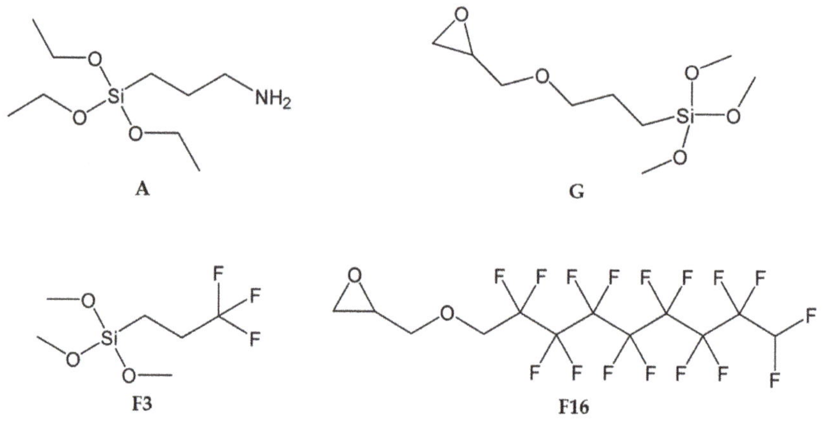

Figure 2. Chemical structure of the employed functional alkoxysilane sol–gel precursors together with the adopted acronyms.

In this way, we would rather try to simulate the lotus effect, with well-separated long chain brushes, in whose cavities (i.e., in correspondence of F3 chains) air may be entrapped, thus preventing the penetration of water and surface wettability.

All coated surfaces were characterized by different chemical–physical, morphological, and rheological techniques, and the good antibacterial and antifouling properties were assessed by the biological test. In particular, biological tests aimed, from one side, to evaluate if the coatings could have released some antibacterial compounds in the liquid medium (biocide-release), while the antifouling effect was evaluated via the reductions of the number of adhering cells on the effective surfaces as compared to the controls.

These results will open the way to the development of eco-friendly, economical, durable, and easy-to-apply matrices that show also interesting fouling release and antibacterial activities, that may find useful applications in blue growth and buildings, as well as for cultural heritage protection (Figure 3).

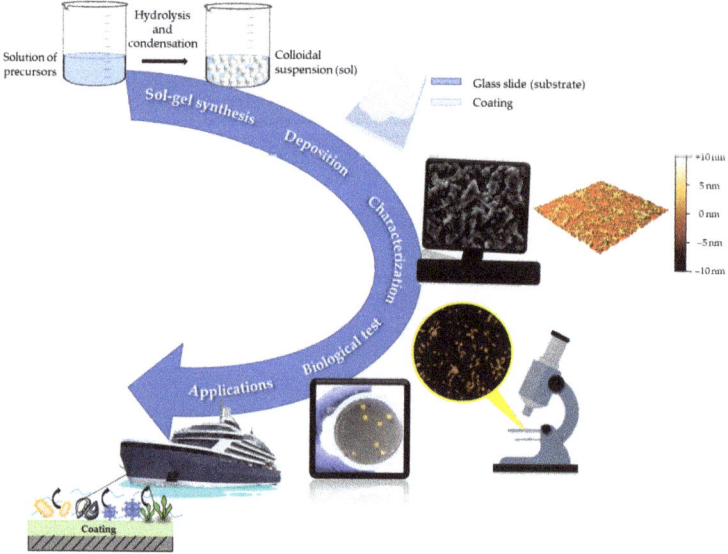

Figure 3. Overall aim of the work, from the FR sol–gel synthesis to blue growth applications.

2. Results and Discussion

The proper design and development of fluorinated, biocide-free amphiphilic sol–gel-based coating has been run, following the synthetic procedures reported in Figures 4 and 5.

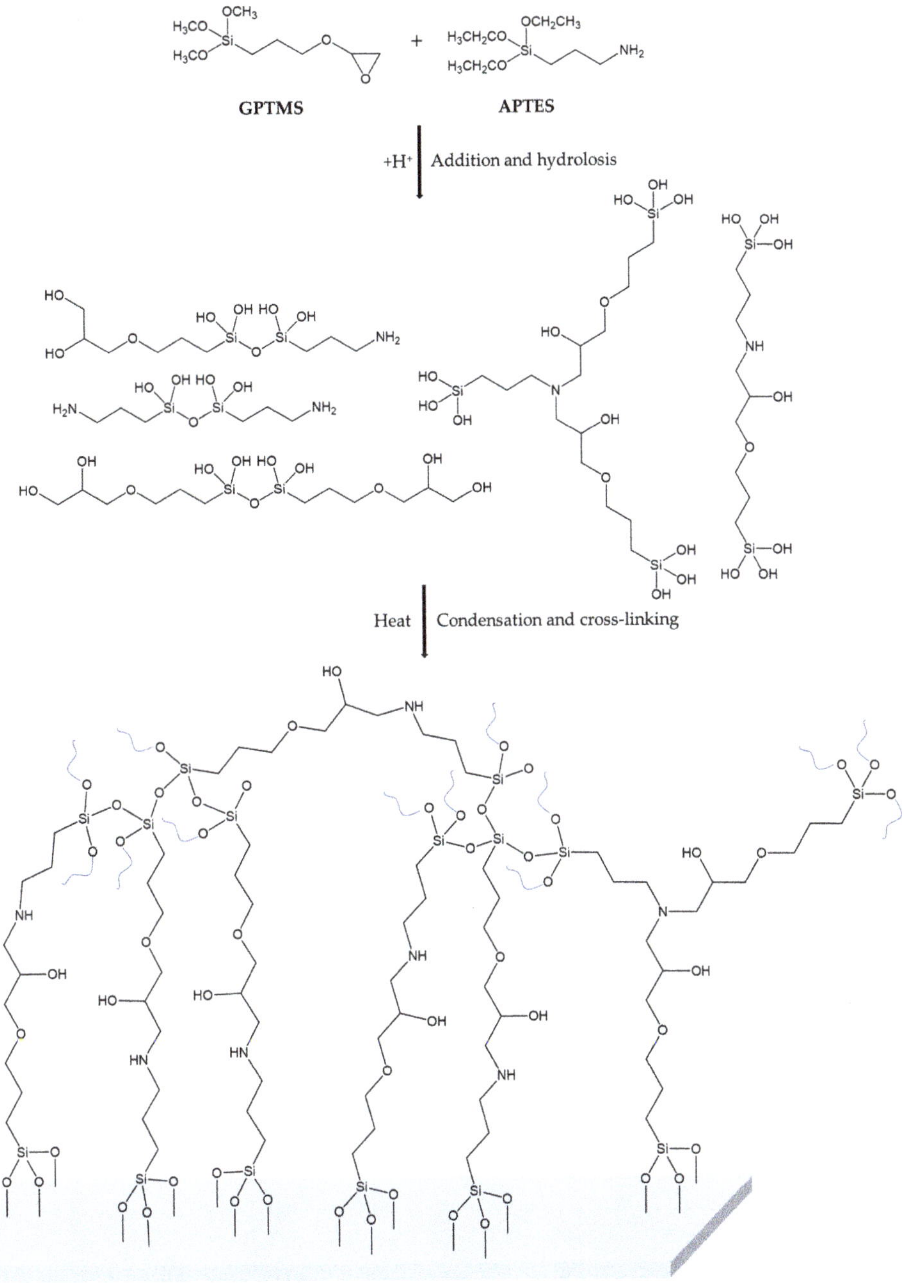

Figure 4. Sol–gel synthesis towards the formation of the cross-linked PEA-PEO-based G_A coating by reaction of G/A (2:1) [61].

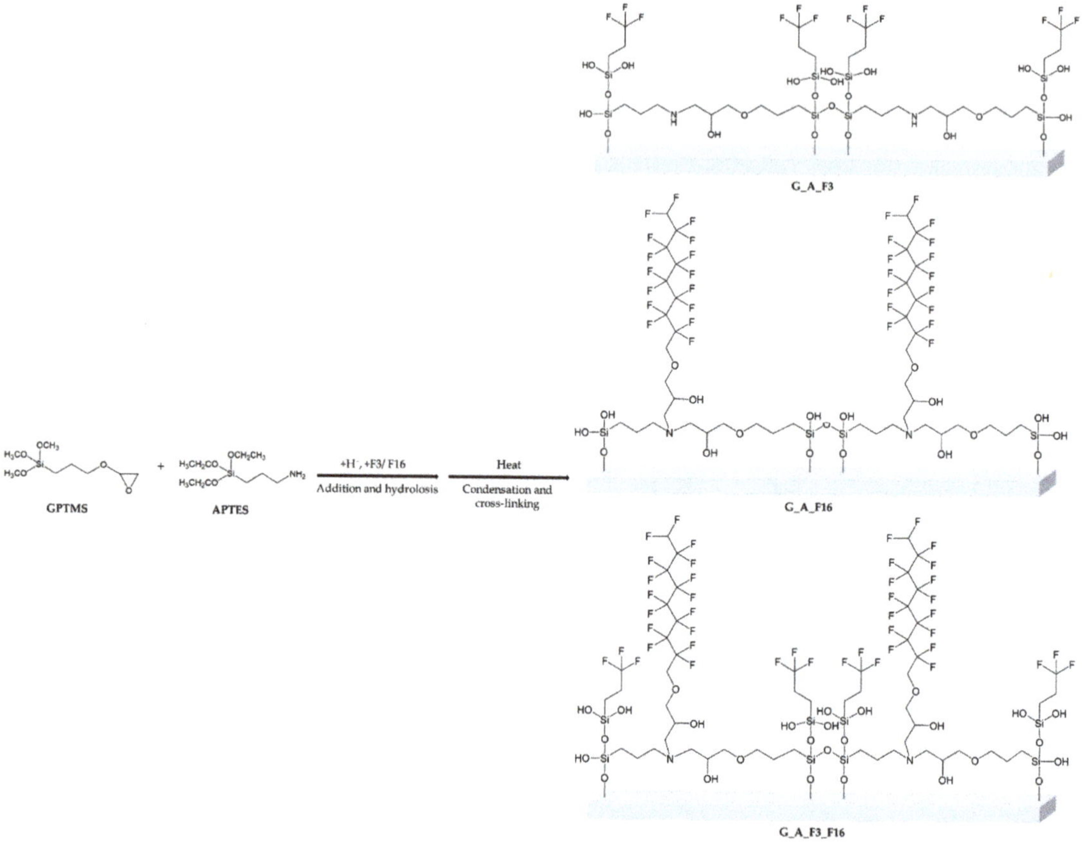

Figure 5. Sol–gel synthesis towards the formation of the three functional G_A_F3, G_A_F16, G_A_F3_F16 coating by reaction of GPTMS/APTES (2:1), and F3 and F16 (overall 5 wt%).

In particular, the amine group is able to react first of all with one or two epoxy group of the G moiety. Reversely the alkoxysilane ends, after a first acidic hydrolysis step, may statistically bound each other or after the application on a glass surface, in the condensation step may bond stably to the glass or either cross link each other, giving rise to a diffuse polar polyether amine (PEA) containing polyethylene oxide (PEO) in the G_A sol–gel-based matrix [61].

Moreover, the other three designed sol–gel functional fluorinated mixtures were also prepared by the addition of an overall 5% of F3, F16, or both F3 and F16, i.e., fluorinated functional long or short chain sol–gel precursors. In this way, the overall sol–gel technique will allow the development of a hybrid functional coating, bearing both polar/hydrophilic components related to the amino, ether, and hydroxyl groups, and nonpolar/hydrophobic –CF$_3$ (F3) and –C$_8$HF$_{16}$ (F16) components, whose FR amphiphilic properties will be tested by mechanical and biological tests.

2.1. Characterization of Sol–Gel Coated Glass Slides

2.1.1. FT-IR Analysis

To investigate the nature of the coatings and to confirm their successful deposition, ATR FT-IR spectra of silane xerogel coatings applied and annealed on glass slides were registered and investigated. The frequencies of major absorption bands are shown in Figure 6 and Table 1, respectively [62].

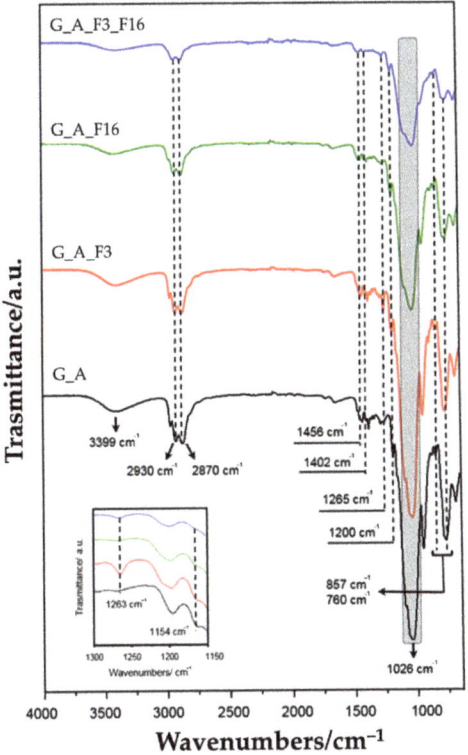

Figure 6. FT-IR spectra of glass slides treated with G_A, G_A F3, G_A F16, and G_A F3_F16 sol.

Table 1. Main vibration modes ascribable to xerogels.

Wavenumbers (cm^{-1})		Reference	Vibrational Modes
On the glass	From literature		
3399	3450–3261	[63,64]	ν (N-H)
2930–2870	2980–2800	[65,66]	
1456	1450		ν (C-H)
1250	1263	[67]	ν (C-F) in CF$_3$
1206	1200	[27,65]	ν (Si-O)
1150	1154	[66,67]	ν (C-F) in CF$_2$
1026	1080	[68,69]	(Si-O-Si)
958	950	[69,70]	(Si-OH)
856	816–847	[27,65]	ν (Si-O-Si)
760	786–749	[27,65]	ν_s (Si-O-Si)

The nature of the hybrid structure silica precursors based on either epoxy or amino groups is strongly influenced by the epoxy ring opening that can follow different well-known subsequent reaction pathways: (a) hydrolysis with formation of diol; (b) alcoholysis with the formation of ethyl ether terminal groups; (c) consecutive polymerization steps to give oligo- or poly(ethylene)oxide groups or a hybrid 3D network; (d) reaction with primary/secondary amino group, thus transformed in secondary/tertiary. Indeed, during the final thermal curing step, each silanol group obtained by sol–gel precursors hydrolysis, can react with each other to form stable siloxane bonds (Si-O-Si). At the same time, further bonds can be formed by glycidyloxy groups able to react both with themselves and with hydroxyl groups of hydrolyzed precursors. Consequently, the polyaddition reaction of the opened epoxy groups, with the formation of Si-O-C bonds, can increase the flexibility of the network, favoring a great homogeneity between the organic and inorganic components of the so-obtained network.

In the coating (G_A) obtained by the combination of GPTMS and APTES, a broad peak located at 3399 cm^{-1} is dominated by -NH stretching of -NH$_2$ group, while some characteristic bands at 2930 cm^{-1} and 2870 cm^{-1} were assigned to C-H symmetric stretching of CH$_2$ in the propyl chain present in both precursors. The peak at 1456 cm^{-1} was characterized as a C-H deformation in alkyl chains, while bands at 1026 cm^{-1}, 857 cm^{-1} (bending), and 790 cm^{-1} (stretching) were due to Si-O-Si stretching and Si-O-Si bending modes, confirming the formation of an inorganic SiO$_x$ matrix. Further bands at 1402 and 760 cm^{-1} were ascribed to the C-N stretching and the N-H out-of-plane bonding, respectively. The coatings realized by adding to the GPTMS/APTES combination two fluorinated precursors, both individually and in combination, show the same hybrid structure. In the coating (G_A) obtained by the combination of GPTMS and APTES, a broad peak located at 3399 cm^{-1} is dominated by -NH stretching of the -NH$_2$ group, while some characteristic bands at 2930 cm^{-1} and 2870 cm^{-1} were assigned to C-H symmetric stretching of CH$_2$ in the propyl chain presents in both precursors. The peak at 1456 cm^{-1} was characterized as a C-H deformation in alkyl chains, while bands at 1026 cm^{-1}, 857 cm^{-1} (bending), and 790 cm^{-1} (stretching) were due to Si-O-Si stretching and Si-O-Si bending modes, confirming the formation of an inorganic SiOx matrix. Further bands at 1402 and 760 cm^{-1} were ascribed to the C-N stretching and the N-H out-of-plane bonding, respectively. The coatings realized by adding to the GPTMS/APTES combination two fluorinated precursors, both individually and in combination, show the same hybrid structure. In addition to the already characterized sol–gel coatings infrared bands, in G_A_F3, G_A_F16, and G_A_F3_F16 samples the presence of fluorine was confirmed by the variation of spectra in the range 1130–1260 cm^{-1}, due to the presence of the peaks ascribable to fluorinate groups. Even if partially overlapped by other bands, the shoulder at 1154 and the peak at 1263 cm^{-1} were related to CF stretching in CF2 and CF3 groups. In particular, to confirm this interpretation, it should be noted that the band assigned to CF2 does not appear in the G_A_F3 coating spectrum, since it does not exist in the chain of the 3,3,3-trifluoropropyltrimethoxy-silane precursor.

2.1.2. Morphological and Topography Characterization

The nanoscale morphology of functionalized glass substrates was investigated via atomic force spectroscopy (AFM).

Bare glass slides were found to have a relatively featureless, homogeneous appearance with a root mean square surface roughness (Sq) of 1.3 ± 0.1 nm. The surface of all G_A-functionalized substrates evidenced numerous ridge-like protrusions with an average length of 62 ± 12 nm (Figure 7). Successive functionalization with F3 and F16 had a marginal impact on the substrates' surface morphology from both a qualitative and a quantitative point of view, as attested by the fact that the Sq of G_A, G_AF3, G_A F16, and G_A F3_F16 substrates fell within the 1.8 ± 0.3 nm range. Roughness values recorded on the second series of samples were generally lower, with an average Sq of 0.7 ± 0.2, except for sample G_A_F3_F16 which showed a higher Sq of 1.7 ± 0.2. Interestingly, the characteristic ridge-like features of the previous sample set were not visible, while occasional depressions and holes were observed. These observations are compatible with the presence of an additional coating layer of amorphous material, masking the underlying surface, on the second sample set.

Figure 8 shows the surface profile of the antifouling coatings, measured with a surface profilometer.

The surface roughness parameters of these coatings (Ra) are listed in Table 2. Functionalization of G and A chains with fluorinated organic compounds should lead to an increase in surface roughness compared to an unmodified matrix, but this roughness decreases in particular for the sample G_A_F3_F16. In fact, all functionalized samples show a decrease in roughness from an Ra value of 0.53 μm for G_A_F16 to an Ra value of 0.26 μm for

G_A_F3_F16. In order to fully understand the chemical structure influence of the coating, wettability and surface roughness need to be measured at the same sample spot, and the results evaluated according to Equation (2) to separate the effect of the roughness from the wettability. Figure 9 and Table 2 show the contact angle (θ) values given by Wenzel's equation and Young's equation concerning the sol–gel coatings.

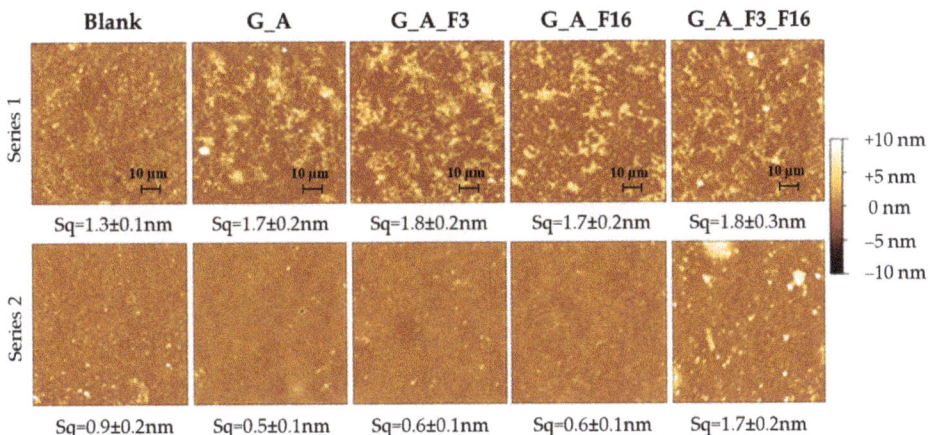

Figure 7. Representative AFM images of bare and functionalized glass substrates (all images are 2 × 2 μm, series 2 differs from series 1 for the presence of an additional applied coating layer).

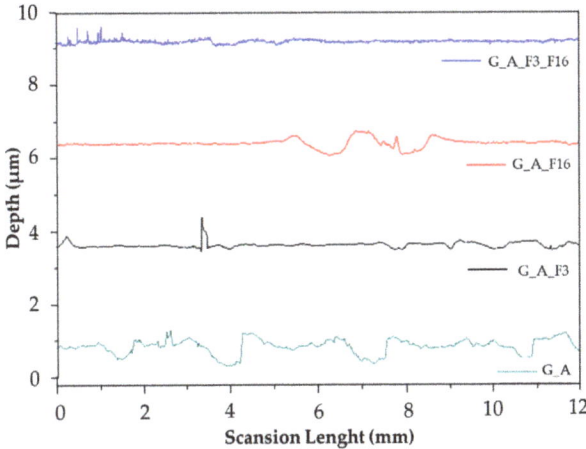

Figure 8. Surface roughness profiles of G_A, G_A_F3, G_A_F16, and G_A_F3_F16 samples.

Table 2. Roughness values of Ra and comparison of the contact angles of Wenzel (θ_W) and Young (θ_Y) of the coatings.

Name	Ra [μm]	θ_W [°]	θ_Y [°/μm]
G_A	1.40 ± 0.01	81.84 ± 0.85	84.18 ± 0.85
G_A_F3	0.41 ± 0.03	80.52 ± 0.85	66.31 ± 0.95
G_A_F16	0.53 ± 0.03	81.44 ± 0.85	73.69 ± 0.85
G_A_F3_F16	0.26 ± 0.003	75.80 ± 0.95	21.69 ± 0.95

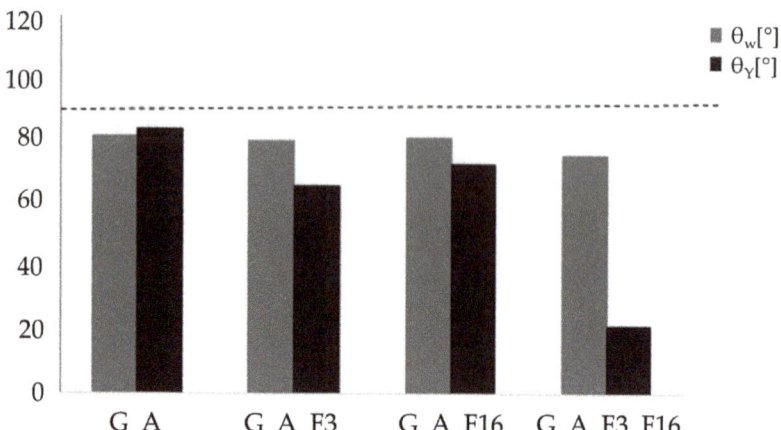

Figure 9. Comparison of contact angles θ_w and θ_Y of the coatings.

Considering that, for hydrophilicity, there is a low contact angle (θ < 90°), while for a hydrophobic situation there is a high contact angle (θ > 90°), the functionalized coatings as such have hydrophilic characteristics, even more than the G_A matrix. The comparison between the contact angles θ_w and θ_Y of the coatings shows an ideal decrease in the value of the contact angle θ_Y with respect to that θ_w.

The roughness values of the functional coated surface are in the range expected for the coated glass surface, i.e., 0.53–0.26 μm. On the other hand, the roughness of the coated surface (that may be influenced by the manual "doctor blade" method, as described in the material and method section) affects Young's contact angle values, since an increase in the roughness will decrease Young's contact angle value (by applying the Equations (1) and (2) in Section 4.2). Anyway, it is worthwhile to remark that all the developed coatings have both Young and Wenzel contact angles lower than 90 °C, even by employing functional alkoxysilanes featuring fluorinated alkyl chains by increasing length, leading us to conclude that we are dealing with hydrophilic coated surfaces.

2.2. Characterization of Sol–Gel Colloidal Solutions

In the present study, it was observed that shear-thinning behavior varies as a function of the type of fluorinated compound used for the functionalization. Figure 10a shows the viscosity changes as a function of the shear rate for the different coatings.

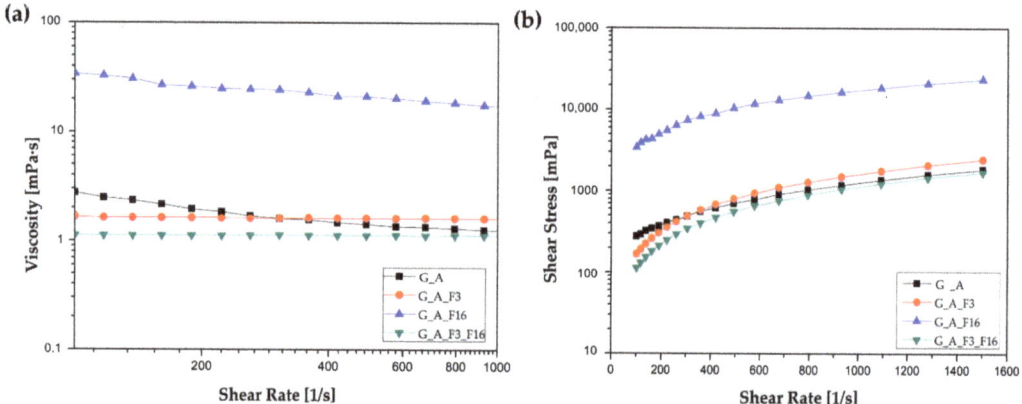

Figure 10. (a) Viscosity vs shear rate and (b) flow curves of the coatings at room temperature.

Viscosity generally represents the flow resistance of materials based on the interaction between the components. The results revealed that the viscosity was almost dependent on the shear rate, behaving like a Newtonian fluid, and it gradually decreased with increasing shear rate, indicating shear thinning properties [71]. Across the entire measuring range, the coating G_A_F16 exhibited a higher viscosity of η = 22.94 mPa·s within the shear rate range shown (100 s^{-1} to 1000 s^{-1}), while the coatings G_A, G_A_F3 and G_A_F3_F16 remains constant, thus showing ideally viscous flow behavior with η = 1.68 mPa·s, η = 1.60 mPa·s and η = 1.12 mPa·s, respectively. The measuring results of viscosity curves are presented as a diagram with shear rate plotted on the x-axis and viscosity plotted on the y-axis; both axes are presented on a logarithmic scale. The relationship between the shear stress and shear rate of the coatings is shown in Figure 10b.

2.3. Evaluation of Antifouling Properties and Characterization

The microscopic analysis carried out on untreated (control) and treated glasses showed significant differences in adhered cells between untreated and treated glasses.

2.3.1. LM and EM

Figure 11 summarizes all the results of the percentage of adhesion of the three microorganisms tested as compared to the untreated control (=100%).

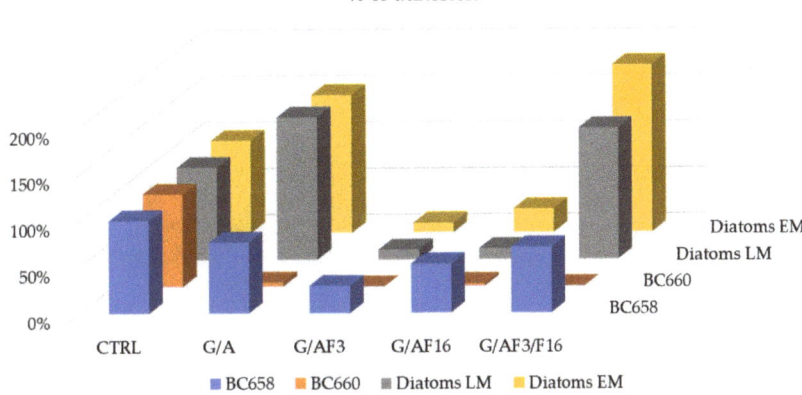

Figure 11. Histogram of the percentage of adhesion on the treated surface of glass slides compared to the adhering cells on the untreated glass slides. For diatoms were reported the results obtained by using two different microscopes (LM and EM).

Each set of experiments carried out with the three microorganisms is shown in the following figures. Figure 12 shows the results obtained with the Gram-negative strain *S. maltophilia* (BC 658).

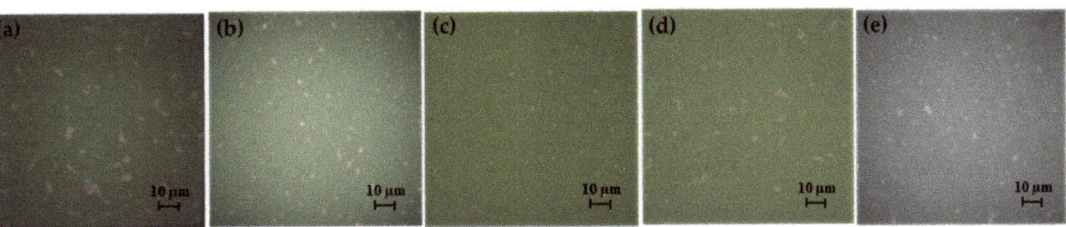

Figure 12. Images of the Gram-negative strain *S. maltophilia* BC 658 observed with an epifluorescence microscope. (**a**) Control 63×; (**b**) G_A 40×; (**c**) G_A_F3 40×; (**d**) G_A_F16 40×; (**e**) G_A_F3_F16 40×.

All glass slides treated with the four sols determined a significant decrease in adhering cells with respect to the control. G_A covered glass slide reduced the adhered cells number to 76.5% with respect to the control. G_A_F3 and G_A_F16 reduced significantly the adhering cells number, respectively, to 29.5% and 53%. G_A_F3_F16 showed also an FR activity decreasing the adhering cells number to 71% with respect to the control.

G_A_F3 coated glass slide showed the best FR attitude against the Gram-negative strain.

Figure 13 shows the results obtained with the strain *Rossellomorea aquimaris* strain BC 660. On the control glass slides, the strain produced abundant biofilm that contributed to the coverage of glass slides, despite the number of cells covering only ~12.5% of the area considered. G_A_F3 and G_A_F3_F16 resulted very effectively against the Gram-positive strain and did not allow bacterial adhesion. A significant decrease in adhering cells was observed.

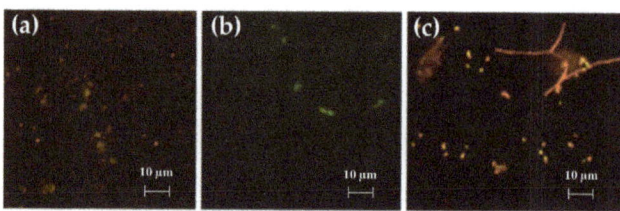

Figure 13. Images of the spore-positive Gram-positive strain *R. aquimaris* BC 660 observed under an epifluorescence microscope. (**a**) control 40×; (**b**) G_A 40×; (**c**) G_A_F16 40×.

The results obtained with the strain of *Navicula* sp. are shown in Figure 14.

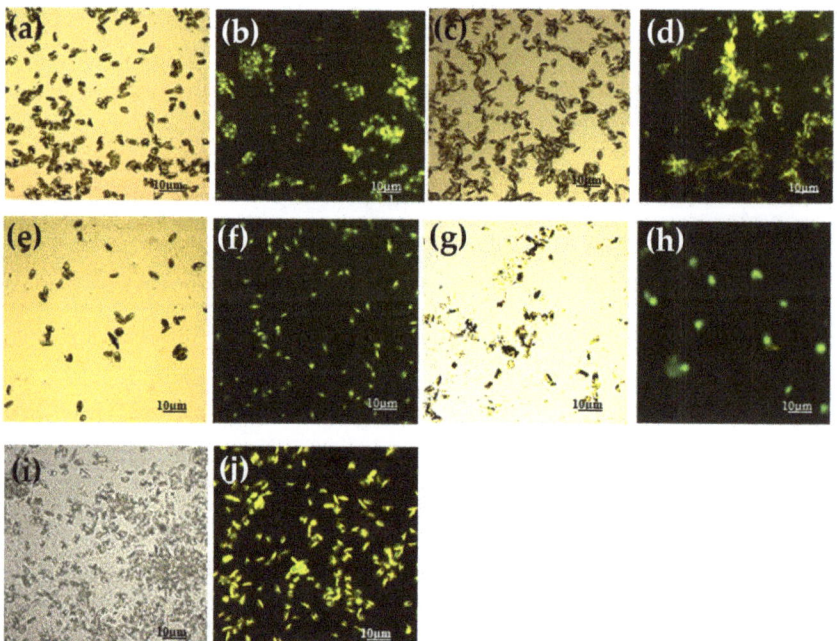

Figure 14. Images of the slides treated and not in contact with *Navicula* sp. obtained with the light microscope and with the epifluorescence microscope 20× magnification. (**a**,**b**) Control; (**c**,**d**) G_A; (**e**,**f**) G_A_F3; (**g**,**h**) G_A_F16; (**i**,**j**) G_A_F3_F16. Note that figures g and h (G_A_F16) showed suffering and ghost cells.

The coating alone G_A increased the adhesion of the diatom with an increase in the number of 154% (seen under LM) and 149% (for EM) than control. A diminishing number of adhering cells (regarding G_A_F3 covered glass slide) was observed (10.8%) with LM, whereas with a fluorescence microscope number of cells seemed to be similar to the control one (105%). However, in more careful observation of the images, we clearly pointed out that while the cells of the control tend to aggregate into clusters, on the G_A_F3 covered glass slides the cells are spread and not aggregated (Figure 14f). G_A_F3_F16 shows a behavior similar to G_A with a cell number increasing considerably higher with respect to the control (141.7%, 181% in epifluorescence).

2.3.2. Morphological Characterization: Scanning Electron Microscopy (SEM)

The influence of different glass functionalization procedures on organism surface adhesion was investigated by SEM imaging (Figure 15). All images showed clearly resolved organisms in different amounts. Individual diatoms were found to measure 80 ± 20 μm^2 and bacteria 1.4 ± 0.3 μm^2 when adhered to substrates. The surface densities of organisms and the percentage of surface area covered by them were determined by quantitative SEM image analysis for all substrates.

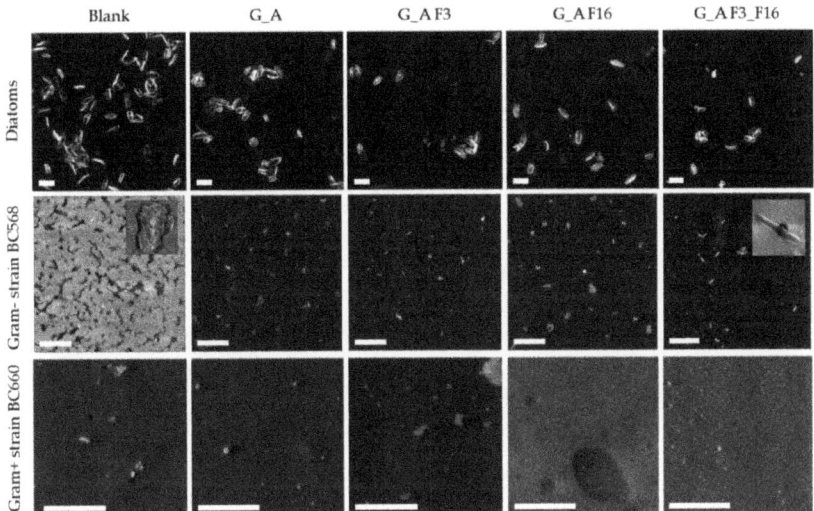

Figure 15. Representative SEM micrographs of bare and functionalized glass substrates after exposition to different microbial suspensions. All scale bars are 20 µm; insets are 20 by 20 µm (on the left, showing clustered bacteria) and 2 by 2 µm (on the right, showing an isolated bacterium).

Bare glass substrates were found to be the most prone to the adhesion of both diatoms ($6.9 \pm 3.9\%$ surface coverage) and bacteria ($1.63 \pm 0.29\%$), whereas functionalized surfaces inhibited their adhesion to different extents. Functionalization with G_A alone was sufficient to drastically drop the surface coverage of diatoms (to $3.3 \pm 1.8\%$) with respect to bare glass substrates, while bacteria were largely unaffected ($1.55 \pm 0.22\%$).

Functionalization of G_A treated substrate with F3 and F16, or their combination further depressed diatom surface coverage percentages to, respectively, $2.4 \pm 1.0\%$, $2.1 \pm 0.6\%$, and $1.8 \pm 0.1\%$, thus reaching an almost four-fold decrease in surface adhesion for the G_A_F3_F16 combination. Bacteria were instead mostly influenced by the presence of F3, showing a surface coverage of $0.22 \pm 0.02\%$ for the substrate treated with G_A_F3 and of $0.17 \pm 0.02\%$ for the G_A_F3_F16 combination. Functionalization with F16 of the G_A treated substrate showed a comparatively milder effect, only reducing bacteria surface coverage to $0.45 \pm 0.06\%$.

The main factor influencing surface coverage of bacteria was observed to be the tendency to form large clusters on certain substrates.

While the number of adhering bacteria clusters varied only slightly across all substrates, the number of individual bacteria found within those clusters varied considerably. Three substrates showed large clusters (see left insert in Figure 15) of adjoining bacteria containing 10+ individuals (bare glass, G_A, and G_A_F16). Conversely, substrates containing F3 (G_A_F3 and G_A_F3_F16) instead only showed individual, discrete bacteria (see right inset in Figure 15).

Figure 16 shows the percentage of surface covered by adhered organisms (i.e., diatoms and the Gram-negative bacteria) as estimated by SEM imaging.

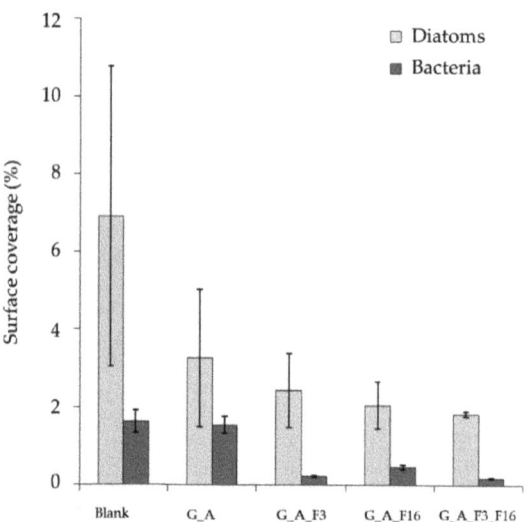

Figure 16. Percentage of area covered by adhered diatoms and Gram-negative bacteria as estimated by SEM imaging.

The sample surfaces with the Gram-positive bacteria are apparently empty due to the biofilm production induced by the Gram-positive strain itself.

In the SEM images few rod-shaped cells were visible and several spores, for this reason, a statistical analysis of the adhesion of surfaces was not carried out for this strain.

To support the hypothesis of a biofilm on the surface of the samples treated with the Gram-positive strain, occasionally holes were seen, and in addition, the surface was charged with excess electrons much more than the corresponding series containing the Gram-negative strain.

2.4. Evaluation of Toxicity of the Coating against Bacteria and Diatoms

No biocide-release effect was observed against both the Gram-positive and Gram-negative bacteria. In fact, the OD_{550} measured before and after the experiment showed an increment of turbidity of 10-fold (from OD_{550} 0.17–0.18 to an average of OD_{550} 1.88 to 1.93, respectively, for BC658 and BC660 (Figure 17).

These values were coherent to the increased number of cell/mL from the initial value of 1×10^8 cell/mL to a value more than 10-fold higher (more than 2×10^9 cell/mL, Figures 18 and 19).

No biocide-release activity was observed against *Navicula* strain as shown in Figure 20. Additionally, in this case, the number of diatom cells after 6 days presented slight differences among suspensions in contact with untreated and treated glasses.

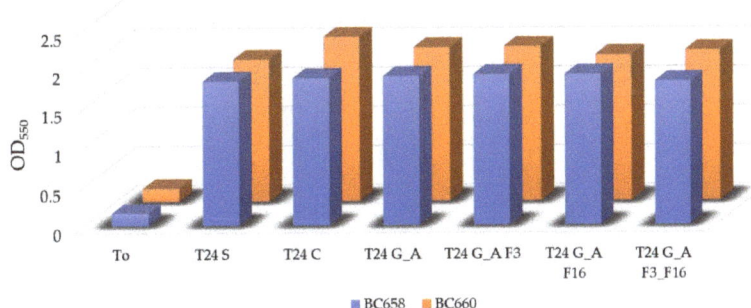

Figure 17. Histogram showing the increase in OD_{550} of the bacterial suspensions before the experiment and after. No significant differences were found between the control suspensions (no glasses, T_{24S}, and untreated glasses T_{24C}) and the microbial suspensions coming from the Petri dishes with treated glasses.

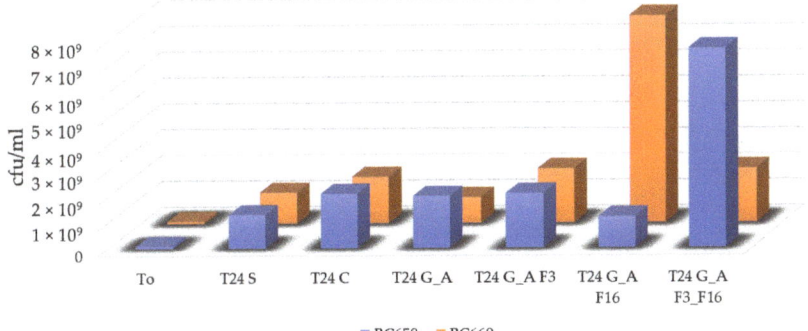

Figure 18. Histogram showing the increase in number of vital bacteria (cfu/mL) after 24 h of incubation in the inoculated liquid medium in presence of untreated and treated glasses. Results were coherent to the OD measurement, except for BC658 in presence of F3/F16 glass slides and for BC660 in the medium in presence of glass slides treated with F16, where a higher number of bacteria was recovered.

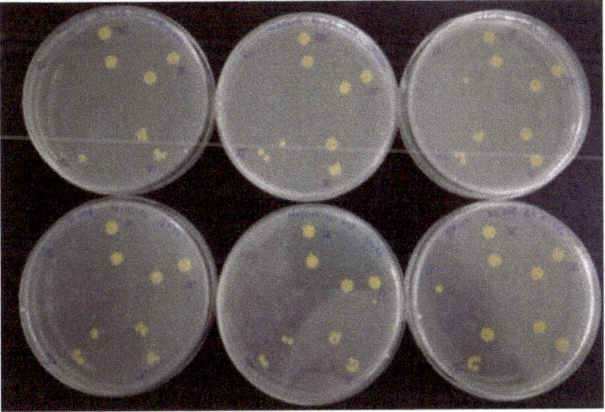

Figure 19. Petri dishes containing marine agar inoculated with the spot technique. Separate colonies were counted only at the higher dilutions.

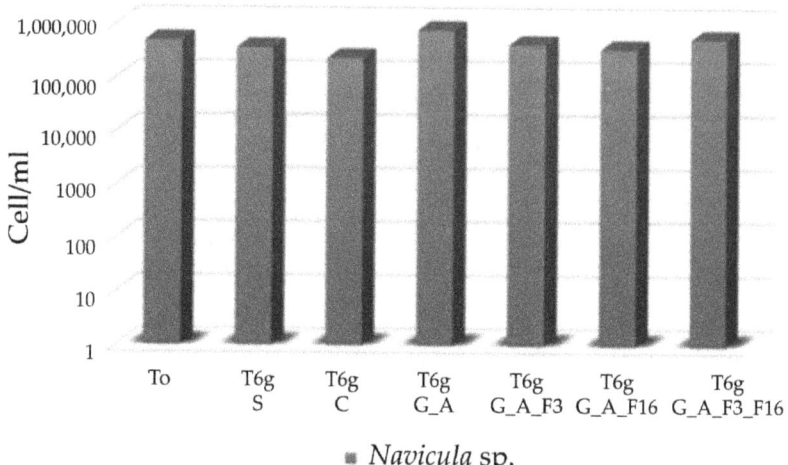

Figure 20. Histogram showing the absence of toxicity against Navicula in contact to untreated and treated glasses as measured through direct count of cells under microscope in Burker counting chamber.

Bacterial Adhesion Tests in Simulation Experiment

Data obtained after 60 days of exposure evidenced, through DAPI staining and fluorescent direct count, are shown in Figures 21 and 22. In untreated glasses (control) and in the G_A system the adhesion of marine bacteria and the consequent biofilm formation is maximal (almost 100%); on the other hand, in alternative systems (treaded glasses) the rate of adhering cells decreased with values of about 47 and 53% for systems G_A_F3 and G_A_F16, respectively. Values of 87% have been registered for G_A_F3_F16 glasses.

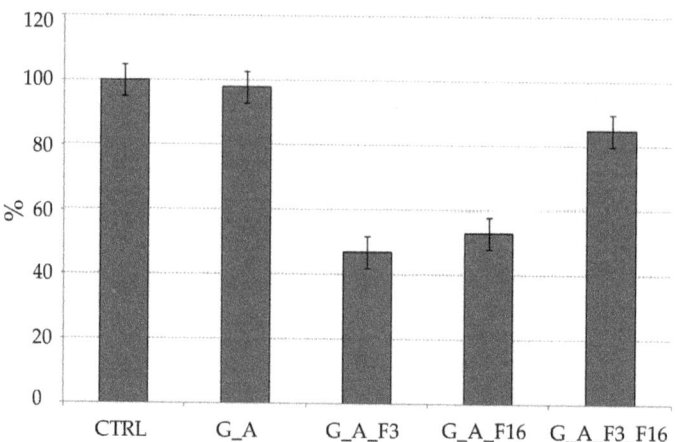

Figure 21. Percentage of bacterial adhesion in contact with untreated (control, CTRL) and treated glasses (G_A, G_A_F3, G_A_F16, and G_A_F3_F16). Area covered by adhering organisms has been estimated by staining with DAPI and counting under epifluorescent microscope.

Even if all developed fluorinated coated cannot be classified as hydrophobic, as evidenced by wettability tests, all experimental findings led us to conclude that the best FR activity is shown by G_A_F3 coating, or to some extent by the G_A_F3_F16 mixed coating.

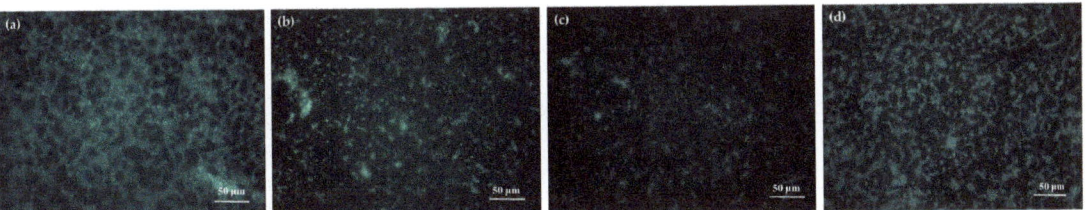

Figure 22. Images of microbial biofilm present on coating surfaces after 60 days of experimentation seen after DAPI staining under fluorescent microscopy (EM); (**a**) G_A, (**b**) G_A_F3, (**c**) G_A_F16, and (**d**) G_A_F3_F16.

Rheological analyses run on the mixed G_A_F3_F16 coating show its lowest roughness values among the developed coatings. This behavior may be ascribed to asymmetric hyperbranched developed coated surface in which most probably F3 and F16 collapse each other and flatten after the curing step in a mushroom-like form for the F16 long chain (Figure 23) [72].

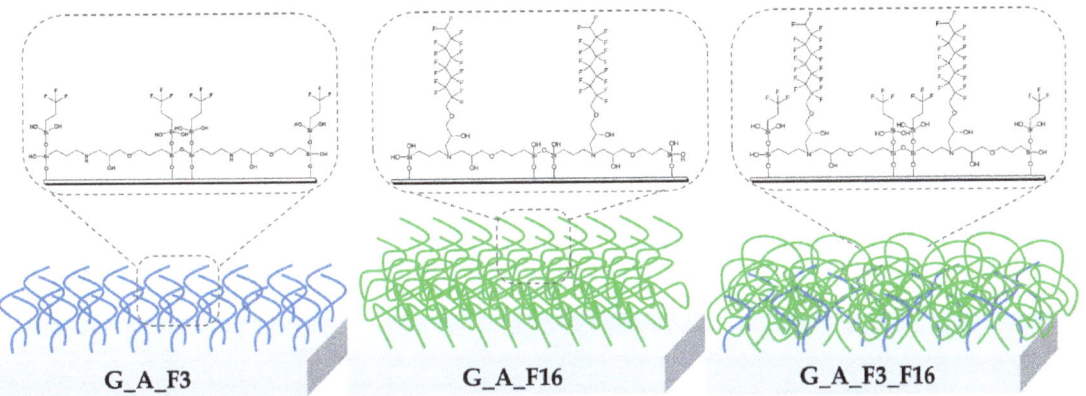

Figure 23. Sketches of F3 and F16 brushes distribution on fluorinated coated glasses.

As evidenced by SEM analysis, the different extent of diatom/bacteria adhesion inhibition by specific substrates hints at the different adhesion mechanisms employed by the model organisms, which also differ by almost two orders of magnitude in size.

In this respect, it is not surprising that the change in surface roughness evidenced by AFM imaging has a considerably stronger impact on the larger diatoms with respect to bacteria. As a matter of fact, it is already well-known that a reduction in roughness on the coated or nanocomposite surface can significantly reduce bacterial adhesion [73].

In particular, diatoms adhesion differs from bacteria one, not only because of surface roughness, but also for their nano-porous architecture that enhances their adhesion [74] on hydrophobic surfaces and fouling release coatings [75].

In both cases, the best inhibition of organism adhesion by SEM microscopy was observed on the G_A_F3_F16 substrates. These results led us to conclude that, as shown before [72], in this latter case in the presence of water, there are swelling phenomena taking place at the water/coating interface, leading to an elongation of the F16 nonpolar brushes that re-build the fouling unfavorable and FR asymmetry of the G_A_F3_F16 coating (Figure 24).

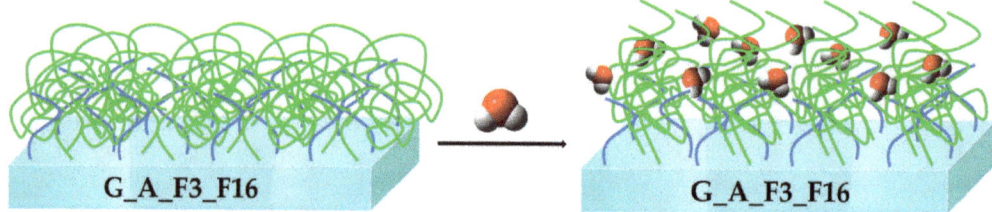

Figure 24. Swelling and elongation phenomena of F16 nonpolar long chain branches taking place at the water/coating interface.

3. Conclusions

Functional hybrid fluorinated and biocide-free formulations are diffusely employed in order to develop antifouling (AF), fouling release (FR), amphiphilic or hydrophobic marine coatings, based on a physical principle. In particular, this work presents an easy procedure for manufacturing an anti-biofouling/fouling release sol–gel-based polymeric hybrid coating featuring a short- or long-alkyl fluorinated chain. The effective persistence of the amphiphilic character with respect to the replacement of the fluorinate silanes by the alkoxysilanes was ensured by the measurement of the static water contact angle.

The efficiency of the antifouling/foul release properties was assessed through testing against the adhesion and deposition of selected marine Gram-positive/Gram-negative bacteria and diatoms and by natural marine microbial population experimental microcosm. Chemical–physical and morphological characterization of the coated and uncoated surface before and after microbial adhesion tests have been performed. Microbiological experiments carried out in two different conditions (laboratory and microcosm), were useful to demonstrate that systems F3 and F16, alone or in combination with each other, largely reduce the rate of adhesion through the fouling release mechanism of the newly prepared coatings, while the matrix alone showed behavior similar of the untreated control and in the case of diatoms even a higher rate of cell adhesion. Slight differences were noticed in the different groups of bacteria. In particular, the Gram-positive strain formed on the glass slides control an abundant biofilm that was not observed on the treated glass slides, Further, the products including the system G_A alone did not show toxicity on the tested microorganisms as they were able to grow in the planktonic state.

All the performed tests indicated the high anti-fouling/fouling release performance of the F3-containing coating (either in combination with F16), in order to prevent microbial biofilm settlement and adhesion.

The results of this work will open the way to further research studies that should be devoted to better sustainable FR coatings, i.e., durable, reliable, stable, easy to be applied on different surfaces (i.e., metals and steel used in ship construction, buildings, cultural heritages) and cost-effective, and last but not least, able to be scaled-up and employed in a large scale.

We firmly believe that this goal may be reached by the use of nanotechnological improved multifunctional and multicomponent sol–gel-based hybrid materials.

4. Materials and Methods

4.1. Sol–Gel Synthesis and Application

(3-aminopropyl)triethoxysilane (APTES, hereafter A), (3-Glycidyloxypropyl) trimethoxysilane (GPTMS, hereafter G), 3,3,3-trifluoropropyl-trimethoxysilane (F3), glycidyl-2,2,3,3,4,4,5,5,6,6,7,7,8,8,9,9-hexadecafluorononyl ether (F16), absolute Ethanol (HPLC grade) were purchased from Sigma-Aldrich and used without further purification. Before treatment, glass slides were cleaned with a concentrate sulfuric acid/potassium permanganate solution, then washed several times with ultrapure water and dried in an oven at 80 °C for 24 h prior to all experiments.

Four separate solutions (named G_A, G_A_F3, G_A_F16, G_A_F3_F16) all containing GPTMS and APTES in 2:1 = [G]:[A], and F3 and/or F16 at a total 0.5 wt.%, were prepared. In a typical procedure, 2.013 g of GPTMS were mixed with 0.943 g of APTES in 37.48 g of ethanol under stirring. Then 0.2 g of F3 was added to the clear ethanol solution and left at room temperature under stirring for 24 h. The same reactions were also carried out with the other functional alkoxysilane. The squeegee method, also known as "doctor blade", was chosen as the deposition technique for forming the different coatings. A cylindrical glass rod with a diameter of about 0.5 cm is used to spread about 1 mL of precursor on a glass substrate. The substrate used for coating deposition is always a microscope slide 1 mm thick and 76 × 26 mm in size. Before coating deposition, the slides were pretreated with a piranha solution that could clean the surface of any organic residues and, at the same time, make the glass hydrophilic by hydroxylating the surface. Next, the slides were washed with distilled water and left in the oven to dry for several hours. To ensure greater protection, two layers of each component were deposited, adequately respecting the drying times of the layers. Once the deposition is complete, heat treatment follows to consolidate the gel coating. The heat treatment consists of a controlled heating ramp at 20 °C/min, followed by a holding phase at 180 °C for 10 min.

4.2. Characterization

The four sols (G_A, G_A F3, G_A F16, and G_AF3_F16) were fully investigated through FT-IR spectroscopy. Coated glass slides were characterized by scanning electron microscopy (SEM), and atomic force microscopy (AFM). To realize xerogels and investigate their chemical structure by FT-IR spectroscopy, small amounts of each sol were applied on glass slides, the solvent was removed at 80 °C for 2 h, and the thin coatings were cured at 120 °C for 1 h. FT-IR spectra of the combined, hydrolyzed, and cured silane precursors as solid residue removed by glass slides were acquired by means of a Thermo Avatar 370, equipped with attenuated total reflection (ATR) accessory. A diamond crystal was used as an internal reflectance element on the ATR accessory. Spectra were recorded, at room temperature, in the range from 4000 to 650 cm^{-1}, with 32 scans and a resolution of 4 cm^{-1}. AFM imaging was performed on a Bruker Multimode 8 (Bruker, Billerica, MA, USA) equipped with a Nanoscope V controller and a type J piezoelectric scanner. Micrographs were recorded in PeakForce mode in air using Bruker SNL-A probes (Bruker, Billerica, MA, USA) with a nominal spring constant of 0.4 N/m. Background subtraction and image analysis were performed with Gwyddion v2.48. Surface roughness parameters were calculated as the average value of five distinct 1 µm^2-sized areas on each sample, using the standard deviation as a measure of the error. The wettability of the sol–gel coating on the microscope slides was evaluated by measuring ten times the height h (mm) and the base diameter d (mm) of 1 µL drop of deionized water on the horizontal surface of the sample, by means of a microlithic syringe (Hamilton, 10 µL). For each material, 10 measurements of the contact angle of deionized water are typically performed, of which the average value with standard deviation was calculated. Wenzel's contact angle θ_W and Young's contact angle θ_Y, have been evaluated by the sessile drops method (ASTM D7334) [76–78] and they were derived from Equations (1) and (2):

$$\theta_W = 2 \text{arctg} \left(\frac{2h}{d} \right) \quad (1)$$

$$\theta_Y = \arccos \left(\frac{\cos \theta_W}{r} \right) \quad (2)$$

where d is the diameter and h the height (both in mm) of the drop, θ_W is the Wenzel angle apparent dependent on the roughness of the surface, r is the surface roughness (Ra), and θ_Y is Young's contact angle of equilibrium on a perfectly smooth surface.

The surface roughness (Ra) of the coatings was calculated by using a roughness tester, Surftest SJ-210- Series 178 (Mitutoyo, Milan, Italy) using Equation (3), in which Ra is calculated as the arithmetic mean of the absolute values of the deviations of the evaluation profile (*Yi*) from the mean line:

$$Ra = \frac{1}{N} \sum_{i=1}^{n} |Yi| \quad (3)$$

The roughness analysis is carried out using a diamond tip that touches the surface of the sample in order to follow its profile. The measurement conditions of the instrument have been set according to the JIS2001 roughness standard: the roughness R profile for compliance, λs = 2.5 μm, λc = 0.8 μm, five sampling lengths and a stylus translation speed of 0.5 mm/s. On average, n. 3 roughness profiles per type of sample were performed and then an average profile was obtained. The viscosities of the coatings were measured by using a modular compact rheometer MCR-502 (RheoCompass Software, Anton Paar Italia S.r.l, Rivoli, Italy). The coatings were tested by using a cone/plate system at a temperature of T = +25 °C. The shear-rate-controlled test was carried out with 18 measuring points using ascending logarithmic steps. The duration for each measuring point was decreased continuously with increasing shear rates, starting at a shear rate of 100 s^{-1} and ending at a shear rate of 1500 s^{-1}. Each test was carried out three times.

4.3. Evaluation of Antifouling Properties and Toxicity of the Coatings

The evaluation of both antifouling and toxic properties of the proposed coatings was performed in laboratory and microcosm conditions as explained below [79].

4.3.1. Strains and Culture Media

A Gram-negative strain *Stenotrophomonas maltophilia* BC658 (*Xanthomonadaceae*; *Xanthomonadales*; Gamma Proteobacteria) and a Gram-positive spore forming strain *Rossellomorea aquimaris* BC 660 (*Bacillaceae*; *Bacillales*, Bacteria) and a marine diatom strain *Navicula* sp. (*Naviculaceae*; *Naviculales*; Eukarya) were used in all laboratory experiments. All strains (isolated previously from marine habitat) were kept in the strains collection of the Dept. of Chemical, Biological Pharmaceutical and Environmental Sciences (CHIBIOFARAM) of University of Messina, Italy [80–82]. Bacterial strains were cultivated and maintained in Tryptone Soy Agar (TSA; Oxoid Limited, Basingstoke, Hampshire, UK) for the AF test they were grown in marine broth MB (Difco) and in marine agar MA (Difco). The diatom *Navicula* sp. strain was kept and grown in F/2 medium [83].

4.3.2. Microbial Suspensions

Bacterial suspensions were prepared from fresh cultures of bacteria grown on TSA Petri dishes (24 h at 28 ± 1 °C) by picking 2–3 colonies and suspending them in phosphate buffer saline (PBS) 0.2 M (130 mM NaCl, 10 mM NaH$_2$PO$_4$, pH 7.2); bacteria were washed 3 times in PBS 0.2 M, centrifuged at 10,000 rpm (5417R Eppendorf Centrifuge) at 25 °C per 10 min. At the end of this procedure, the resulting pellets were suspended in MB in order to reach an OD550 nm of about 1.0 corresponding to a concentration of about 10^{10} cell/mL. After it, for the AF tests, the final concentration of bacteria was adjusted at about 1×10^8 cells/mL in MB liquid medium. Bacterial suspension density was measured through a Shimadzu UV mini 1240 UV Vis spectrophotometer. For measurement of the number of diatoms, after 7 days of growth in F/2 at 4000 lux and 30 °C, diatoms cells were directly counted through a Bürker chamber and adjusted to a final concentration of about 5×10^5 cells/mL in fresh F/2 medium [84].

4.3.3. Antifouling and Biocide-Release Assessment

Short-term antifouling properties of the different coatings were determined through the evaluation of microorganism's adhesion to the untreated and treated glasses after 24 h, observed under light microscopy, (LM), epifluorescence microscopy (EM), and scanning electron microscopy (SEM). The toxicity released by the coating to the environment was assessed by the measurement of the suspension density and/or by counting the cells (cfu/mL) before and after the incubations in the liquid medium. Different sets of glasses were used, prepared as described in par. 2.1 (control, G_A, G_A F3, G_A F16, G_A F3_F16 and 4 replicates) were placed in separate sterile glass Petri dishes (Ø 120 mm). Twenty ml of each microbial suspension were then added into the Petri dishes and these latter placed in a horizontal shaker and incubated for 24 h at room temperature set at 25 ± 1 °C for bacteria, and for 6 days in the light at 30 °C for the diatoms. Before and after the incubation time, the density of bacterial suspension inside the Petri dishes was assessed through OD_{550} measurement (as specified in the previous paragraph) and verified by counting the colony forming units (CFU)/mL on marine agar (MA) by using the spots method of inoculation (10 µL of each decimal dilution of the suspension in double). Diatoms number was directly counted as previously described. After the incubation time, glasses were rinsed with sterile PBS, and treated differently: (i) one set was air dried and fixed by placing the glass directly to the Bunsen flame for 30 s observation was carried out in LM after methylene blue staining; (ii) one set was prepared for epifluorescent microscopy and fixed in a 3:1 freshly prepared solution consisting of 3 parts of fixation buffer (4% paraformaldehyde in PBS, pH 7.2) and 1 part of PBS, incubated for 90 min at 4 ± 1 °C. Bacterial cells were then stained with Acridine orange (AO) (0.1 mg/mL) diluted in sterile distilled water 1:2 (v/v) for 3–4 min and (iii) the last set of glasses was prepared for SEM microscopy by fixing in glutaraldehyde 2.5% in phosphate buffer (PB) 0.1M (pH 7.4) for 6 h at 4 °C and then air dried [85].

4.3.4. Microbial Adhesion Assessment

Adhesion to the untreated and treated glass slides was observed under three different microscopes as it follows:

(a) For LM and EM were performed in a LEICA DM RE equipped with a video camera (LEICA DC 300 F). For each slide, 3 different images of different fields were acquired through the software Leica QWin Color (RGB). After, images were cut to obtain a final dimension of 602 × 602 µm and treated by using the plugin threshold [86] of Image J. 1.52c. In this way, both the number and percentage of the covered surface were determined for each field. The number of cells counted on the untreated glasses was considered as 100% of cells that could adhere to the glass substrate and thus the other cells counting adhering to the different coatings could be more similar or less than 100%, meaning, respectively, an attractive, similar or a repulsive action toward the microorganisms.

(b) SEM images were obtained on a Zeiss EVO LS10 SEM equipped with a LaB6 thermionic electron source and a variable pressure secondary electron detector. No additional treatment was performed prior to SEM observation of samples, which were kept at 3.0×10^{-1} Torr during measurements. Images obtained under both microscope observations were analyzed with Image J 1.52c. Surface densities were estimated by counting individual adhered organisms in at least five randomly selected 100 × 100 µm fields on each sample. The average area of adhered diatoms and bacteria was estimated by measuring 20 individual organisms of each type. In each case, standard deviation was used as the error.

4.4. Bacterial Adhesion Tests in Microcosm Experiment

To test adhesion of natural marine microbial population (in the glass slides covered with different coatings) experimental microcosm was carried out. As reported in Figure 25, the experiment was performed in glass tank (100 × 30 × 40 cm, volume 120 L) filled in continuous (8 L h^{-1}) with seawater (salinity 37–38‰) collected directly from the station "Mare Sicilia" (38°12.23′ N, 15°33.10′ E; Messina, Italy) by a pipeline from the sea, in order to ensure approximately two complete water turnover daily. Before introduction to the experimental system, natural seawater was filtered through a 300 µm nylon mesh to remove large metazoans and detritus. To ensure a constant level of water, each microcosm was equipped with a relief valve connected by vertical conduct (PVC-u pn10, 200 mm Ø) placed laterally of the tank to continuously discharge the excess seawater. Water within each microcosm was gently mixed in a continuous mode with a pump (35 L h^{-1}) placed close to the entrance of each tank to provide more homogenous conditions within each microcosm. The measurement of pH and temperature, as performed through a multi-parametric probe Waterproof CyberScan PCD 650 (Eutech Instruments, Breda, The Netherlands), and reveal values of temperature of 19.5–20.5 °C (daily temperature fluctuations not exceeded 1 °C) and approximatively constant pH values (around 8). Different sets of glasses (control, G_A, G_A_F3, G_A_F16, G_A_ F3_F16) were inserted inside the experimental microcosm and incubated for 60 days. The glass slides were immersed in a vertical orientation according to the orientation of the other biological tests. The liquid in which the slides were immersed in the previous tests is moved from left to right and right to left (horizontal movement of the shaker), while in the microcosm a similar movement is created by the pump. After the incubation time, the density of bacterial adhesion (biofilm formation) to different support has been evaluated by epifluorescence microscopy (EM). The total bacterial cell counts were performed by DAPI (4′,6-diamidino-2-phenylindole 2HCl, Sigma-Aldrich, Milan, Italy) staining on samples fixed with formaldehyde (2% final concentration), according to Porter and Feig (1980) [87]. Slides were examined by epifluorescence with an Axioplan 2 Imaging (Zeiss) microscope (Carl Zeiss, Thornwood, NY, USA). Results were expressed as number of cells mL^{-1}.

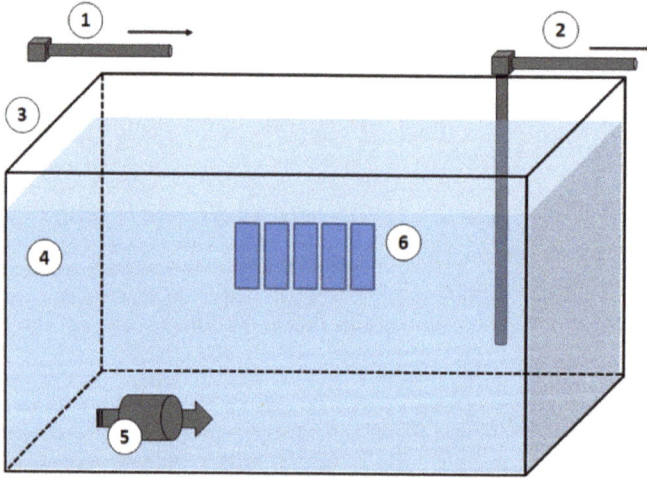

Figure 25. Schematic representation of the microcosm used in the study. Legend: (❶) inlet seawater system, (❷) outlet "too full" seawater system; (❸) experimental tank, (❹) seawater, (❺) internal pump, and (❻) experimental glass.

Author Contributions: Conceptualization, M.R.P., G.R. and S.S.; methodology, M.G., M.B., G.R., S.S., P.C., C.U., V.T., A.V. and M.R.P.; software, F.D.L. and M.G.; validation, S.S.; investigation, M.B., G.R., S.S., C.U., A.V. and M.R.P.; resources, M.R.P., A.V. and C.U.; data curation, G.R., S.S., M.B., S.C., F.D.L., V.T., M.R.P., A.V. and C.U.; writing—original draft preparation, S.S., I.I., G.R., M.B., V.T., M.R.P., A.V. and C.U.; writing—review and editing S.S., G.R., M.B., V.T., M.R.P., A.V. and C.U.; supervision, M.R.P., A.V. and C.U. All authors have read and agreed to the published version of the manuscript.

Funding: The research is supported by PON MUR 2014–2020 "THALASSA—TecHnology And materials for safe Low consumption And low life cycle cost veSSels And crafts" project-Area of specialization "BLUE GROWTH". CUP ARS01_00293. This work was conducted within the framework of the doctoral program of S.S. as financed by Confindustria (Noxosorkem Group S.r.l.) and CNR. G.R. acknowledge the financial support of the PON-MUR "Ricerca e Innovazione 2014–2020" RESTART project. MIUR and CNR are gratefully acknowledged for financial support.

Institutional Review Board Statement: Not applicable.

Informed Consent Statement: Not applicable.

Data Availability Statement: Not applicable.

Acknowledgments: All authors wish to thank Marianna Barbalinardo (CNR-ISMN, Bologna) and Franco Corticelli (CNR-IMM, Bologna) for help in performing SEM imaging. We thank the SPM@ISMN facility for support in AFM imaging. Authors are grateful to Marco Nicoló for supplying the *Navicula* sp. strain and for the useful suggestions and discussions. Salvatore Romeo, Giuseppe Napoli and Francesco Giordano are gratefully acknowledged for technical and informatics assistance during all instrumentation set-up and data fitting.

Conflicts of Interest: The authors declare no conflict of interest.

References

1. Qiu, H.; Feng, K.; Gapeeva, A.; Meurisch, K.; Kaps, S.; Li, X.; Yu, L.; Mishra, Y.K.; Adelung, R.; Baum, M. Functional polymer materials for modern marine biofouling control. *Prog. Polym. Sci.* **2022**, *127*, 101516.
2. Xu, S.J.; Chen, G.E.; Xu, Z.L. Excellent anti-fouling performance of PVDF polymeric membrane modified by enhanced CaA gel-layer. *J. Ind. Eng. Chem.* **2018**, *58*, 179–188.
3. Ielo, I.; Giacobello, F.; Sfameni, S.; Rando, G.; Galletta, M.; Trovato, V.; Rosace, G.; Plutino, M.R. Nanostructured Surface Finishing and Coatings: Functional Properties and Applications. *Materials* **2021**, *14*, 2733.
4. Ielo, I.; Giacobello, F.; Castellano, A.; Sfameni, S.; Rando, G.; Plutino, M.R. Development of Antibacterial and Antifouling Innovative and Eco-Sustainable Sol–Gel Based Materials: From Marine Areas Protection to Healthcare Applications. *Gels* **2022**, *8*, 26.
5. Gao, X.; Yan, R.; Xu, L.; Ma, H. Effect of amorphous phytic acid nanoparticles on the corrosion mitigation performance and stability of sol-gel coatings on cold-rolled steel substrates. *J. Alloy. Compd.* **2018**, *747*, 747–754.
6. Kwanyoung, K.; YoungJae, Y.; Min-Ji, K.; Jihyang, K.; Haegeun, C. Improvement in fouling resistance of silver-graphene oxide coated polyvinylidene fluoride membrane prepared by pressurized filtratio. *Sep. Purif. Technol.* **2018**, *194*, 161–169.
7. Liu, Y.; Zhao, Y.J.; Teng, J.L.; Wang, J.H.; Wu, L.S.; Zheng, Y.L. Research progress of nano self-cleaning anti-fouling coatings. *IOP Conf. Ser. Mater. Sci. Eng.* **2017**, *284*, 012016.
8. Amini, S.; Kolle, S.; Petrone, L.; Ahanotu, O.; Sunny, S.; Sutanto, C.N.; Hoon, S.; Cohen, L.; Weaver, J.C.; Aizenberg, J.; et al. Preventing mussel adhesion using lubricant-infused materials. *Science* **2017**, *357*, 668–673.
9. Silva, E.; Tulcidas, A.V.; Ferreira, O.; Bayón, R.; Igartua, A.; Mendoza, G.; Mergulhão, F.J.M.; Faria, S.I.; Gomes, L.C.; Carvalho, S.; et al. Assessment of the environmental compatibility and antifouling performance of an innovative biocidal and foul-release multifunctional marine coating. *Environ. Res.* **2021**, *198*, 111219.
10. Gu, Y.; Yu, L.; Mou, J.; Wu, D.; Xu, M.; Zhou, P.; Ren, Y. Research Strategies to Develop Environmentally Friendly Marine Antifouling Coatings. *Mar. Drugs* **2020**, *18*, 371.
11. Silva, E.R.; Ferreira, O.; Ramalho, P.A.; Azevedo, N.F.; Bayón, R.; Igartua, A.; Bordado, J.C.; Calhorda, M.G. Eco-friendly non-biocide-release coatings for marine biofouling prevention. *Sci. Total Environ.* **2019**, *650*, 2499–2511.
12. Ulaeto, S.B.; Pancrecious, J.K.; Rajan, T.P.D.; Pai, B.C. Smart Coatings. *Noble Met-Met. Oxide Hybrid Nanopart.* **2019**, *17*, 341–372.
13. Nurioglu, A.G.; Carvalho Esteves, A.C.; With, G. Non-toxic, non-biocide-release antifouling coatings based on molecular structure design for marine applications. *J. Mater. Chem. B* **2015**, *3*, 6547.
14. Salta, M.; Wharton, J.A.; Stoodley, P.; Dennington, S.P.; Goodes, L.R.; Werwinski, S.; Mart, U.; Wood, R.J.K.; Stokes, K.R. Designing biomimetic antifouling surfaces. *Phil. Trans. R. Soc. A* **2010**, *368*, 4729–4754.
15. Nisol, B.; Lerouge, S.; Watson, S.; Meunier, A.; Wertheimer, M.R. Energetics of reactions in a dielectric barrier discharge with argon carrier gas: VI PEG-like coatings. *Plasma Process Polym.* **2017**, *15*, 1700132.
16. Zouaghi, S.; Six, T.; Bellayera, S.; Coffinier, Y.; Abdallah, M.; Chihib, N.E.; Andréa, C.; Delaplace, G.; Jimenez, M. Atmospheric pressure plasma spraying of silane-based coatings targeting whey protein fouling and bacterial adhesion management. *Appl. Surf. Sci.* **2018**, *455*, 392–402.
17. Dong, W.; Qian, F.; Li, Q.; Tang, G.; Xiang, T.; Chun, T.; Lu, J.; Han, Y.; Xia, Y.; Hu, J. Fabrication of superhydrophobic PET filter material with fluorinated SiO2 nanoparticles via simple sol–gel process. *J. Sol-Gel Sci. Technol.* **2021**, *98*, 224–237.
18. Tang, Y.; Zhang, Q.; Zhan, X.; Chen, F. Superhydrophobic and anti-icing properties at overcooled temperature of a fluorinated hybrid surface prepared via a sol-gel process. *Soft Matter* **2015**, *11*, 4540–4550.
19. Mielczarski, J.A.; Mielczarski, E.; Galli, G.; Morelli, A.; Martinelli, E.; Chiellini, E. The Surface-Segregated Nanostructure of Fluorinated Copolymer-Poly(dimethylsiloxane) Blend Films. *Langmuir* **2010**, *26*, 2871–2876.
20. Kim, J.; Kim, C.; Baek, Y.; Hong, S.P.; Kim, H.J.; Lee, J.C.; Yoon, J. Facile surface modification of a polyamide reverse osmosis membrane using a TiO_2 sol-gel derived spray coating method to enhance the anti-fouling property. *Desalin. Water Treat.* **2018**, *102*, 9–15.
21. Bai, X.; Li, J.; Zhu, L.; Wang, L. Effect of Cu content on microstructure, mechanical and anti-fouling properties of TiSiN-Cu coating deposited by multi-arc ion plating. *Appl. Surf. Sci.* **2018**, *427*, 444–451.

22. Xin, L.; Jiansheng, L.; Bart, V.; Xiuyun, S.; Jinyou, S.; Weiqing, H.; Lianjun, W. Fouling behavior of polyethersulfone ultrafiltration membranes functionalized with sol–gel formed ZnO nanoparticles. *RSC Adv.* **2015**, *5*, 50711.
23. Munch, A.S.; Wolk, M.; Malanin, M.; Eichhorn, K.J.; Simon, F.; Uhlmann, P. Smart functional polymer coatings for paper with anti-fouling properties. *J. Mater. Chem. B* **2018**, *6*, 830.
24. Reverdy, C.; Belgacem, N.; Moghaddam, M.S.; Sundin, M.; Swerin, A.; Bras, J. One-step superhydrophobic coating using hydrophobized cellulose nanofibrils. *Colloids Surf. A Physicochem. Eng. Asp.* **2018**, *544*, 152–158.
25. Pieper, R.J.; Ekin, A.; Webster, D.C.; Casse, F.; Callow, J.A.; Callow, M.E. Combinatorial approach to study the effect of acrylic polyol composition on the properties of crosslinked siloxane-polyurethane fouling-release coatings. *J. Coat. Technol. Res.* **2017**, *4*, 453–461.
26. Lin, X.; Yang, M.; Jeong, H.; Chang, M.; Hong, J. Durable superhydrophilic coatings formed for anti-biofouling and oil–water separation. *J. Memb. Sci.* **2016**, *506*, 22–30.
27. Plutino, M.R.; Colleoni, C.; Donelli, I.; Freddi, G.; Guido, E.; Maschi, O.; Mezzi, A.; Rosace, G. Sol-gel 3-glycidoxypropyltriethoxysilane finishing on different fabrics: The role of precursor concentration and catalyst on the textile performances and cytotoxic activity. *J. Colloid Interface Sci.* **2017**, *506*, 504–517.
28. Abou Elmaaty, T.; Elsisi, H.G.; Elsayad, H.H.; Elhadad, H.H.; Sayed-Ahmed, K.; Plutino, M.R. Fabrication of New Multifunctional Cotton/Lycra Composites Protective Textiles through Deposition of Nano Silica Coating. *Polymers* **2021**, *13*, 2888.
29. Puoci, F.; Saturnino, C.; Trovato, V.; Iacopetta, D.; Piperopoulos, E.; Triolo, C.; Bonomo, M.G.; Drommi, D.; Parisi, O.I.; Milone, C.; et al. Sol–Gel Treatment of Textiles for the Entrapping of an Antioxidant/Anti-Inflammatory Molecule: Functional Coating Morphological Characterization and Drug Release Evaluation. *Appl. Sci.* **2020**, *10*, 2287.
30. Trovato, V.; Mezzi, A.; Brucale, M.; Rosace, G.; Plutino, M.R. Alizarin-functionalized organic-inorganic silane coatings for the development of wearable textile sensors. *J. Colloid Interface Sci.* **2022**, *617*, 463–477.
31. Trovato, V.; Mezzi, A.; Brucale, M.; Abdeh, H.; Drommi, D.; Rosace, G.; Plutino, M.R. Sol-Gel Assisted Immobilization of Alizarin Red S on Polyester Fabrics for Developing Stimuli-Responsive Wearable Sensors. *Polymers* **2022**, *14*, 2788. [CrossRef]
32. Chobba, M.B.; Weththimuni, M.L.; Messaoud, M.; Urzì, C.; Bouaziz, J.; De Leo, F.; Licchelli, M. Ag-TiO2/PDMS nanocomposite protective coatings: Synthesis, characterization, and use as a self-cleaning and antimicrobial agent. *Prog. Org. Coat.* **2021**, *158*, 106342.
33. Iannazzo, D.; Pistone, A.; Visco, A.; Galtieri, G.; Giofrè, S.V.; Romeo, R.; Romeo, G.; Cappello, S.; Bonsignore, M.; Denaro, R. 1,2,3-Triazole/MWCNT Conjugates as Filler for Gelcoat Nanocomposites: New Active Antibiofouling Coatings for Marine. *Appl. Mater. Res. Express* **2015**, *2*, 115001.
34. Pistone, A.; Scolaro, C.; Visco, A. Mechanical Properties of Protective Coatings against Marine Fouling: A Review. *Polymer* **2021**, *13*, 173.
35. Pistone, A.; Visco, A.; Galtieri, G.; Iannazzo, D.; Espro, C.; Marino Merlo, F.; Urzì, C.; De Leo, F. Polyester resin and carbon nanotubes based nanocomposite as new-generation coating to prevent biofilm formation. *Int. J. Polym. Anal. Charact.* **2016**, *21*, 327–336.
36. Celesti, C.; Gervasi, T.; Cicero, N.; Giofrè, S.V.; Espro, C.; Piperopoulos, E.; Gabriele, B.; Mancuso, R.; Lo Vecchio, G.; Iannazzo, D. Titanium Surface Modification for Implantable Medical Devices with Anti-Bacterial Adhesion Properties. *Materials* **2022**, *15*, 3283.
37. Krzak, J.; Szczurek, A.; Babiarczuk, B.; Gąsiorek, J.; Borak, B. Chapter 5—Sol–gel surface functionalization regardless of form and type of substrate. In *Handbook of Nanomaterials for Manufacturing Applications*; Hussain, C.M., Ed.; Micro and Nano Technologies; Elsevier: Amsterdam, The Netherlands, 2020; pp. 111–147. ISBN 978-0-12-821381-0.
38. Fotovvati, B.; Namdari, N.; Dehghanghadikolaei, A. On Coating Techniques for Surface Protection: A Review. *J. Manuf. Mater. Process.* **2019**, *3*, 28.
39. Figueira, R.B. Hybrid Sol gel Coatings for Corrosion Mitigation: A Critical Review. *Polymers* **2020**, *12*, 689.
40. Rando, G.; Sfameni, S.; Galletta, M.; Drommi, D.; Cappello, S.; Plutino, M.R. Functional Nanohybrids and Nanocomposites Development for the Removal of Environmental Pollutants and Bioremediation. *Molecules* **2022**, *27*, 4856.
41. Ielo, I.; Galletta, M.; Rando, G.; Sfameni, S.; Cardiano, P.; Sabatino, G.; Drommi, D.; Rosace, G.; Plutino, M.R. Design, synthesis and characterization of hybrid coatings suitable for geopolymeric-based supports for the restoration of cultural heritage. *IOP Conf. Ser. Mater. Sci. Eng.* **2020**, *777*, 012003.
42. Trovato, V.; Rosace, G.; Colleoni, C.; Sfameni, S.; Migani, V.; Plutino, M.R. Sol-gel based coatings for the protection of cultural heritage textiles. *IOP Conf. Ser. Mater. Sci. Eng.* **2020**, *777*, 012007.
43. Rosace, G.; Guido, E.; Colleoni, C.; Brucale, M.; Piperopoulos, E.; Milone, C.; Plutino, M.R. Halochromic resorufin-GPTMS hybrid sol-gel: Chemical-physical properties and use as pH sensor fabric coating. *Sens. Actuators B Chem.* **2017**, *241*, 85–95. [CrossRef]
44. Trovato, V.; Colleoni, C.; Castellano, A.; Plutino, M.R. The key role of 3-glycidoxypropyltrimethoxysilane sol–gel precursor in the development of wearable sensors for health monitoring. *J. Sol-Gel Sci. Technol.* **2018**, *87*, 27–40. [CrossRef]
45. Giacobello, F.; Ielo, I.; Belhamdi, H.; Plutino, M.R. Geopolymers and Functionalization Strategies for the Development of Sustainable Materials in Construction Industry and Cultural Heritage Applications: A Review. *Materials* **2022**, *15*, 1725. [CrossRef]
46. Cardiano, P.; Schiavo, S.L.; Piraino, P. Hydrorepellent properties of organic–inorganic hybrid materials. *J. Non-Cryst. Solids* **2010**, *356*, 917–926. [CrossRef]
47. Cardiano, P. Hydrophobic properties of new epoxy-silica hybrids. *J. Appl. Polym. Sci.* **2008**, *108*, 3380–3387. [CrossRef]

48. Pistone, A.; Scolaro, C.; Celesti, C.; Visco, A. Study of Protective Layers Based on Crosslinked Glutaraldehyde/3-aminopropyltriethoxysilane. *Polymers* **2022**, *14*, 801.
49. Galiano, F.; Schmidt, S.A.; Ye, X.; Kumar, R.; Mancuso, R.; Curcio, E.; Gabriele, B.; Hoinkis, J.; Figoli, A. UV-LED induced bicontinuous microemulsions polymerisation for surface modification of commercial membranes—Enhancing the antifouling properties. *Sep. Purif. Technol.* **2018**, *194*, 149–160.
50. Dett, M.R.; Ciriminna, R.; Bright, F.V.; Pagliaro, M. Environmentally Benign Sol-Gel Antifouling and Foul-Releasing Coatings. *Acc. Chem. Res.* **2014**, *47*, 678–687.
51. Donnelly, B.; Sammut, K.; Tang, Y. Materials Selection for Antifouling Systems in Marine Structures. *Molecules* **2022**, *27*, 3408.
52. Chen, J.; Jian, R.; Yang, K.; Bai, W.; Huang, C.; Lin, Y.; Zheng, B.; Wei, F.; Lin, Q.; Xu, Y. Urushiol-based benzoxazine copper polymer with low surface energy, strong substrate adhesion and antibacterial for marine antifouling application. *J. Clean. Prod.* **2021**, *318*, 128527.
53. Sfameni, S.; Rando, G.; Marchetta, A.; Scolaro, C.; Cappello, S.; Urzì, C.; Visco, A.; Plutino, M.R. Development of eco-friendly hydrophobic and fouling-release coatings for blue-growth environmental applications: Synthesis, mechanical characterization and biological activity. *Gels* **2022**, *8*, 528. [CrossRef]
54. Callow, J.A.; Callow, M.E. Trends in the development of environmentally friendly fouling-resistant marine coatings. *Nat. Commun.* **2011**, *2*, 244.
55. Callow, M.E.; Fletcher, R.L. The influence of low surface energy materials on bioadhesion—A review. *Int. Biodeterior. Biodegrad.* **1994**, *34*, 333–348.
56. Maan, A.M.C.; Hofman, A.H.; de Vos, W.M.; Kamperman, M. Recent Developments and Practical Feasibility of Polymer-Based Antifouling Coatings. *Adv. Funct. Mater.* **2020**, *30*, 2000936.
57. Barletta, M.; Aversa, C.; Pizzi, E.; Puopolo, M.; Vesco, S. Design, manufacturing and testing of anti-fouling/foul-release (AF/FR) amphiphilic coatings. *Prog. Org. Coat.* **2018**, *123*, 267–281.
58. Zhou, Z.; Calabrese, D.R.; Taylor, W.; Finlay, J.A.; Callow, M.E.; Callow, J.A.; Fischer, D.; Kramer, E.J.; Ober, C.K. Amphiphilic triblock copolymers with PEGylated hydrocarbon structures as environmentally friendly marine antifouling and fouling-release coatings. *Biofouling* **2014**, *30*, 589–604.
59. Rodríguez-Hernández, J. Chapter 11—Nano/Micro and Hierarchical Structured Surfaces in Polymer Blends. In *Nanostructured Polymer Blends*; Thomas, S., Shanks, R., Chandrasekharakurup, S.B.T.-N.P.B., Eds.; William Andrew Publishing: Oxford, UK, 2014; pp. 357–421. ISBN 978-1-4557-3159-6.
60. Galli, G.; Martinelli, E. Amphiphilic Polymer Platforms: Surface Engineering of Films for Marine Antibiofouling. *Macromol. Rapid Commun.* **2017**, *38*, 1600704.
61. Karnati, S.; Oldham, D.; Fini, E.H.; Zhang, L. Application of surface-modified silica nanoparticles with dual silane coupling agents in bitumen for performance enhancement. *Constr. Build. Mater.* **2020**, *244*, 118324.
62. Nečas, D.; Klapetek, P. Gwyddion: An open-source software for SPM data analysis. *Cent. Eur. J. Phys.* **2012**, *10*, 181–188.
63. Langer, R. Drug delivery and targeting. *Nature* **1998**, *392*, 5–10.
64. Brouwers, J.R.B.J. Advanced and controlled drug delivery systems in clinical disease management. *Pharm. World Sci.* **1996**, *18*, 153–162.
65. Uhrich, K.E.; Cannizzaro, S.M.; Langer, R.S.; Shakesheff, K.M. Polymeric Systems for Controlled Drug Release. *Chem. Rev.* **1999**, *99*, 3181–3198.
66. Ferrero, F.; Periolatto, M. Application of fluorinated compounds to cotton fabrics via sol-gel. *Appl. Surf. Sci.* **2013**, *275*, 201–207.
67. Ten Breteler, M.R.; Nierstrasz, V.A.; Warmoeskerken, M.M.C.G. Textile Slow-Release Systems with Medical Applications. *AUTEX Res. J.* **2002**, *2*, 175–189.
68. Brannon-Peppas, L. *Controlled Release in the Food and Cosmetics Industries*; ACS Publications: Washington, DC, USA, 1993; pp. 42–52.
69. Mansur, H.S.; Oréfice, R.L.; Vasconcelos, W.L.; Lobato, Z.P.; Machado, L.J.C. Biomaterial with chemically engineered surface for protein immobilization. *J. Mater. Sci. Mater. Med.* **2005**, *16*, 333–340.
70. Amiri, A.; Sharifian, P.; Soltanizadeh, N. Application of ultrasound treatment for improving the physicochemical, functional and rheological properties of myofibrillar proteins. *Int. J. Biol. Macromol.* **2018**, *111*, 139–147.
71. Hu, Z.; Haruna, M.; Gao, H.; Nourafkan, E.; Wen, D. Rheological properties of partially hydrolyzed polyacrylamide seeded by nanoparticles. *Ind. Eng. Chem. Res.* **2017**, *56*, 3456–3463.
72. Wang, Z.; Zuilhof, H. Antifouling Properties of Fluoropolymer Brushes toward Organic Polymers: The Influence of Composition, Thickness, Brush Architecture, and Annealing. *Langmuir* **2016**, *32*, 6571–6581. [CrossRef]
73. Lu, A.; Gao, Y.; Jin, T.; Luo, X.; Zeng, Q.; Shang, Z. Effects of surface roughness and texture on the bacterial adhesion on the bearing surface of bio-ceramic joint implants: An in vitro study. *Ceram. Int.* **2020**, *46*, 6550–6559.
74. Khan, M.J.; Singh, R.; Shewani, K.; Shukla, P.; Bhaskar, P.V.; Joshi, K.B.; Vinayak, V. Exopolysaccharides directed embellishment of diatoms triggered on plastics and other marine litter. *Sci. Rep.* **2020**, *10*, 18448.
75. Scardino, A.J.; Zhang, H.; Cookson, D.J.; Lamb, R.N.; Nys, R.D. The role of nano-roughness in antifouling. *Biofouling* **2009**, *25*, 757–767.
76. Marmur, A.A. Guide to the Equilibrium Contact Angles Maze. In *Contact angle, Wettability and Adhesion*; CRC Press: Boca Raton, FL, USA, 2009; pp. 1–18.

77. Wenzel, R.N. Resistance of solid surfaces to wetting by water. *Ind. Eng. Chem.* **1936**, *28*, 988–994.
78. Scurria, A.; Scolaro, C.; Sfameni, S.; Di Carlo, G.; Pagliaro, M.; Visco, A.; Ciriminna, R. Towards AquaSun practical utilization: Strong adhesion and lack of ecotoxicity of solar-driven antifouling sol-gel coating. *Prog. Org. Coat.* **2022**, *165*, 106771.
79. Cappello, S.; Calogero, R.; Santisi, S.; Genovese, M.; Denaro, R.; Genovese, L.; Giuliano, L.; Mancini, G.; Yakimov, M.M. Bioremediation of oil polluted marine sediments: A bio-engineering treatment. *Int. Microbiol.* **2015**, *18*, 127–134.
80. Urzì, C.; De Leo, F. Sampling with adhesive tape strips: An easy and rapid method to monitor microbial colonization on monument surfaces. *J. Microbiol. Methods* **2001**, *44*, 1–11.
81. Urzì, C.; Albertano, P. Studying phototrophic and heterotrophic microbial communities on stone monuments. *Methods Enzymol.* **2001**, *336*, 340–355.
82. Dufour, A.P.; Strickland, E.R.; Cabelli, V.J. Membrane filter method for enumerating *Escherichia coli*. *Appl. Environ. Microbiol.* **1981**, *41*, 1152–1158.
83. Urzì, C.; De Leo, F.; Krakova, L.; Pangallo, D.; Bruno, L. Effects of biocide treatments on the biofilm community in Domitilla's catacombs in Rome. *Sci. Total Environ.* **2016**, *572*, 252–262.
84. Urzì, C.; De Leo, F. Evaluation of the efficiency of water-repellent and biocide compounds against microbial colonization of mortars. *Int. Biodeterior. Biodegrad.* **2007**, *60*, 25–34.
85. La Cono, V.; Urzì, C. Fluorescent in situ hybridization applied on samples taken with adhesive tape strips. *J. Microbiol. Methods* **2003**, *55*, 65–71.
86. Scandura, G.; Ciriminna, R.; Ozer, L.Y.; Meneguzzo, F.; Palmisano, G.; Pagliaro, M. Antifouling and Photocatalytic Antibacterial Activity of the AquaSun Coating in Seawater and Related Media. *ACS Omega* **2017**, *2*, 7568–7575.
87. Porter, K.G.; Feig, Y.S. The use of DAPI for identifying and counting aquatic microflora. *Limnol. Oceanogr.* **1980**, *25*, 943–948.

Article

Synthesis and CO$_2$ Capture of Porous Hydrogel Particles Consisting of Hyperbranched Poly(amidoamine)s

Hojung Choi [1,†], Sanghwa Lee [1,†], SeongUk Jeong [1], Yeon Ki Hong [2] and Sang Youl Kim [1,*]

[1] Department of Chemistry, Korea Advanced Institute of Science and Technology (KAIST), Daejeon 34141, Korea
[2] School of Chemical and Materials Engineering, Korea National University of Transportation, Chungju-si 27469, Korea
* Correspondence: kimsy@kaist.ac.kr; Tel.: +82-42-350-2834
† These authors contributed equally to this work.

Abstract: We successfully synthesized new macroporous hydrogel particles consisting of hyperbranched poly(amidoamine)s (HPAMAM) using the Oil-in-Water-in-Oil (O/W/O) suspension polymerization method at both the 50 mL flask scale and the 5 L reactor scale. The pore sizes and particle sizes were easily tuned by controlling the agitation speeds during the polymerization reaction. Since O/W/O suspension polymerization gives porous architecture to the microparticles, synthesized hydrogel particles having abundant amine groups inside polymers exhibited a high CO$_2$ absorption capacity (104 mg/g) and a fast absorption rate in a packed-column test.

Keywords: carbon dioxide capture; poly(amidoamine)s; hyperbranched polymers; macroporous polymers; suspension polymerization

Citation: Choi, H.; Lee, S.; Jeong, S.; Hong, Y.K.; Kim, S.Y. Synthesis and CO$_2$ Capture of Porous Hydrogel Particles Consisting of Hyperbranched Poly(amidoamine)s. *Gels* 2022, *8*, 500. https://doi.org/10.3390/gels8080500

Academic Editors: Luca Burratti, Paolo Prosposito and Iole Venditti

Received: 15 July 2022
Accepted: 9 August 2022
Published: 11 August 2022

Publisher's Note: MDPI stays neutral with regard to jurisdictional claims in published maps and institutional affiliations.

Copyright: © 2022 by the authors. Licensee MDPI, Basel, Switzerland. This article is an open access article distributed under the terms and conditions of the Creative Commons Attribution (CC BY) license (https://creativecommons.org/licenses/by/4.0/).

1. Introduction

A major anthropogenic greenhouse gas in the atmosphere, CO$_2$ gas has been emitted in large quantities, mainly due to the burning of fossil fuels and other chemical processes [1,2]. Despite the global economic slowdown caused by the COVID-19 pandemic, the atmospheric concentration of CO$_2$ increased to 421 ppm in May 2022, from 405 ppm in 2017 and 340 ppm in the 1980s [3,4], and the estimated amount of annual CO$_2$ capture necessary to maintain the current level of atmospheric CO$_2$ concentrations has reached 3.7 gigatonnes/year [5]. However, fossil fuels are the major source of energy for human activities [6,7], and this trend is expected to continue over the coming decades [8]. Therefore, capture and storage of CO$_2$ gas from power plants is an urgent and important issue with respect to reducing CO$_2$ emissions.

The current CO$_2$ capture technology from flue gases involves the passage of streams of CO$_2$-containing flue gases through an aqueous amine solution, for example, monoethanolamine (MEA), in the absorber column. The amine solution is then regenerated by heating it in a stripping tower to liberate the CO$_2$. This 'wet-scrubbing' method with an alkanolamine solution has been employed industrially since the 1980s and is suitable for large-scale removal of CO$_2$ [8]. However, the regeneration of the total solution requires considerable energy consumption, and the amine concentration of the solution is limited due to viscosity and foaming [9]. Furthermore, careful control of the alkanolamine solution is necessary to avoid corrosion and oxidative degradation [2,10].

In efforts to overcome these problems, recently many materials have been developed as dry type solid CO$_2$ adsorbents. In the early stage of research, porous inorganic materials drew interest as solid type adsorbents for their high surface area [11–13]. However, industrial flue gases contain a considerable amount of water vapor [14], and these types of adsorbents are vulnerable to moisture, resulting in diminution of their sorption capacity and durability [15,16]. To address these drawbacks, amine-containing CO$_2$ adsorbents that tolerate moisture have been developed, including amine-modified mesoporous silica [17–25], Metal-Organic Frameworks (MOFs) [26–31], and amine-containing polymeric adsorbents [32–36]. However, implementing amines in solid supports disrupts the pore structure and reduces the surface area [25,37,38]. Furthermore, nonuniform dispersion of the amine and the complicated synthetic process of amine-implemented solid CO_2 adsorbents are hurdles toward practical applications.

To exploit the strengths of wet-scrubbing and a dry type adsorbent, we designed a hydrogel absorbent that works in an aqueous phase. A hydrogel is introduced because hydrogels are widely used as adsorbents for various substances such as heavy metal ions, dyes, and carbon dioxide due to their high adsorption performance and ease of use in an aqueous solution [39–46]. Because amines are verified carbon dioxide absorbers in wet-scrubbing processes by forming carbamate and ammonium bicarbonate with CO_2, we adopted poly(amidoamine) (PAMAM) as our key functional material. PAMAM has a high content of N-heteroatom groups. Abundant secondary and hindered amine groups of PAMAM are expected to provide high CO_2 sorption capacity [10,47]. Previously, we reported a facile one-step synthesis and application of micro-sized hyperbranched PAMAM (HPAMAM) particles [48]. Compared with PAMAM dendrimers, HPAMAM particles offer advantages of simplicity in the synthetic process and ease of use as packed columns without an ultrafiltration step for separation. In this work, we synthesize porous HPAMAM particles and investigate the CO_2 sorption ability of the porous HPAMAM hydrogel microbeads in packed columns (Scheme 1). The amine groups of HPAMAM particles in the packed column resemble the environment of amines in conventional wet scrubbing. Emulsion templating including double emulsions is generally used to produce porous materials. We adopted the Oil-in-Water-in-Oil (O/W/O) suspension polymerization method to provide a porous feature to HPAMAM particles that resembles the structural pores of solid type adsorbents [49–52]. The macropores inside of the hydrogel particles facilitate mass transfer and diffusion and thereby accelerate the sorption rate.

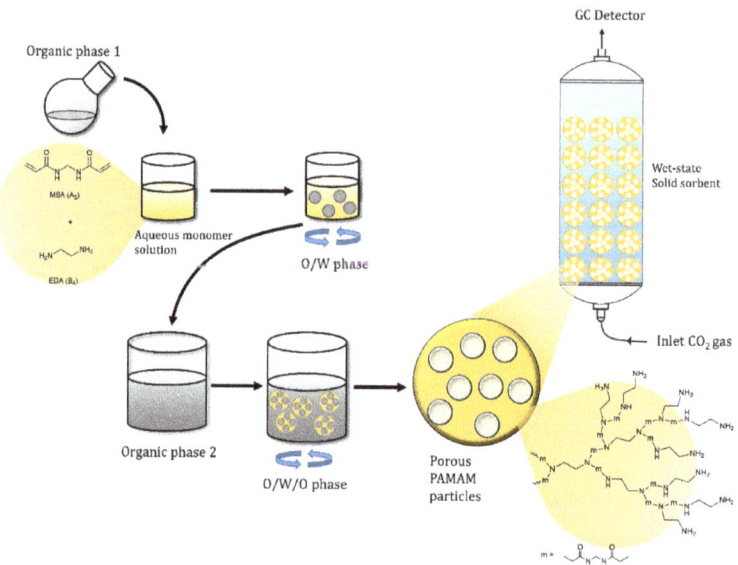

Scheme 1. Schematic illustration of synthesis and carbon dioxide capture of porous HPAMAM.

2. Results and Discussion

The preparation of porous HPAMAM particles was carried out via the O/W/O suspension polymerization method. We prepared three solutions: (1) an inner phase organic solution (O_1, solution 1) used to create interior pores; (2) an aqueous solution (W, solution 2) in which a polymerization reaction actually takes place; and (3) an outer phase organic solution (O_2, solution 3) for developing spherical particles.

A_2 (N,N'-methylenebisacrylamide, MBA) and B_4 (ethylenediamine, EDA) monomers and surfactant (poly(vinylalcohol), PVA) were mixed together in aqueous solution W until complete dissolution of the solids to induce aza-michael addition between amine and acrylamide. The O_1 solution, containing a cosurfactant (Span 60®) and toluene, was added to aqueous solution W with vigorous agitation to make an O/W suspension. The O/W mixture was then suspended to the O_2 solution. Since the polymerization of hydrophilic monomers occurs in the aqueous phase, the gelated polymer forms a spherical shape with the inner dispersion of organic solutions. After the polymerization, during the workup, the inner organic phase was removed from the particles, leaving macropores. Samples of porous HPAMAM particles are designated as **P-MxAyTz**, where x, y and z denote the molar feed ratio of MBA to EDA, the agitation speed (rpm) of the O/W suspension, and the volume ratio of solution 1 to solution 2, respectively (Scheme 1).

Three kinds of porous HPAMAM particles, **P-$M_{1.5}A_{1400}T_{0.6}$**, **P-$M_{1.5}A_{3000}T_{0.6}$**, and **P-$M_{1.5}A_{6000}T_{0.6}$**, were synthesized via O/W/O suspension polymerization with different agitation speeds to verify that the pore size can be controlled by the agitation speed of the O/W phase. To verify the chemical structures of the product, we carried out ATR-FTIR analysis for a representative sample of **P-$M_{1.5}A_{1400}T_{0.6}$** (Figure 1). ATR-FTIR spectra of MBA, EDA, and PAMAM polymers showed the peaks of N-H stretching (3000~3350 cm^{-1}), C=O stretching (1650 cm^{-1}), and CONH stretching (1520 cm^{-1}) in the PAMAM product, suggesting the successful synthesis of poly(amidoamine). The C=CH_2 stretching peak (3020 cm^{-1}) of MBA disappeared in the PAMAM product. The 3080 cm^{-1} peak is an overtone of 1540 cm^{-1} (N-H bending), which commonly appears in amide compounds. Energy Dispersive X-ray Spectroscopy (EDS) was performed to obtain additional information on the monomer ratio (Figure S1). The morphology of the synthesized porous HPAMAM particles was investigated with an optical microscope (OM) and a scanning electron microscope (SEM), as shown in Figure 2. Particle sizes obtained with a slower agitation speed (1000 rpm) were smaller than the particle sizes obtained at a faster agitation speed (1400 rpm) (Figure S2). Since the size of the pores was in the range of the micrometer scale, we analyzed the pore sizes through SEM images rather than with porosimetry such as BET, which is suitable for micro or mesopores. The average pore size changed from 2.9 μm to 2.3 μm to 1.3 μm with each agitation speed of 1400 rpm, 3000 rpm, and 6000 rpm, respectively (Figure 2c and Figure S3). As expected, the pore size decreased with the increase of the O/W phase agitation speed. However, the size of the particles did not change regardless of the O/W phase agitation speed (50–300 μm). The proposed synthetic route for porous HPAMAM hydrogel particles via O/W/O suspension polymerization is an eco-friendly process. It does not produce any byproducts and does not require catalysts or pore-generating reagents (porogens). Scale-up synthesis of the porous HPAMAM hydrogel particles was attempted with O/W/O suspension polymerization. The 2 L scale O/W/O suspension polymerization reaction of MBA and EDA monomers proceeded successfully and a product (**P-$M_{1.5}A_{1400}T_{0.6}$-2L**) with 50–300 μm particle diameters was obtained (Figure 3). The porous HPAMAM hydrogel particles exhibited no differences compared to the particles obtained with the smaller scale reaction. The synthesized porous PAMAM hydrogel particles contain a large amount of amine functional groups that are known to interact with carbon dioxide to form carbamates (Figure S4). On the basis of the feasibility of large-scale production through an eco-friendly process and the high reactivity of the amines in synthesized polymers to carbon dioxide [10,47,53,54], the porous HPAMAM hydrogel particles are suitable CO_2 absorbing materials for flue gas.

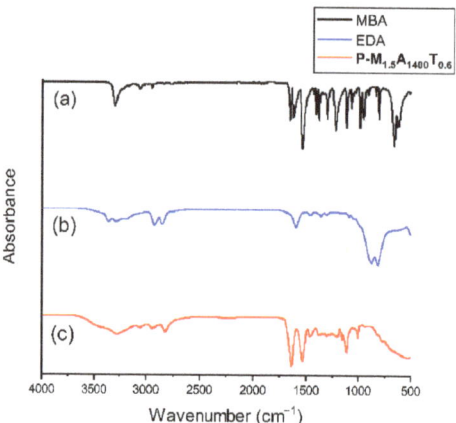

Figure 1. ATR-FTIR spectra of methylenebiscrylamide (a), ethylenediamine (b), and poly(amidoamine) particles (**P-M$_{1.5}$A$_{1400}$T$_{0.6}$**) (c).

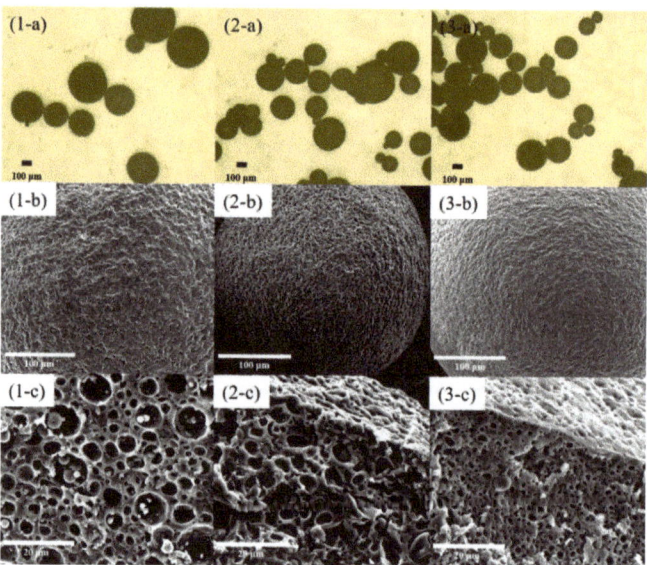

Figure 2. Optical images (**a**), SEM image (**b**), and cross-sectional SEM image of (**1-a**, **1-b** and **1-c**) **P-M$_{1.5}$A$_{1400}$T$_{0.6}$**, (**2-a**, **2-b** and **2-c**) **P-M$_{1.5}$A$_{3000}$T$_{0.6}$**, and (**3-a**, **3-b** and **3-c**) **P-M$_{1.5}$A$_{6000}$T$_{0.6}$**. Magnification is ×1000 (**b**) and ×5000 (**c**).

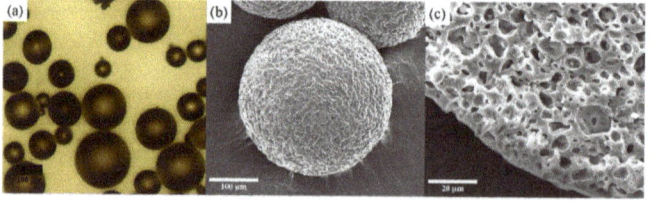

Figure 3. Optical images of **P-M$_{1.5}$A$_{1400}$T$_{0.6}$-2L** (**a**) and cross-sectional SEM images of **P-M$_{1.5}$A$_{1400}$T$_{0.6}$-2L** (**b**) ×1000 and (**c**) ×5000.

To measure the CO$_2$ absorption capacity of the hydrogel particles, water-soaked porous HPAMAM particles (**P-M$_{1.5}$A$_{1400}$T$_{0.6}$-2L**) packed in a 60 cm column were used with a continuous flow of CO$_2$. A schematic illustration of the column and a photograph of the experimental setup are shown in Figure 4. We used a column with a pocket where water can be circulated to control the temperature of the column at 50 °C. For the measurement, 45 g of porous PAMAM particles (**P-M$_{1.5}$A$_{1400}$T$_{0.6}$-2L**) was used at various CO$_2$ concentrations (6, 9, 12, and 15 wt%) with a gas flow rate of 150 mL/min (Figure 5). The concentration of outlet CO$_2$ gas was measured by gas chromatography at 25 °C. The porous HPAMAM hydrogel particles exhibited 2.374 mol/kg of CO$_2$ sorption (104 mg/g) within 60 min for saturation at 15 wt% CO$_2$ concentration while the sorption value was 2.150 mol/kg and 100 min for saturation at 9 wt% CO$_2$. These values are comparable to the reported values of amine-modified mesoporous silica (1.36–2.59 mol/kg) [35,55,56].

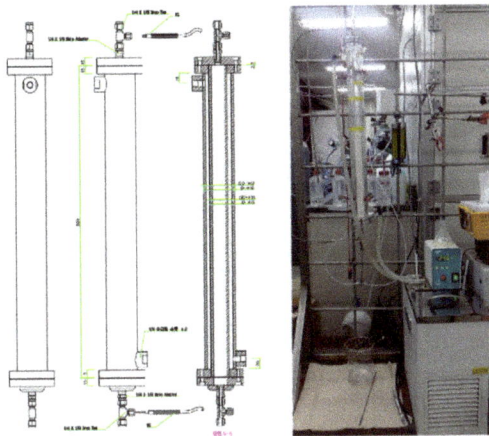

Figure 4. Schematic illustration of the column and a photograph of the experimental setup.

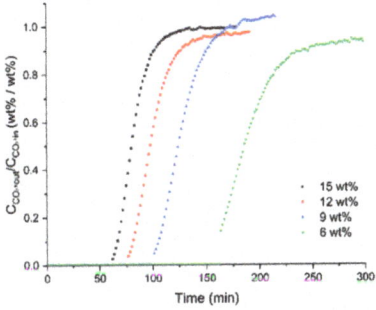

Figure 5. CO$_2$ sorption properties of porous PAMAM hydrogel particles (**P-M$_{1.5}$A$_{1400}$T$_{0.6}$-2L**) with various CO$_2$ concentrations of input gas.

3. Conclusions

We synthesized a new type of CO$_2$ absorbing porous hydrogel particle consisting of hyperbranched poly(amidoamine)s by Oil-in-Water-in-Oil (O/W/O) suspension polymerization of N,N'-methylenebisacrylamide (MBA) and ethylenediamine (EDA) monomers. By introducing the double emulsion method to the conventional inverse suspension polymerization, synthesis of macroporous hydrogel particles was achieved successfully. The particle sizes and the size of macropores were controlled by simply changing the agitation speed of Oil-in-Water-in-Oil (O/W/O) suspension polymerization, demonstrating the high applicability of the synthetic method in a large scale. The synthesized porous PAMAM hydrogel particles contain many amine functional groups that are known to interact with carbon dioxide. A CO_2 sorption test revealed the CO_2 absorption capacity of the porous hydrogel particles was 104.4 mg/g, which is comparable to the values of high performance dry-state adsorbents. Easy scale-up of the O/W/O suspension polymerization of MBA and EDA without any byproducts supports the practical usability of porous hydrogel particles for CO_2 capture from flue gas.

4. Materials and Methods

4.1. General Information

Ethylenediamine (EDA), N,N'-methylenebisacrylamide (MBA), poly(vinylalcohol) (PVA) (Mw = 89,000–98,000 g/mol), and Span 60® were purchased from Sigma-Aldrich (Boston, MA, USA). Methanol (MeOH), toluene, and cyclohexane were purchased from Junsei (Chuo-ku, Tokyo, Japan). All reagents were used as received without further purification. A 5L Chemglass ChemRxnHub (Vineland, NJ, USA) Process system was used in the scale up process. Images of the particles were obtained by Nikon ECLIPSE ME600 (Tokyo, Japan) optical microscopy (OM), FEI Nova230 (Hillsboro, OR, USA), and JEOL JSM IT800 (Akishima, Tokyo, Japan) scanning electron microscopy (SEM). EDS analysis was conducted by a Bruker Quantax ESPRIT (Billerica, MA, USA) attachment. CO_2 was measured by DS6200 gas chromatography (DS Science Inc.; Gwangju-si, Gyeonggi-do, Korea).

4.2. Various Agitation Speeds of Aqueous Phase

Three solutions, solution 1 (Span60 (0.128 g, 5 wt%) in toluene (3 mL)), solution 2 (MBA (2.024 g), EDA (0.526 g), PVA (0.026 g, 1 wt%) in water (5.1 mL)), and solution 3 (Span60 (0.013 g, 0.5 wt%) in toluene (20.4 mL)) were prepared. Solution 3 was stirred vigorously at 60 °C. MBA, EDA, PVA, and water in solution 2 were mixed together until complete dissolution of the solids at 45 °C. After solution 1 was added to solution 2, it was agitated at 1400 rpm (stirring bar), 3000 rpm (homogenizer), and 6000 rpm (homogenizer) for one minute. After that, solution 2 was added to solution 3, and agitated at 1400 rpm at 60 °C. After 6 hr of polymerization, particles were filtered and washed with methanol and acetone, and dried under vacuum at 70 °C for 24 hr.

4.3. Scale up Process

Three solutions, solution 1 (Span60 (25 g, 0.5 wt%) in toluene (600 mL)), solution 2 (MBA (398.6 g), EDA (103.2 g), PVA (5 g, 1 wt%) in water (1 L)), and solution 3 (Span60 (1.5 g, 0.5 wt%) in toluene (2 L)) were prepared. Solution 3 was stirred vigorously at 60 °C. MBA, EDA, PVA, and water in solution 2 were mixed together until dissolution of the solids at 45 °C. After solution 1 was added to solution 2, it was agitated at 5000 rpm (homogenizer), for one minute. Solution 2 was then added to solution 3, and it was agitated at 1400 rpm at 60 °C. After 6 hr of polymerization, particles were filtered and washed with methanol and acetone, and dried under vacuum at 70 °C for 24 h.

4.4. CO_2 Absorption Test

Water was poured to 45 g of each type of PAMAM particle (P1.5, P1.8, P2.0, and P1.5 porosity) in a beaker. After 90 min, wet PAMAM particles were placed in an acryl column (diameter: 7 cm, height 60 cm). Removing the moisture in column, N_2 gas was flowed for 20 min. CO_2 absorption was then carried out using 15 wt%, 12 wt%, 9 wt% and 6 wt% of CO_2 (The rest of the gas was N_2) at a flow rate of 150 mL/min at 50 °C. The concentration of CO_2 in the outlet gas was measured by gas chromatography.

Supplementary Materials: The following supporting information can be downloaded at: https://www.mdpi.com/article/10.3390/gels8080500/s1, Figure S1: SEM and EDS analysis of P-$M_{1.5}A_{1400}T_{0.6}$; Figure S2: SEM images of P-$M_{1.5}A_{3000}$, W/O agitation speed of 1400 rpm (a,b) and P-M1.5A3000, W/O agitation speed of 1000 rpm (c,d); Figure S3: Pore size distribution of various O1/W agitation speeds, 1400 rpm (a), 3000 rpm (b), 6000 rpm (c), and merged graph (d), based on SEM images of Figure 2; Figure S4: Reactions of amine and carbon dioxide [1–4].

Author Contributions: Conceptualization, S.L. and S.Y.K.; methodology, S.L.; software, S.L.; validation, S.L. and H.C.; formal analysis, S.L.; investigation, H.C.; resources, H.C.; writing—original draft preparation, H.C. and S.J.; writing—review and editing, Y.K.H. and S.Y.K.; visualization, H.C.; supervision, S.Y.K. All authors have read and agreed to the published version of the manuscript.

Funding: This work was supported by the National Research Foundation (NRF) of Korea (NRF-2021R1A4A1052070) and by a Korea Institute of Energy Technology Evaluation and Planning (KETEP) grant funded by the Korean government (MOTIE) (20191510301070).

Institutional Review Board Statement: Not applicable.

Informed Consent Statement: Not applicable.

Data Availability Statement: The data presented in this study are available on request from the corresponding author.

Conflicts of Interest: The authors declare no conflict of interest.

References

1. Carapellucci, R.; Milazzo, A. Membrane systems for CO_2 capture and their integration with gas turbine plants. *Proc. Inst. Mech. Eng. Part A J. Power Energy* **2003**, *217*, 505–517. [CrossRef]
2. Yang, H.; Xu, Z.; Fan, M.; Gupta, R.; Slimane, R.B.; Bland, A.E.; Wright, I. Progress in carbon dioxide separation and capture: A review. *J. Environ. Sci.* **2008**, *20*, 14–27. [CrossRef]
3. Ziemke, J.R.; Cooper, O.R. Tropospheric Ozone, in "State of the Climate in 2017". *Bull. Am. Meteorol. Soc.* **2018**, *99*, S56–S59.
4. Carbon Dioxide Now More than 50% Higher than Pre-Industrial Levels. Available online: https://www.noaa.gov/news-release/carbon-dioxide-now-more-than-50-higher-than-pre-industrial-levels (accessed on 25 June 2022).
5. International Energy Agency (IEA). *Energy Technology Systems Analysis Program (IEA-ETSAP) Technology Roadmap*; International Energy Agency (IEA): Paris, France, 2013; Volume 59.
6. Li, J.-R.; Ma, Y.; McCarthy, M.C.; Sculley, J.; Yu, J.; Jeong, H.-K.; Balbuena, P.B.; Zhou, H.-C. Carbon dioxide capture-related gas adsorption and separation in metal-organic frameworks. *Coord. Chem. Rev.* **2011**, *255*, 1791–1823. [CrossRef]
7. Kenarsari, S.D.; Yang, D.; Jiang, G.; Zhang, S.; Wang, J.; Russell, A.G.; Wei, Q.; Fan, M. Review of recent advances in carbon dioxide separation and capture. *RSC Adv.* **2013**, *3*, 22739–22773. [CrossRef]
8. Rochelle, G.T. Amine Scrubbing for CO_2 Capture. *Science* **2009**, *325*, 1652–1654. [CrossRef] [PubMed]
9. Franchi, R.S.; Harlick, P.J.E.; Sayari, A. Applications of Pore-Expanded Mesoporous Silica. 2. Development of a High-Capacity, Water-Tolerant Adsorbent for CO_2. *Ind. Eng. Chem. Res.* **2005**, *44*, 8007–8013. [CrossRef]
10. D'Alessandro, D.M.; Smit, B.; Long, J.R. Carbon Dioxide Capture: Prospects for New Materials. *Angew. Chem. Int. Ed.* **2010**, *49*, 6058–6082. [CrossRef] [PubMed]
11. Pourhakkak, P.; Taghizadeh, M.; Taghizadeh, A.; Ghaedi, M. *Adsorption: Fundamental Processes and Applications*, 1st ed.; Ghaedi, M., Ed.; Academic Press: Cambridge, MA, USA, 2021; Volume 33, pp. 71–210.
12. Schumacher, C.; Gonzalez, J.; Pérez-Mendoza, M.; Wright, P.A.; Seaton, N.A. Design of Hybrid Organic/Inorganic Adsorbents Based on Periodic Mesoporous Silica. *Ind. Eng. Chem. Res.* **2006**, *45*, 5586–5597. [CrossRef]
13. González, A.; Plaza, M.; Rubiera, F.; Pevida, C. Sustainable biomass-based carbon adsorbents for post-combustion CO_2 capture. *Chem. Eng. J.* **2013**, *230*, 456–465. [CrossRef]
14. Aouini, I.; Ledoux, A.; Estel, L.; Mary, S. Étude du captage du CO_2 dans des gaz de combustion d'un incinérateur de déchets à l'aide d'un pilote utilisant un solvant à base de MEA. *Oil Gas Sci. Technol.* **2014**, *69*, 1091–1104. [CrossRef]
15. Burtch, N.C.; Jasuja, H.; Walton, K.S. Water Stability and Adsorption in Metal-Organic Frameworks. *Chem. Rev.* **2014**, *114*, 10575–10612. [CrossRef] [PubMed]
16. DeCoste, J.B.; Peterson, G.W.; Schindler, B.J.; Killops, K.L.; Browe, M.A.; Mahle, J.J. The effect of water adsorption on the structure of the carboxylate containing metal-organic frameworks Cu-BTC, Mg-MOF-74, and UiO-66. *J. Mater. Chem. A* **2013**, *1*, 11922–11932. [CrossRef]
17. Wang, L.; Yao, M.; Hu, X.; Hu, G.; Lu, J.; Luo, M.; Fan, M. Amine-modified ordered mesoporous silica: The effect of pore size on CO_2 capture performance. *Appl. Surf. Sci.* **2014**, *324*, 286–292. [CrossRef]
18. Kishor, R.; Ghoshal, A.K. Amine-Modified Mesoporous Silica for CO_2 Adsorption: The Role of Structural Parameters. *Ind. Eng. Chem. Res.* **2017**, *56*, 6078–6087. [CrossRef]
19. Qian, Z.; Wei, L.; Mingyue, W.; Guansheng, Q. Application of amine-modified porous materials for CO_2 adsorption in mine confined spaces. *Colloids Surf. A Physicochem. Eng. Asp.* **2021**, *629*, 127483. [CrossRef]

20. Vilarrasa-Garcia, E.; Moya, E.O.; Cecilia, J.; Cavalcante, C.; Jiménez-Jiménez, J.; Azevedo, D.; Rodríguez-Castellón, E. CO_2 adsorption on amine modified mesoporous silicas: Effect of the progressive disorder of the honeycomb arrangement. *Microporous Mesoporous Mater.* **2015**, *209*, 172–183. [CrossRef]
21. Anyanwu, J.-T.; Wang, Y.; Yang, R.T. Amine-Grafted Silica Gels for CO_2 Capture Including Direct Air Capture. *Ind. Eng. Chem. Res.* **2020**, *59*, 7072–7079. [CrossRef]
22. Garip, M.; Gizli, N. Ionic liquid containing amine-based silica aerogels for CO_2 capture by fixed bed adsorption. *J. Mol. Liq.* **2020**, *310*, 113227. [CrossRef]
23. Fayaz, M.; Sayari, A. Long-Term Effect of Steam Exposure on CO_2 Capture Performance of Amine-Grafted Silica. *ACS Appl. Mater. Interfaces* **2017**, *9*, 43747–43754. [CrossRef]
24. Ojeda, M.; Mazaj, M.; Garcia, S.; Xuan, J.; Maroto-Valer, M.; Logar, N.Z. Novel Amine-impregnated Mesostructured Silica Materials for CO_2 Capture. *Energy Procedia* **2017**, *114*, 2252–2258. [CrossRef]
25. Hou, X.; Zhuang, L.; Ma, B.; Chen, S.; He, H.; Yin, F. Silanol-rich platelet silica modified with branched amine for efficient CO_2 capture. *Chem. Eng. Sci.* **2018**, *181*, 315–325. [CrossRef]
26. Demessence, A.; D'Alessandro, D.M.; Foo, M.L.; Long, J.R. Strong CO_2 Binding in a Water-Stable, Triazolate-Bridged Metal-Organic Framework Functionalized with Ethylenediamine. *J. Am. Chem. Soc.* **2009**, *131*, 8784–8786. [CrossRef] [PubMed]
27. Cui, X.; Chen, K.; Xing, H.; Yang, Q.; Krishna, R.; Bao, Z.; Wu, H.; Zhou, W.; Dong, X.; Han, Y.; et al. Pore chemistry and size control in hybrid porous materials for acetylene capture from ethylene. *Science* **2016**, *353*, 141–144. [CrossRef] [PubMed]
28. Martínez, F.; Sanz, R.; Orcajo, G.; Briones, D.; Yángüez, V. Amino-impregnated MOF materials for CO_2 capture at post-combustion conditions. *Chem. Eng. Sci.* **2016**, *142*, 55–61. [CrossRef]
29. Zhang, G.; Wei, G.; Liu, Z.; Oliver, S.R.J.; Fei, H. A Robust Sulfonate-Based Metal-Organic Framework with Permanent Porosity for Efficient CO_2 Capture and Conversion. *Chem. Mater.* **2016**, *28*, 6276–6281. [CrossRef]
30. Ghalei, B.; Sakurai, K.; Kinoshita, Y.; Wakimoto, K.; Isfahani, A.P.; Song, Q.; Doitomi, K.; Furukawa, S.; Hirao, H.; Kusuda, H.; et al. Enhanced selectivity in mixed matrix membranes for CO_2 capture through efficient dispersion of amine-functionalized MOF nanoparticles. *Nat. Energy* **2017**, *2*, 17086. [CrossRef]
31. Molavi, H.; Eskandari, A.; Shojaei, A.; Mousavi, S.A. Enhancing CO_2/N_2 adsorption selectivity via post-synthetic modification of NH2-UiO-66(Zr). *Microporous Mesoporous Mater.* **2018**, *257*, 193–201. [CrossRef]
32. Min, K.; Choi, W.; Kim, C.; Choi, M. Rational Design of the Polymeric Amines in Solid Adsorbents for Postcombustion Carbon Dioxide Capture. *ACS Appl. Mater. Interfaces* **2018**, *10*, 23825–23833. [CrossRef]
33. Min, K.; Choi, W.; Kim, C.; Choi, M. Oxidation-stable amine-containing adsorbents for carbon dioxide capture. *Nat. Commun.* **2018**, *9*, 726. [CrossRef]
34. Park, S.; Kim, J.; Won, Y.-J.; Kim, C.; Choi, M.; Jung, W.; Lee, K.S.; Na, J.-G.; Cho, S.-H.; Lee, S.Y.; et al. Epoxide-Functionalized, Poly(ethylenimine)-Confined Silica/Polymer Module Affording Sustainable CO_2 Capture in Rapid Thermal Swing Adsorption. *Ind. Eng. Chem. Res.* **2018**, *57*, 13923–13931. [CrossRef]
35. Park, S.; Choi, K.; Yu, H.J.; Won, Y.-J.; Kim, C.; Choi, M.; Cho, S.-H.; Lee, J.-H.; Lee, S.Y.; Lee, J.S. Thermal Stability Enhanced Tetraethylenepentamine/Silica Adsorbents for High Performance CO_2 Capture. *Ind. Eng. Chem. Res.* **2018**, *57*, 4632–4639. [CrossRef]
36. Choi, W.; Park, J.; Kim, C.; Choi, M. Structural effects of amine polymers on stability and energy efficiency of adsorbents in post-combustion CO_2 capture. *Chem. Eng. J.* **2021**, *408*, 127289. [CrossRef]
37. Sujan, A.R.; Pang, S.H.; Zhu, G.; Jones, C.W.; Lively, R.P. Direct CO_2 Capture from Air using Poly(ethylenimine)-Loaded Polymer/Silica Fiber Sorbents. *ACS Sustain. Chem. Eng.* **2019**, *7*, 5264–5273. [CrossRef]
38. Zhang, G.; Zhao, P.; Hao, L.; Xu, Y. Amine-modified SBA-15(P): A promising adsorbent for CO_2 capture. *J. CO2 Util.* **2018**, *24*, 22–33. [CrossRef]
39. Verduzco-Navarro, I.P.; Mendizábal, E.; Mayorga, J.A.R.; Rentería-Urquiza, M.; Gonzalez-Alvarez, A.; Rios-Donato, N. Arsenate Removal from Aqueous Media Using Chitosan-Magnetite Hydrogel by Batch and Fixed-Bed Columns. *Gels* **2022**, *8*, 186. [CrossRef]
40. Guo, S.; Su, K.; Yang, H.; Zheng, W.; Zhang, Z.; Ang, S.; Zhang, K.; Wu, P. Novel Natural Glycyrrhetinic Acid-Derived Super Metal Gel and Its Highly Selective Dyes Removal. *Gels* **2022**, *8*, 188. [CrossRef]
41. Choi, H.; Kim, T.; Kim, S.Y. Poly (Amidehydrazide) Hydrogel Particles for Removal of Cu^{2+} and Cd^{2+} Ions from Water. *Gels* **2021**, *7*, 121. [CrossRef]
42. Choi, H.; Eom, Y.; Lee, S.; Kim, S.Y. Copper Ions Removal from Water using A_2B_3 Type Hyperbranched Poly(amidoamine) Hydrogel Particles. *Molecules* **2019**, *24*, 3866. [CrossRef]
43. Xu, X.; Heath, C.; Pejcic, B.; Wood, C.D. CO_2 capture by amine infused hydrogels (AIHs). *J. Mater. Chem. A* **2018**, *6*, 4829–4838. [CrossRef]
44. Xu, X.; Pejcic, B.; Heath, C.; Wood, C.D. Carbon capture with polyethylenimine hydrogel beads (PEI HBs). *J. Mater. Chem. A* **2018**, *6*, 21468–21474. [CrossRef]
45. Kang, D.W.; Lee, W.; Ahn, Y.-H. Superabsorbent polymer for improved CO_2 hydrate formation under a quiescent system. *J. CO2 Util.* **2022**, *61*, 102005. [CrossRef]
46. Sun, M.-T.; Song, F.-P.; Zhang, G.-D.; Li, J.-Z.; Wang, F. Polymeric superabsorbent hydrogel-based kinetic promotion for gas hydrate formation. *Fuel* **2021**, *288*, 119676. [CrossRef]

47. Taniguchi, I.; Kinugasa, K.; Toyoda, M.; Minezaki, K. Effect of amine structure on CO_2 capture by polymeric membranes. *Sci. Technol. Adv. Mater.* **2017**, *18*, 950–958. [CrossRef]
48. Lee, S.; Eom, Y.; Park, J.; Lee, J.; Kim, S.Y. Micro-hydrogel Particles Consisting of Hyperbranched Polyamidoamine for the Removal of Heavy Metal Ions from Water. *Sci. Rep.* **2017**, *7*, 10012. [CrossRef] [PubMed]
49. Zhu, Y.; Zheng, Y.; Wang, F.; Wang, A. Fabrication of magnetic porous microspheres via $(O_1/W)/O_2$ double emulsion for fast removal of Cu^{2+} and Pb^{2+}. *J. Taiwan Inst. Chem. Eng.* **2016**, *67*, 505–510. [CrossRef]
50. Zhang, T.; Sanguramath, R.A.; Israel, S.; Silverstein, M.S. Emulsion Templating: Porous Polymers and Beyond. *Macromolecules* **2019**, *52*, 5445–5479. [CrossRef]
51. Blin, J.-L.; Jacoby, J.; Kim, S.; Stébé, M.-J.; Canilho, N.; Pasc, A. A meso-macro compartmentalized bioreactor obtained through silicalization of "green" double emulsions: W/O/W and W/SLNs/W. *Chem. Commun.* **2014**, *50*, 11871–11874. [CrossRef]
52. Jeong, W.-C.; Choi, M.; Lim, C.H.; Yang, S.-M. Microfluidic synthesis of atto-liter scale double emulsions toward ultrafine hollow silica spheres with hierarchical pore networks. *Lab Chip* **2012**, *12*, 5262–5271. [CrossRef]
53. Henao, W.; Jaramillo, L.Y.; López, D.; Romero-Sáez, M.; Buitrago-Sierra, W.A.H. Insights into the CO_2 capture over amine-functionalized mesoporous silica adsorbents derived from rice husk ash. *J. Environ. Chem. Eng.* **2020**, *8*, 104362. [CrossRef]
54. Varghese, A.M.; Karanikolos, G.N. CO_2 capture adsorbents functionalized by amine-bearing polymers: A review. *Int. J. Greenh. Gas Control* **2020**, *96*, 103005. [CrossRef]
55. Wilfong, W.C.; Kail, B.W.; Jones, C.W.; Pacheco, C.; Gray, M.L. Spectroscopic Investigation of the Mechanisms Responsible for the Superior Stability of Hybrid Class 1/Class 2 CO_2 Sorbents: A New Class 4 Category. *ACS Appl. Mater. Interfaces* **2016**, *8*, 12780–12791. [CrossRef] [PubMed]
56. Kishor, R.; Ghoshal, A.K. N^1-(3-Trimethoxysilylpropyl)diethylenetriamine grafted KIT-6 for CO_2/N_2 selective separation. *RSC Adv.* **2016**, *6*, 898–909. [CrossRef]

Review

Hydrogel-Based Adsorbent Material for the Effective Removal of Heavy Metals from Wastewater: A Comprehensive Review

Zenab Darban [1], Syed Shahabuddin [1,*], Rama Gaur [1,*], Irfan Ahmad [2] and Nanthini Sridewi [3,*]

1. Department of Chemistry, School of Technology, Pandit Deendayal Energy University, Raisan 382426, India; darbanzenab05@gmail.com
2. Department of Clinical Laboratory Sciences, College of Applied Medical Sciences, King Khalid University, Abha 61421, Saudi Arabia; irfancsmmu@gmail.com
3. Department of Maritime Science and Technology, Faculty of Defence Science and Technology, National Defence University of Malaysia, Kuala Lumpur 57000, Malaysia
* Correspondence: syedshahab.hyd@gmail.com or syed.shahabuddin@sot.pdpu.ac.in (S.S.); rama.gaur@sot.pdpu.ac.in (R.G.); nanthini@upnm.edu.my (N.S.); Tel.: +91-8585932338 (S.S.); +91-8266907756 (R.G.); +60-124-675-320 (N.S.)

Citation: Darban, Z.; Shahabuddin, S.; Gaur, R.; Ahmad, I.; Sridewi, N. Hydrogel-Based Adsorbent Material for the Effective Removal of Heavy Metals from Wastewater: A Comprehensive Review. *Gels* 2022, *8*, 263. https://doi.org/10.3390/gels8050263

Academic Editors: Luca Burratti, Paolo Prosposito and Iole Venditti

Received: 30 March 2022
Accepted: 19 April 2022
Published: 22 April 2022

Publisher's Note: MDPI stays neutral with regard to jurisdictional claims in published maps and institutional affiliations.

Copyright: © 2022 by the authors. Licensee MDPI, Basel, Switzerland. This article is an open access article distributed under the terms and conditions of the Creative Commons Attribution (CC BY) license (https://creativecommons.org/licenses/by/4.0/).

Abstract: Water is a vital resource that is required for social and economic development. A rapid increase in industrialization and numerous anthropogenic activities have resulted in severe water contamination. In particular, the contamination caused by heavy metal discharge has a negative impact on human health and the aquatic environment due to the non-biodegradability, toxicity, and carcinogenic effects of heavy metals. Thus, there is an immediate need to recycle wastewater before releasing heavy metals into water bodies. Hydrogels, as potent adsorbent materials, are a good contenders for treating toxic heavy metals in wastewater. Hydrogels are a soft matter formed via the cross-linking of natural or synthetic polymers to develop a three-dimensional mesh structure. The inherent properties of hydrogels, such as biodegradability, swell-ability, and functionalization, have made them superior applications for heavy metal removal. In this review, we have emphasized the recent development in the synthesis of hydrogel-based adsorbent materials. The review starts with a discussion on the methods used for recycling wastewater. The discussion then shifts to properties, classification based on various criteria, and surface functionality. In addition, the synthesis and adsorption mechanisms are explained in detail with the understanding of the regeneration, recovery, and reuse of hydrogel-based adsorbent materials. Therefore, the cost-effective, facile, easy to modify and biodegradable hydrogel may provide a long-term solution for heavy metal removal.

Keywords: hydrogels; heavy metals removal; wastewater

1. Introduction

1.1. Problem Statement

Water is essential for all living organisms on the planet. Although, it occupies 71% of the total surface area of the earth, only 3% of water is available as freshwater and less than 1% is potable. The remaining percentage of water is inaccessible in different forms such as ice, glaciers, and snow on the south and north poles [1]. Water plays a significant part in the hydrological cycle, food-processing industries, chemical weathering, domestic usage, agricultural irrigation, and so on. Therefore, there is an increasing need for freshwater, but the availability is limited. Freshwater is contaminated by discarding waste in various water bodies in the form of marine dumping, oil leakage, industrial waste, sewage waste, etc. Different pollutants present in wastewater are summarized in Figure 1. Among them, heavy metals are found to be the most common pollutant in contaminated water, which deteriorates the sustainable environment. Water contamination by heavy metals has harmed human health all around the world due to the fast development in industries, economics, and population [2].

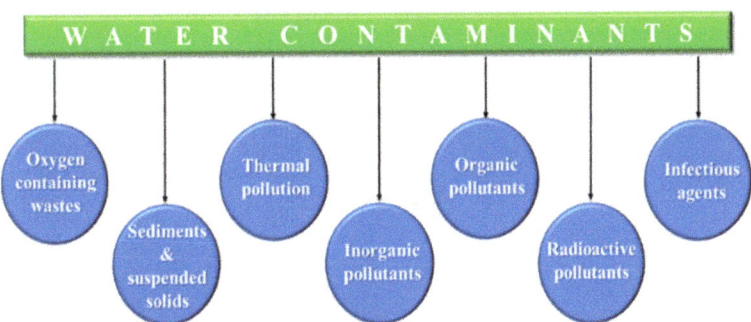

Figure 1. Different pollutants in contaminated water.

1.2. Heavy Metals and Their Hazardous Effect

Heavy metals are referred to as metals with a density of 5 gm/cm^3 and are poisonous, toxic, and hazardous even at very low concentrations. The sources of heavy metal contamination into water are categorized in two ways: (1) natural ways like soil erosion, rainfall, dissolution of soluble salts, etc., and (2) artificial ways like industrial waste, and urban wastewater [3]. Heavy metals include mercury (Hg), zinc (Zn), arsenic (As), cadmium (Cd), silver (Ag), iron (Fe), lead (Pb), tin (Sn), and the platinum group of metals. Heavy metals are non-biodegradable elements [4] that cause detrimental effects on the natural ecosystem and human health when their concentration goes beyond permissible limits. For instance, persistent intake of inorganic arsenic causes lung, bladder, skin, and kidney cancer in humans via consumption of drinking water [5]. Mercury accumulation in the food chain shows a negative impact on human health such as kidney and pulmonary function impairment, chest pain, and damage to the central nervous system [6]. Some other examples are listed in Table 1.

Heavy metals are not degraded by natural mechanisms and hence persist in the environment for a long duration of time. They may be converted into insoluble compounds or other forms. Water, air, and soil are the three key environmental compartments that get affected by heavy metal contamination (Table 1). Runoffs from cities, villages, towns, and factories transport the heavy metals that accumulate in a flowing stream. Even if a low concentration is transferred to water streams, it is extremely harmful to humans and the natural ecosystem [7]. Air pollution is caused by dust and particulate matters such as PM$_{2.5}$ and PM$_{10}$ which are discharged by various natural and anthropogenic processes. Soil erosion, dust storms, rock weathering, and volcanic eruptions are examples of natural processes that release particulate matter in the air, whereas anthropogenic activities are mainly transport-related and industrial. These particulate matters cause corrosion, haze, and eutropication, and lead to the formation of acid rains [8]. Heavy metals pollute soil by damping wastes like animal manures, pesticides, fertilizers, sewage sludge, spillage of petroleum distillates, etc. The use of this untreated waste has resulted in a high concentration of heavy metals in agricultural fields, which affects the entire biosphere. They are directly absorbed by plants, causing a risk to the plant and the food chain that consumes it. They affect soil qualities such as color, pH, and porosity and also pollute water [9]. Therefore, it is urgent and necessary to remove toxic heavy metals from contaminated wastewater. A variety of wastewater recycling techniques have been developed, which are further discussed in this review article.

Table 1. Toxic effects of different heavy metals on human health [10,11].

Heavy Metals	Leading Source	Path of Entry	Toxic Effects on Human Health	Environmental Hazards	MCL (mg/L)
Lead (Pb)	Mining, automobile emissions, smoking, pesticide, paint, burning of coal	Ingestion and inhalation	Damages the central nervous system, fetal brain, kidney, reproductive system, liver, basic cellular processes, and causes diseases, namely, anemia, nephrite syndrome, hepatitis, etc.	Soil and water pollution	0.015
Cadmium (Cd)	Pesticide fertilizer, electroplating, Cd-Ni batteries, welding	Ingestion and inhalation	Irritation of respiratory system, damages liver, kidney, and lungs	Soil and water pollution	0.005
Nickel (Ni)	Electrochemical industries	Inhalation	Causes lung, kidney, and gastronomical pain, renal edema, pulmonary fibrosis, and skin dermatitis	Soil and water pollution	0.1
Zinc (Zn)	Plumbing, refineries, metal plating, brass manufacture	Ingestion, inhalation, and through skin	Vomiting, pain in the stomach, skin irritation, nausea, and anemia	Soil and water pollution	0.8

2. Methods Used for Recycling Wastewater

Over the years, numerous methods have been developed to remediate heavy metal-contaminated wastewater before discharging it into the environment. Heavy metals can be removed from wastewater by using a variety of methods, including ion exchange, coagulation-flocculation, flotation, membrane filtration, chemical precipitation, and adsorption (Figure 2). However, each method has its own set of advantages and disadvantages (Summarized in Table 2).

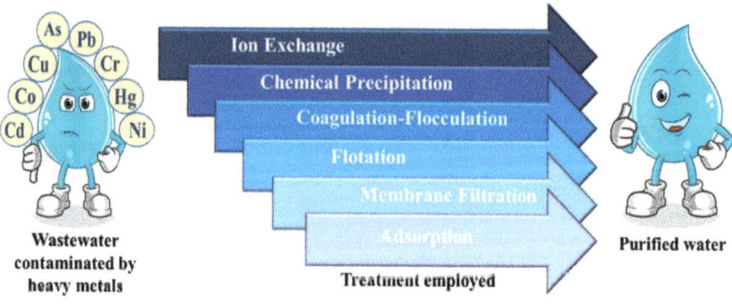

Figure 2. Wastewater recycling methods.

Among the presented methods in Figure 2, adsorption is considered to be one of the most efficient, low-cost, and simple-to-operate methods to remove heavy metals from contaminated water when compared with other methods [12]. Moreover, the adsorption process is technologically feasible and attractive, as the adsorbent material can be reused and regenerated. In this process, no secondary waste is generated during the removal of heavy metals [13]. The process also has the advantage of removing low concentrations of heavy metals from the solution with low energy consumption [14,15]. Adsorption is a mass transfer surface phenomenon that leads to the binding of molecules from liquid bulk (adsorbate) onto the solid surface (adsorbent) [16]. This binding occurs due to the presence of residual imbalance forces that attracts and retains molecules on the surface of the solid or liquid phase [17]. The adsorbent adsorbs the adsorbate via bonding interactions like a covalent bond or Van der Waals forces [18].

Over the years, researchers have developed adsorbent materials for the removal of toxic heavy metals from wastewater like rice husk bio-char, sugar beet pulp, TiO_2, activated carbon, clay, etc. [19–23]. However, these adsorbent materials suffer from certain disadvantages such as their difficulty separating from the water after the decontamination process, higher production cost, economic unsustainability for large-scale applications, and many other reasons [24]. This calls for the immediate development of an adsorbent material that is cost-effective, easy to handle, biodegradable, and biocompatible adsorbent material to purify contaminated water. In recent years, hydrogels have gained tremendous attention as the potential adsorbent material owing to their excellent water affinity, controllable swelling behavior, high porosity, better mechanical properties, and easy handling; these are the main factors for the reuse of adsorbent material. Hydrogel-based material has shown substantial attention for applications in different fields, such as biomedicine, agriculture, food additives, drug delivery, wound dressing, regenerative medicine, and cosmetics. Even though hydrogel-based adsorbents have a long history in a few of the above-mentioned fields, their application for contaminant removal from wastewater has been only reported over the last decade. Our emphasis will be on the application of hydrogel-based material for the removal of toxic and hazardous heavy metals from wastewater.

Table 2. Advantages and disadvantages of various methods used for recycling wastewater.

Methods	Advantages	Disadvantages	References
Ion exchange	Does not produce a large amount of sludge, easy regeneration of resins	High operational cost, selective towards certain metal ions	[25]
Chemical precipitation	Low capital cost, simple process	Produces a large amount of sludge, ineffective in treating low concentration of heavy metal ions	[26]
Coagulation-flocculation	Easy to employ, inexpensive, low energy consumption	Complete removal of heavy metals is difficult, generation of a large quantity of sludge	[27]
Flotation	Economically efficient	Low elimination efficiency, complex process, high operational expense due to membrane fouling	[28,29]
Membrane filtration	Small space requirement, high efficiency, high separation selectivity		[30]
Adsorption	Technologically feasible, effective, low-cost adsorbent, no waste generation, easy operation conditions	Low selectivity	[3,13]

3. Hydrogels for Removal of Heavy Metals

In 1894, the term "hydrogel" was first coined by Bemmelen to explain colloidal gels [31]. DuPont scientists reported the first synthetic hydrogel, poly (2-hydroxyethyl methacrylate) (PHEMA), in 1936 [32]. Witcher and Lim were the first to report the use of PHEMA for the application of contact lenses in 1960 [33]. Since then, hydrogels have been an intriguing topic for researchers, and they now represent a developing and active research field aimed at providing better solutions for various needs in many applicative fields. The use of hydrogels for the extraction of heavy metals from wastewater is becoming more popular, as they can capture and store different heavy metals found in wastewater within their network structure. Hydrogels are regarded as hydrophilic gels, which consists of chemically reactive functional groups and physically distinct three-dimensional (3D) network [34]. The porous three-dimensional network of hydrogel allows absorption and retention of a large volume of water without dissolving [35]. The hydrophilic groups in the polymeric network enable the formation of a flexible structure, which allows easy diffusion of solute into the three-dimensional framework of gels and forms a stable complex with the functional group present on a long polymeric chain [36]. Due to distinct properties like hydrophilicity, biocompatibility, biodegradability, viscoelasticity, and superabsorbancy, hydrogel adsorbents can play a prime role in the capture of heavy metals from contaminated water (Figure 3a) and can discharge these hazardous pollutants upon changes in the external environment (change in pH, temperature, etc.) (Figure 3b) [34,37].

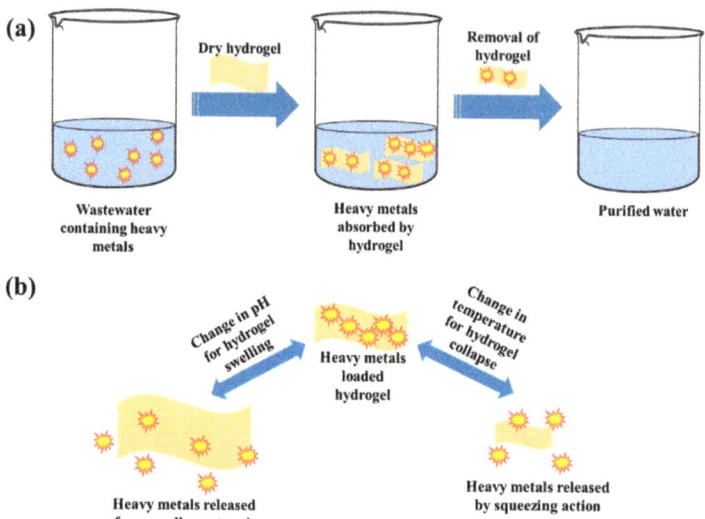

Figure 3. (a) Schematic illustration for removal of heavy metals from wastewater by using hydrogel-based adsorbent material. (b) Change in structure on applying external stimuli like pH and temperature.

In recent years, hydrogel adsorbents have shown a high potential for the effective elimination of heavy metals. Hydrogel absorbs heavy metals in a three-dimensional interstitial structure ensuring more sites per unit volume [38]. Unlike other adsorbents, hydrogels adsorb heavy metals in a three dimensional, highly porous network that leads to high adsorption efficiency [37]. Adsorption or desorption of heavy metals is mainly due to the surface chemistry and presence of hydrophilic functional groups (–COOH, –NH_2, –OH, –SO_3H, etc.) that act as a complexing agent for heavy metal removal from aqueous media [36,37]. Moreover, hydrogels can be modified by the addition of new functional groups or the preparation of composites with natural or synthetic sources to enhance heavy metal absorption capacities [39]. The swelling behavior of hydrogels is associated to an extent with hydrophilic functional groups present in the polymer backbone, the degree of cross-linking, the elasticity of the network, and the porosity of the polymer [40]. The hydrophilic polymers in hydrogel can swell up to several times their original volume in aqueous media and hold large content of water about 400 times its original weight [41]. Hydrogels are insoluble in water which leads to easy regeneration because of the presence of chemically cross-linked polymers, which enhance their mechanical strength as well as decreases the swelling ratio. Therefore, it is necessary to balance the amount of crosslinking and swelling ratio to obtain a stronger hydrogel [42]. Some other advantages of hydrogels are that they can be synthesized with the desired charges, controllable sizes, and functional groups [43]. The three most important parameters on which the capacity of cross-linked hydrogel synthesized depend are: (a) polymer volume fraction of hydrogel in swollen shape, which determines the quantity of fluid absorbed into hydrogel network, (b) the average molecular weight between two cross-links, which determines the degree of cross-linking for the prepared hydrogel, and (c) the network mesh size, which determines the degradability, mechanical strength, and diffusivity of releasing components into hydrogel structure [44].

An ideal hydrogel should have the following characteristics to be widely applied for the effective removal of heavy metals from polluted wastewater [24].

- Cost-effective
- High adsorption capacity to absorb heavy metals from wastewater
- High adsorption rate (determined by porosity and particle size)

- Biodegradable
- Easy to modify
- The low content of the unreacted residual monomer
- High stability and durability during swelling and storage
- Non-toxic, colorless, and odorless
- pH neutrality after swelling in aqueous media
- Deswelling capabilities and re-watering (able to release back the stored water)

4. Properties of Hydrogel

An ideal hydrogel has distinctive characteristic properties such as swelling or deswelling in the presence of external stimuli, responsiveness to the change in temperature, pH, light, etc., and biodegradability.

4.1. Swelling or Deswelling of Hydrogels

Hydrogels are the cross-linked polymer that swells and imbibes water when immersed in the aqueous media. These three-dimensional structures can swell up to several times their dry weight [45]. Hydrogels can respond to external stimuli (temperature, pH, light, salt, magnetic fields, biomolecules, and ionic strength) by shrinking, swelling, and discoloration [46]. Factors affecting swelling kinetics and equilibrium are the chemical structure of the polymers, cross-linking ratio, synthesis state, and ionic media. Chemical structure affects the swelling of hydrogel; hydrophilic groups swell more in comparison to hydrophobic groups. Cross-linking also has a significant effect on swelling behavior, as a highly cross-linked polymer network will show less swelling and vice versa. The swelling behavior of hydrogels is also affected by temperature and pH [47]. Temperature-sensitive hydrogels undergo swelling or de-swelling (change in volume) with the change in the temperature. These hydrogels swell below the low critical solution temperature and shrink above the low critical solution temperature [48]. pH-sensitive hydrogels undergo swelling or deswelling by varying pH levels. These hydrogels consist of ionizable acidic and basic groups connected to the polymer backbone that can add or release protons by varying pH levels. At the high pH value, the acidic groups on polymer chains deprotonate whereas at low pH values, the basic groups get protonated [49].

Swelling in hydrogels takes place in three steps:—(a) the diffusion of water into the three-dimensional network of hydrogel, (b) loosening of the polymeric chains, and (c) expansion of hydrogel structure. A hydrogel in swollen form is referred to as the rubbery state and the dry form as the glassy state. When dry hydrogel comes in contact with the solvent, the free space in a polymer network permits the solvent to enter the hydrogel matrix easily. This transforms the dry or glassy state into a swollen or rubbery state. The de-swelling of hydrogel occurs when water is removed from a hydrogel matrix [47]. Experimentally, the swelling ratio can be calculated by the following formula [50].

$$\text{Swelling ratio} = [(W_s - W_d) \div W_d] \times 100$$

where, W_s and W_d are the weight of the swollen and dry hydrogel, respectively.

4.2. Stimuli-Responsive Hydrogels

Gels that respond to changes in the external environment (temperature, pH, magnetic field, electric field, etc.) are termed stimuli-responsive hydrogels. These hydrogels have characteristics to transform their shape (from solution to gel) based on the application [51]. Furthermore, stimuli-responsive hydrogels are categorized into three classes: chemical, physical, and biological. pH, solvency, ionic strength, and electrochemical field are typical chemical stimuli. Physical stimuli include temperature, magnetic field, electric field, light, mechanical force, and ultrasound. Biological stimuli include enzymes, glucose, antigen, ligands, etc. (Figure 4) [52]. Multi-responsive hydrogels are the kind of hydrogels that respond to two or more stimuli [53].

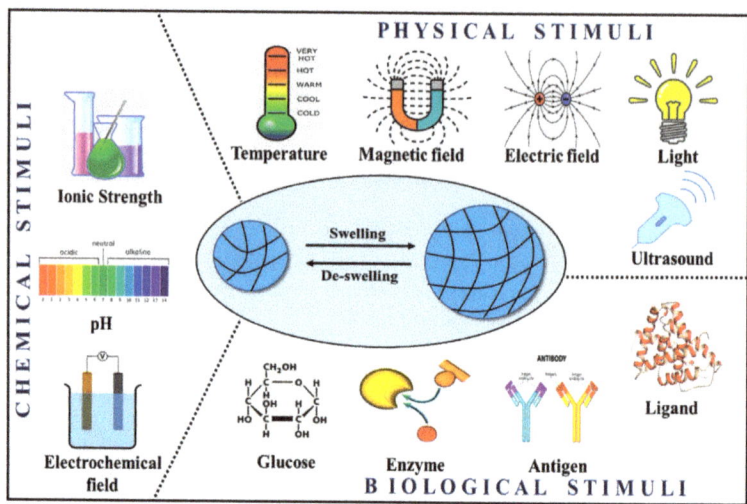

Figure 4. Response of hydrogel to different stimuli.

4.3. Biodegradable

Biodegradability refers to the ability of the hydrogel to break down into harmless and non-toxic end products by bacteria or other organisms. Hydrogels' biodegradability is determined by the functional groups present in the system as well as the method of synthesis. The degradation process involves solubilization and hydrolysis of biological entities of hydrogel into safer end products. Biodegradable polymers include a wide range of hydrophilic synthetic and natural polymers. Due to diffusion, these polymers absorb an ample amount of water and expand to a large extent. The breakdown of these polymers is influenced by various parameters such as molecular weight, hydrophilicity, and the interaction of the polymer with water. Other environmental conditions like temperature and pH also influence the breakdown of polymers via solubilization.

Chemical hydrolysis can be used to degrade a variety of polymers that cannot be destroyed by simple hydrolysis. These polymers do not produce hydrogel; rather they mix with hydrogel to form a hydrophilic monomer, which is then combined to form a biodegradable hydrogel. The formed hydrogel undergoes degradation via chemical hydrolysis via ester bonds. Furthermore, hydrogels can be degraded by enzyme hydrolysis and this category of hydrogels involves polymers such as proteins, polysaccharides, and synthetic polypeptides. Enzyme hydrolysis takes place by a set of hydrolases that catalyze the hydrolysis of C-N, C-O, and C-C bonds. Peptidases and proteinases are hydrolases that degrade polypeptide and protein hydrogels, respectively. Moreover, glycosidase is the sole enzyme that degrades polysaccharide hydrogels [47].

5. Classification of Hydrogel

Hydrogels are classified depending on their source, nature of cross-linking, chain composition, ionic charge, response to external stimuli, configuration, and size. The most important parameters for the classification of hydrogels are depicted in Figure 5.

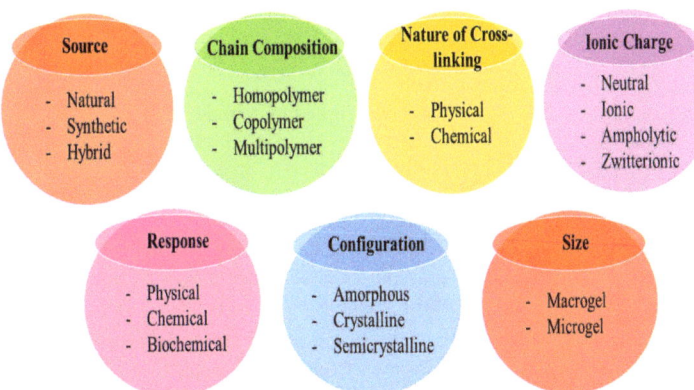

Figure 5. Classification of hydrogels.

5.1. Based on Source

Hydrogels can be classified as natural, synthetic, and hybrid.

(a) Natural hydrogels are synthesized by using natural sources including chitosan, agar-agar, cellulose, lignin, gelatin, alginate, dextran, collagen, and many other materials [54].
(b) Synthetic hydrogels are prepared by using synthetic polymers namely hydroxy methyl methacrylate (HEMA), acrylic acid (AA), vinyl acetate (VAc), ethylene glycol (EG), ethyleneglycoldimethacrylate (EGDMA), methacrylic acid, N-vinyl 2-pyrrolidone (NVP), and many other materials [54].
(c) Hybrid hydrogels are made by combining natural and synthetic sources [55]. Zhang et al. prepared chitosan-g-poly (acrylic acid)/attapulgite/sodium humate hydrogel for effective removal of Pb^{2+} [56].

5.2. Based on the Nature of Chain Composition

Hydrogels can be classified into three principal classes: homo-polymeric hydrogels, co-polymeric hydrogels, and multipolymer hydrogels.

(a) Homo-polymeric hydrogels are cross-linked polymer-network originating from a single type of monomer unit [57]. The structural unit of these hydrogels depends on the type of monomer, cross-linker, and polymerization technique [58].
(b) Co-polymeric hydrogels are composed of two or more types of monomer units with at least one hydrophilic monomer, arranged in a random, block, and irregular structure along the backbone of the polymer network [59]. These hydrogels are prepared by cross-linking or polymerization between both the monomers by using a cross-linker and initiator. An example of such hydrogels is chitosan, k-carrageenan, carboxymethyl cellulose composite hydrogel which is used to remove metal ions.
(c) Multipolymer hydrogels are cross-linked polymer-network prepared by three or more monomer units via cross-linking and polymerization reactions. For example, Kim et al. synthesized chitosan-based multicomponent functional gel comprising multiwall carbon nanotubes, polyaniline, poly (acrylic acid), and poly (4-amino diphenyl amine) [60].
(d) Interpenetrating polymeric hydrogels are comprised of two independent, intertwined polymer networks, having natural and/or synthetic polymer components. In a semi-interpenetrating polymer hydrogel, one polymer has a linear network that diffuses into another cross-linked network. There is no chemical bonding between the polymers [54].

5.3. Based on the Nature of Cross-Linking

Hydrogels can be classified into two categories: physically cross-linked hydrogels and chemically cross-linked hydrogels.

(a) Physically cross-linked hydrogels have a transient junction that arises due to physical interaction such as hydrogen bond, ionic interaction, and hydrophobic interaction.
(b) Chemically cross-linked hydrogels have permanent junctions that arise due to covalent bonds [50].

5.4. Based on the Reaction of Hydrogel with the External Stimulus

Hydrogels can be classified into two distinct categories: traditional hydrogels and environmentally sensitive hydrogels.

(a) Traditional hydrogel is not reactive to environmental changes
(b) An environment-sensitive hydrogel can detect changes caused by chemical (pH, concentration), biochemical (antigen, enzyme, ligand), and physical (temperature, pressure, light) factors [50].

5.5. Based on the Configuration

On the basis of the physical structure and chemical composition, hydrogels can be classified as amorphous, semi-crystalline (mixture of crystalline and amorphous phases), and crystalline [41].

5.6. Based on the Size

Hydrogels can be classified into two classes: macrogel and microgel. The macrogel is further classified as a porous sponge, columnar, membranous, fibrous, and spherical according to its morphology. The prepared microgel also can be classified into nanometer and micron [50].

5.7. Based on Ionic Charge

Hydrogels can be classified into four categories based on the electric charge placed on the cross-linked network: neutral, ionic, ampholytic, and zwitterionic.

(a) Neutral hydrogels are also known as non-ionic hydrogels. These hydrogels contain no charge on side groups or polymer backbone.
(b) Ionic hydrogels are further classified as anionic and cationic. Anionic hydrogels carry negatively charged functional groups like sulfonyl, carboxyl, etc. and at high pH values show an increase in swelling behavior. Cationic hydrogels carry positively charged functional groups like amines, thiol, etc., and at low pH values exhibit an increase in swelling behavior.
(c) Ampholytic or amphoteric hydrogels contain acidic as well as basic groups.
(d) Zwitterionic hydrogels contain cationic and anionic groups in their structure [61,62].

6. Surface Functional Groups of Hydrogel

A group of atoms in a compound responsible for chemical reactions is known as the functional group. Functional groups play a significant role in determining the chemical reactivity of the molecule as well as the type and strength of intermolecular forces. The paramount functional groups incorporated in a three-dimensional network of hydrogels for metal adsorption are classified into three groups:—(a) nitrogen-containing functional groups, (b) oxygen-containing functional groups, and (c) sulfur-containing functional groups [63]. Table 3 summarizes hydrogels containing different functional groups and removed heavy metals.

6.1. Nitrogen-Containing Functional Groups

6.1.1. Amine Group

The amine group contains a nitrogen atom that has a lone pair of electrons, that readily attach to cationic metal ions. The methods used to functionalize the amine group on the hydrogel surface are atom-transfer radical polymerization, formaldehyde treatment, and gamma ray-induced polymerization [63].

6.1.2. Amide Group

The general formula for the amide group is –CONH. In general, amine groups are more often functionalized on the hydrogel surface than amide groups. Moreover, monomers like 2-acrylamido-2-methyl-1-propanesulfonic acid sodium salt have an amide group in the polymer backbone, that complexes with heavy metals Cu(II) and Ni (II) [64].

6.1.3. Quaternary Ammonium Groups

Quaternary ammonium groups [R-N$^+$(CH$_3$)$_3$] show strong attraction toward metal oxyanions (Cr$_2$O$_7^{2-}$, HCrO$_4^-$, AsO$_4^{3-}$ and CrO$_4^{2-}$) [65]. Quaternary ammonium groups are highly stable and unaffected by pH change. Hence, they can captivate oxyanions of metals irrespective of the pH of the medium. Monomers, namely, (vinylbenzyl)trimethyl ammonium chloride, (3-acrylamidopropyl)trimethyl ammonium chloride and 2,3-epoxypropyltrimethylammonium chloride contains quaternary ammonium group as an active functional group in hydrogel preparation [66,67]. By ion exchange, hydrogels that consist of these monomers led to the exchange of the chloride (Cl$^-$) ions with oxyanions of metals.

6.2. Oxygen-Containing Functional Groups

6.2.1. Hydroxyl Group

A hydroxyl (R-OH) group is composed of one oxygen atom bonded to one hydrogen atom. According to the International Union of Pure and Applied Chemistry (IUPAC), the word "hydroxyl" refers to a hydroxyl radical. A hydroxyl group can easily remove the proton to attract metal cations.

6.2.2. Carboxyl Group

A carboxyl (R-COOH) group is composed of an electronegative oxygen atom that is double-bonded to the carbon atom and singly bonded to the –OH group. According to literature, the carboxyl group is found to be the most prominently used group to adsorb heavy metals onto the hydrogel surface. A carboxyl group gets ionized by giving an H$^+$ ion from its R-OH group at alkaline pH, forming RCOO$^-$ ion that readily attracts the divalent metal cations. To functionalize the carboxyl group onto the hydrogel surface, different methods used in post-treatment are surface grafting, etherification, and 2,2,6,6-tetramethylpiperidine-1-oxyl radical (TEMPO)-mediated oxidation, which converts a primary hydroxyl group into a carboxyl group [63].

6.3. Sulfur-Containing Groups

6.3.1. Thiol Group

The thiol (R-SH) group plays an important role in functionalizing the hydrogel surface for heavy metal adsorption [68]. The thiol group acts as a Lewis base and interacts with the heavy metal (Lewis acid) by forming a coordinate bond [69]. From the literature, it is demonstrated that the thiol group shows strong bonding with mercury (Hg). The bonding can be well explained by Hard-Soft-Acid-Base (HSAB) theory, where mercury acts as a soft acid and prefers to bind with the thiol group (soft base) [70]. In an article by Kumar et al. the thiol group prefers to form a stable complex with highly polarizable soft heavy metals like mercury (Hg), gold (Au), and silver (Ag); and to a lesser extent with cadmium (Cd) and zinc (Zn), failing to form a coordinate bond with lighter metals like sodium (Na), calcium (Ca), and magnesium (Mg) [71].

6.3.2. Sulfonic Acid Group

A sulfonic acid ($R-SO_3H$) group is a sulfur-containing functional group that contains an electronegative sulfur atom that is double bonded to two oxygen atoms and single bonded to the –OH group. Sulfonic acid turns into sulfonate group ($R-SO_3^-$) on disassociation of hydrogen atom. Functionalizing a sulfonate group onto the surface of hydrogel makes the surface negatively charged, irrespective of the pH of the medium. A monomer 2-acrylamido-2-methyl propane sulfonic acid (AMPS) has been used to synthesize hydrogel via ^{60}cobalt gamma-ray irradiation for the adsorption of Co^{2+}, Mn^{2+}, Cu^{2+}, and Fe^{3+} [72].

6.4. Other Functional Groups

6.4.1. Amidoxime Group

The general formula for the amidoxime group is ($R-C(NH_2)=N-OH$). The amidoxime group forms stable complexes with many heavy metals such as Co^{2+}, Cu^{2+}, Ni^{2+}, and Pb^{2+} but shows a strong affinity towards uranium. Therefore, hydrogel-based adsorbent material has been functionalized with the amidoxime group for the adsorption of uranium [73,74]. Guibal and coworkers synthesized amidoxime grafted chitosan magnetic hydrogel for sorption of uranium (U^{6+}) and europium (Eu^{3+}) [75]. Zeng et al. prepared amidoxime-modified hydrogel via graft copolymerization for adsorption of Cu^{2+}. The maximum adsorption capacity of 40.7 mg/g at pH 5 was achieved over the contact time of 25 h [76].

6.4.2. Phosphate-Containing Functional Group

Phosphate-containing functional groups, namely phosphine, phosphate, and phosphoramide are used to functionalize hydrogel surface. However, phosphate-based functional groups are more popular for functionalizing hydrogel for biomedical areas rather than metal adsorption. There are very few papers reported for metal adsorption. Liao et al. prepared phosphate functionalized graphene hydrogel for electrosorption of U^{6+}. The maximum electrosorption capacity of 545.7 mg/g at 1.2 V and pH 5 was obtained [77].

6.4.3. Chelating Group

The chelating agent such as aminopolycarboxylic acids (APCAs) helps in enhancing the adsorption affinity of the hydrogel by chelation [78]. Because there are many nitrogen- and oxygen-containing functional groups present in the aminopolycarboxylic acid structure. Especially, the nitrogen-containing functional group shows strong bonding interaction with divalent metal cations [79,80]. The four common aminopolycarboxylic acids for metal adsorption are ethylenediaminetetra acetic acid (EDTA), iminodiacetic acid (IDA), diethylenetriaminepentaacetic acid (DTPA), and nitrilotriacetic acid (NTA). Many biosorbents have been functionalized with EDTA because of their chemical stability, chelating ability, and low price [81]. IDA is a tridentate ligand forming a metal complex by chelation [82]. No studies have been reported for hydrogel surface modification by an NTA chelating agent. DTPA possesses five carboxylate groups, which show a high binding affinity for heavy metals after grafting on the hydrogel surface. For example, Huang et al. prepared DTPA-modified chitosan/alginate hydrogels for removal of Cu^{2+} from electroplating wastewater [83].

Table 3. Different hydrogel-based functionalized adsorbent materials for the removal of heavy metals.

Hydrogel	Active Functional Group	Heavy Metals Removed	References
Graphene oxide-chitosan-poly(acrylic acid) (GO-CS-AA) hydrogel nanocomposite	R-COOH	Pb^{2+}	[84]
Hydrous ferric oxide-Poly(trans-aconitic acid/2-hydroxyethyl acrylate (HFO-P(TAA/HEA)) hydrogel	R-OH	Cu^{2+}, Cd^{2+}, Pb^{2+} and Ni^{2+}	[85]
Chitosan-sodium lignosulfonate-acrylic acid (CS-SLS-AA) hydrogel	$R-NH_2$	Co^{2+} and Cu^{2+}	[86]
Poly(3-acrylamidopropyl) trimethyl ammonium chloride/γ-Fe_2O_3	$R-N^+(CH_3)_3$	Cr^{4+}	[87]
Sulfathiazole-based novel UV-curved hydrogel	R-SH	Hg^{2+}, Cd^{2+} and Zn^{2+}	[88]
Magnetic anionic hydrogel (nFeMAH)	$R-SO_3Na$	Cu^{2+} and Ni^{2+}	[64]
Poly(2-acrylamido-2-methyl-1-propane sulfonic acid) magnetic hydrogel	$R-SO_3H$	Cd^{2+}, Co^{2+}, Fe^{2+}, Pb^{2+}, Cu^{2+}, Cr^{2+} and Ni^{2+}	[39]
Acrylamide/crotonic acid (AAm/CA) hydrogel	R-COOH, and $R-CONH_2$	Hg^{2+}	[89]
Glucan/chitosan hydrogel	R-OH and $R-NH_2$	Co^{2+}, Cu^{2+}, Cd^{2+}, Ni^{2+} and Pb^{2+}	[90]
Malic acid enhanced chitosan hydrogel beads (mCHBs)	R-COOH and $R-NH_2$	Cu^{2+}	[91]
Carboxymethyl cellulose/polyacrylamide (CMC/PAM) composite hydrogel	R-OH, R-COOH and $R-NH_2$	Cd^{2+}, Pb^{2+} and Cu^{2+}	[92]
Chitosan poly(acrylic acid) supermacroporous hydrogel	R-OH, R-COOH and $R-NH_2$	Cu^{2+} and Pb^{2+}	[93]
Lignosulfonate-modified graphene hydrogel	R-C=O, R-OH and R-COOH	Pb^{2+}	[94]
Polyacrylonitrile-chitosan-graphene oxide (PCG) hydrogel composite	$R-C(NH_2)=N-OH$	U^{6+}	[73]

7. Synthesis of Hydrogel

A plethora of issues arising from the overuse of non-biodegradable materials and fossil resources have shifted researchers' focus to renewable and environmentally friendly materials. In the present time, polymers are extensively used in different areas, namely agriculture, biomedical applications, wastewater treatment, and food packaging [95–98]. Similarly, for the removal of toxic heavy metals from wastewater by adsorption, polymeric hydrogels are the most promising adsorbent material due to their increased surface area, good solubility in organic solvents, improved functionality, low-priced, biodegradability, recyclability, enhanced adsorption capacity, and ease of fabrication. In addition, the excellent hydrophilic character makes these hydrogels suitable for wastewater treatment [35,99]. However, the effectiveness of the adsorbent material is highly dependent on the physicochemical properties of the adsorbent [100]. As a result, the first and most important step in developing an effective adsorption process is to synthesize a suitable hydrogel-based adsorbent material with high absorptivity of heavy metals present in wastewater.

The essential chemicals required for the synthesis of the hydrogel are a monomer, an initiator, and a cross-linker. Acryl amide (AAm), polyvinyl alcohol (PVA), polyvinyl pyrrolidone, acrylic acid (AA), 2-dimethylamino ethyl methacrylate (DMAEM), polyethylene glycol methyl ether methacrylate (PEGMEM), (3-Acrylamidopropyl) trimethylammonium chloride (APTMACI), N-isopropylacrylamide, 2-acrylamido-2-methyl-1-propan-sulfonic acid (AMPS), and 4-vinyl pyridine, 2-hydroxyethylmetacrylate are the examples of some monomers used in hydrogel synthesis [38,39,101–106]. Distinct monomers have different properties in terms of adsorption capacity, physical strength, and so on. In the synthesis of hydrogels, researchers were able to develop a solution to overcome the limitation of specific monomers. For example, to reduce the physical weakness of biopolymer chitosan, Sun et al. [107] and Liu et al. [108] used cellulose as the blending polymer in the synthesis of chitosan-based hydrogel for heavy metal adsorption.

Cross-linkers or cross-linking agents play a crucial role in the synthesis of polymeric hydrogels because they help to build up the polymeric three-dimensional network by stabilizing the binding sites amid the functional monomer and adsorption target molecule. Therefore, cross-linkers influence the polymers' hydrophilic or hydrophobic properties, selectivity, mechanical stability, and morphology [109]. A cross-linker of organic or inorganic nature can be used in the synthesis process. Moreover, inorganic cross-linkers are mainly used to synthesize hydrogel adsorbents as organic cross-linkers that have certain disadvantages in terms of lower mechanical strength and thus cannot withstand stressed conditions; additionally, they also have a lower swelling capacity [41,110]. Furthermore, the characteristic properties of a hydrogel can differ depending on whether the cross-linking between the chains is covalent or non-covalent. Permanently cross-linked junctions exist in hydrogels that have been cross-linked with covalent bonds. Hydrogels cross-linked with non-covalent bonds (ionic interaction, hydrophobic interaction, or hydrogen bonding), on the other hand, have transient junctions [41,50].

For polymerization reaction, a cross-linker must have more than one active functional group to help linear polymer chains to join with other chains to form a stable three-dimensional structure. A low degree of cross-linking, in particular, corresponds to a small quantity of cross-linker, resulting in poor mechanical strength of polymeric material. As a result, the three-dimensional structure of hydrogel distorts during the application, and adsorption sites are disrupted, giving rise to a high number of non-specific perforations. When the degree of cross-linking is high, a densely packed three-dimensional mesh structure is generated, having excellent mechanical strength and an unexpectedly high mass transfer number. Resulting in the reduction of adsorption sites, as well as the degree of swelling of the hydrogel, causing heavy metals to barely penetrate the hydrogel surface. Therefore, it is optimal to maintain the quantity of the cross-linker in the specified range. Polymers having a cross-link ratio greater than 80% are generally used [109,111,112].

In the synthesis of hydrogels, an initiator is a chemical that helps to initiate the polymerization process. Table 4 summarizes various hydrogel-based adsorbents and the monomers, initiators, and cross-linkers used in their synthesis process.

Table 4. Different hydrogel adsorbents and associated monomers, cross-linker, and initiators in the synthesis process.

Hydrogel	Monomer	Cross-Linker	Initiator/Accelerator	References
Poly(2-acrylamido-2-methyl-1-propansulfonic acid-co- vinylimidazole) hydrogel	2-acrylamido-2-methyl-1-propansulfonicacid (AMPS), N- vinyl imidazole	N,N' methylenebisacrylamide (MBA)	2,2'-azobis(2-methyl propionamide) (MPA) dihydrochloride	[37]
Cationic hydrogel	(3-acrylamidopropyl) trimethylammonium chloride (APTMCI)	N,N' methylenebisacrylamide (MBA)	Ammoniumpersulfate (APS)/N,N,N',N'-tetramethylenediamine (TEMED)	[38]
Hydrogel biochar composite	Acrylamide (AAm)	N,N' methylenebisacrylamide (MBA)	Ammonium persulfate (APS)	[101]
Fe_2O_3 nanoparticles functionalized polyvinyl alcohol/chitosan magnetic composite hydrogel	Polyvinyl alcohol (PVA)	Glutaraldehyde vapor	Glacial acetic acid	[102]
Methacrylate-based hydrogel	Polyethylene glycol methyl ether methacrylate (PEGMEM), 2-dimethylamino ethyl methacrylate	N,N' methylenebisacrylamide (MBA)	Ammonium persulfate (APS)	[104]

Table 4. Cont.

Hydrogel	Monomer	Cross-Linker	Initiator/Accelerator	References
(p-4-VP-co-HEMA) composite hydrogel	4-vinyl pyridine (4-VP), 2-hydroxyethylmetacrylate (HEMA)	N,N' methylenebisacrylamide (MBA)	Ammonium persulfate (APS), N,N,N',N'-tetramethylenediamine (TEMED)	[106]
Chitosan-cellulose hydrogel	Chitosan	Cellulose	-	[107]
Superabsorbent polymer hydrogels	Acrylic acid (AA), acrylamide (AAm)	N,N' methylenebisacrylamide (MBA)	Ammoniumpersulfate (APS)	[113]
Poly(N-hydroxymethylacrylamide) hydrogel	N-hydroxymethylacrylamide	Polyethylene glycol (400) diacrylate	Ammonium persulfate (APS)/N,N,N',N'-tetramethylenediamine (TEMED)	[114]
EDTA Functionalized Chitosan/Polyacrylamide double network hydrogel	Chitosan, acrylamide	N,N' methylenebisacrylamide (MBA)	Potassium persulfate (KPS)	[115]
N-vinyl-2-pyrrolidone/Itaconic acid hydrogel	Itaconic acid (IA), N-vinyl-2-pyrrolidone	N,N' methylenebisacrylamide (MBA)	Ammoniumpersulfate (APS/N,N,N',N'-tetramethylenediamine (TEMED)	[116]
Polyampholyte hydrogel	Methyl methacrylate (MMA), acrylic acid (AA)	N,N' methylenebisacrylamide (MBA)	Ammonium persulfate (APS)/N,N,N',N'-tetramethylenediamine (TEMED)	[117]
Poly(acrylic acid) hydrogel adsorbent	Acrylic acid (AA)	Calcium hydroxide $(Ca(OH)_2)$ nano-spherulites (CNS)	Ammonium persulfate (APS)/N,N,N',N'-tetramethylenediamine (TEMED)	[118]
Magnetic chitosan hydrogel beads	Chitosan	Glutaraldehyde	-	[119]
Hydrogel-based on novel cross-linker	Chitosan, acrylic acid, glucose	Allyl pentaerythritol(AP)[15]/allyl mannitol (AP)[14]/allyl sorbitol	Potassium persulfate (KPS)	[120]

Hydrogels are synthesized via two routes: Chemical and physical.

7.1. Synthesis via the Chemical Route

Polymer chains in chemically cross-linked hydrogels are formed by covalent bonds. The subsequent sections describe various methods for the synthesis of the hydrogel by chemical modification.

7.1.1. Chemical Route of Cross-Linking via Free Radical Polymerization

Free radical polymerization is one of the well-studied approaches for the synthesis of hydrogel in the presence of cross-linking agent N, N' methylene bisacrylamide (MBA). The method involves three steps namely, initiation, polymeric chain propagation, and termination. In this process, the first step involves the generation of free radicals by using an initiator such as ammonium persulfate (APS), potassium persulfate (KPS), etc. in the vicinity of temperature, light, redox reaction, or ultraviolet or gamma radiation [121,122]. After that, in the second step, the free radical will react with the monomer to produce a radical monomer, which then reacts with the other monomers present in the solution to form polymeric chains. The cross-linker is added during the propagation of the polymeric chain, resulting in the formation of a three-dimensional structure of hydrogel. In the last step, the polymeric chain is terminated via disproportionation or combination reaction. The combination reaction connects two growing chains into one long polymeric chain. However, in the case of disproportionation reaction, a hydrogen atom is abstracted from the end of one growing chain and added to the other growing chain. As a result, a polymer with the unsaturated end group and a saturated end group is obtained. To speed up the process, an accelerating agent such as N,N,N′,N-tetramethylene diamine is added to the reaction mixture [123]. For instance, Shah et al. prepared superabsorbent polymer hydrogels containing acrylamide and acrylic acid as monomers via one-step free-radical polymerization. In this work, they aimed to remove multi-metals (Ni^{2+}, Cd^{2+}, Co^{2+}, and Cu^{2+}) from an aqueous medium [113].

7.1.2. Chemical Route of Cross-Linking via High Energy Irradiation

The hydrogel synthesis via ultraviolet light radiation, electron beams, and ɤ-radiation is carried out at ambient or sub-ambient temperatures without the requirement of initiators, catalysts, or cross-linkers [124]. This synthetic route outperforms chemically initiated processes in regards to one-step hydrogel formation with no waste generation as a byproduct [125]. In this method, the density of cross-linking is estimated by duration and dose of irradiation. Polyethylene glycol (PEG), polyvinyl alcohol (PVA), alginate, chitosan, gelatin, hyaluronic acid (HA), carboxymethyl cellulose (CMC) are among the natural and synthetic polymers proposed for the hydrogel synthesis using this method [126,127]. This cross-linking method is similar to free radical polymerization in terms of three-step hydrogel formation: initiator, propagation of polymeric chain, and termination. When the mixture of the reaction solution is irradiated, a hydroxyl free radical is generated, resulting in the formation of a free-radical monomer. The hydrogel is synthesized when the network has reached the critical stage of gelling [128]. Maziad et al. prepared polyacrylic acid/polyvinyl alcohol based hydrogel to treat water decontamination via gamma radiation. They found that the hydrogel swelled 273%and had removal capacity of 150 mg/g, 155mg/g, and 193 mg/g for Ni^{2+}, Co^{2+}, and Cu^{2+} ions, respectively, at acidic pH 5 and after 24 h [129].

7.1.3. Chemical Route of Cross-Linking via Grafting Reactions

In this method, the hydrophilic functional group like carboxyl (–COOH), sulfonic (–SO_3H), amino (–NH_2), and acylamino (–$CONH_2$) are grafted on the surface of hydrogel [92,130]. Grafting the functional groups helps in improving the adsorption or desorption efficiency, as well as selectivity for specific heavy metals. As a result of this, there is an increase in surface polarity, hydrophilicity, and enhancement in the number of active sorption sites [131]. For example, Qi et al. prepared a new salecan polysaccharide-based hydrogel via graft copolymerization of sodium vinyl sulfonate and acrylamide onto the salecan for the effective decontamination of Pb^{2+} from wastewater [132].

7.1.4. Chemical Route of Cross-Linking via Reaction of Functional Groups

The reaction involves the bond formation between the cross-linker and the functional moieties present in the polymer molecule. Hydrophilic groups such as amine (–NH_2 in chitosan and proteins) and hydroxyl (–OH in cellulose and its derivatives) are bonded with cross-linking agents (such as glutaraldehyde) having an aldehyde functional group resulting in aldol product via covalent interaction. Hydrogel synthesis involving the polymers having hydroxyl groups needs certain specific conditions like methanol as a quencher, high temperature, and low pH. However, in the case of protein-based hydrogel no specific conditions are required [112,133,134]. Polymers with ester functional groups, on the other hand, undergo chemical cross-linking through the condensation process in the presence of a cross-linking agent, resulting in the formation of Schiff bases [130].

7.2. Synthesis via Physical Route

The physical route of cross-linking is highly favorable to synthesize non-toxic and environmentally friendly hydrogel as there is no requirement for chemical-based cross-linking agents [135]. In this process, polymer chains are held by weak interactions like hydrophobic interaction, ionic interaction, hydrogen bonding, Van der Waals forces, and π–π interaction [136]. From the literature, it is noted that polysaccharides like dextran, pullulan, carboxymethyl curdlan, and chitosan are used for the synthesis of hydrogels by this method [137].

7.2.1. Synthesis via Freeze-Thaw

Crystallization via the freeze–thaw method is one of the physical processes used to synthesize hydrogel [138]. In this method, crystallization takes place by freezing low molecular solutes or bulk solvents, which enhances the polymer concentration by decreasing the chain gap and allowing the chains to align and join to create a three-dimensional structure [139]. In hydrogels, freeze–thaw cycles give rise to porous structures due to the space created by melting crystals during the thawing stages [140]. By varying the polymer concentration, the freezing temperature, freeze–thaw time duration, and the number of freezing and thawing cycles, the mechanical characteristics of the freeze–thawed hydrogel may be adjusted [139].

Hydrogels synthesized via the freeze–thaw method have greater elastic characteristics in comparison to those synthesized via chemical methods, attracting widespread interest across the world [141]. For instance, poly (vinyl alcohol) (PVA)/carboxy methyl cellulose (CMC) hydrogels are synthesized via the freeze–thaw method and used to absorb heavy metals such as Ni^{2+}, Cu^{2+}, Zn^{2+} and Ag^{2+} [139].

7.2.2. Synthesis via Self-Assembling

Self-assembled hydrogels are prepared by monomeric units that spontaneously self-assemble by non-covalent interaction into supramolecular fibers [142]. When such fibers retain proper solvation in liquid (water), they efficiently entangle and immobilize solvent flow, resulting in a 3D mesh structure. The non-covalent interaction stabilizes hydrogel structures by making them softer than those generated by covalently cross-linked material [143]. These interactions provide self-assembled hydrogel with advantages such as tolerance to environmental perturbation and self-healing characteristics [144].

7.2.3. Synthesis via Instantaneous Gelation

Another approach for synthesizing hydrogel quickly after a one-step procedure is instantaneous gelation [39,145]. For example, Zhou et al. synthesized novel chitosan-based magnetic hydrogel beads comprised of amine-functionalized magnetite nanoparticles, carboxylated cellulose nanofibrils, and polyvinyl alcohol incorporated chitosan for adsorption of Pb^{2+}. The synthesized hydrogel beads exhibited an adsorption efficiency of 171.0 mg/g and could be regenerated in a weakly acidic solution with an adsorption efficacy of 90% after 4 cycles [146].

7.2.4. Synthesis via Ionotropic Gelation

Hydrogel synthesis by ionotropic gelation allows the formation of microparticles and nanoparticles via electrostatic bonding among the two ionic species under suitable conditions, one of which must be a polymer [147]. For instance, sodium alginate(SA)/hydroxypropyl cellulose (HPC) hydrogel beads were synthesized with different ratios of 50:50, 75:25, and 100:0 for the removal of Pb^{2+}. According to the results obtained, 75:25 showed better adsorption capacity in comparison to 50:50 and 100:0. After three hours of contact time, hydrogel beads showcased adsorption capacity and adsorption percentage of 47.72 mg/g and 95.45%, respectively [148].

7.2.5. Synthesis via Inverse Emulsion Method

In the inverse emulsion method, the term "water-in-oil" describes the phenomenon in which water-soluble monomer is dispersed in the continuous phase oil (paraffin oil) by using an appropriate stabilizing agent, namely non-ionic surfactant Triton X-100, and after that the systems go through the phase inversion in a coagulation bath to release the monomer and precipitate out the porous film. This method has an advantage over other methods such as fine powdered product is obtained and by altering the reaction condition, the desired particle size can be achieved [149]. For example, a superabsorbent polymer-based hydrogel consisting of acrylic acid and carboxymethyl cellulose was synthesized by inverse emulsion polymerization method by using N, N′ methylene bisacrylamide (MBA) as a cross-linking agent and potassium persulfate (KPS) as an initiator. The maximum swelling capacity of 44.0 g/g in 0.9% w/v NaCl solution and 544.95 g/g in deionized water [150].

8. Characterization Techniques of Hydrogel

After the successful synthesis of hydrogel adsorbent, it becomes inevitable to investigate the physical, mechanical, structural, and morphological properties of the hydrogel formed. For this purpose, various characterization techniques such as Fourier transform infrared spectroscopy (FTIR), scanning electron microscopy (SEM), thermogravimetric analysis (TGA), zeta sizer, and energy-dispersive X-ray (EDX) are used to characterize the hydrogel (Table 5).

Table 5. Characterization techniques used for hydrogel adsorbent and information obtained from the characterization tools.

Characterization Techniques	Characteristics
Fourier Transform Infrared Spectroscopy (FTIR)	Functional group
Field Emission-Scanning Electron Microscopy (FE-SEM)	Surface morphology
Thermo Gravimetric Analysis (TGA)	Thermal stability
Zeta Sizer	Surface charge
Energy Dispersive X-ray (EDX)	Elemental composition

8.1. Functional Group Analysis

The surface functional groups such as hydroxyl, carboxyl, amide, amine, thiol, and amidoxime, etc. can be identified by using FTIR. Tang et al. synthesized chitosan/sodium alginate/calcium ion physically cross-linked double network hydrogel (PCDNH) for scavenging heavy metal ions. Analysis of FTIR spectra of hydrogels reveals that the peaks of chitosan at 1591 cm^{-1} and 1649 cm^{-1} (bending vibration of N-H and stretching vibrations for C=O of primary amine) disappear after the synthesis of PCDNH because the $-NH_2$ group is converted to $-NH_3^+$. The symmetric and asymmetric stretching vibration peaks of $-COO^-$ of sodium alginate at 1406 cm^{-1} and 1594 cm^{-1} shift to 1404 cm^{-1} and 1588 cm^{-1}, indicating, the interaction between $-COO^-$ with Ca^{2+} and $-NH_3^+$. A new peak was observed at 1714 cm^{-1} corresponding to the partial protonation of $-COO^-$ after the formation of PCDNH (Figure 6a) [151]. Ablouh et al. investigated the adsorption of Cr^{6+} and Pb^{2+} via FTIR analysis for the preparation of Chitosan/Sodium alginate (CSM-SA) hybrid hydrogel beads. It was noticed that the peak of $-NH_2$ or $-OH$ at around 3250 cm^{-1} shifted to 3245 cm^{-1}, indicating hydrogen bonding among the H atoms in $-NH_2$ groups and the O atoms of oxyanions of Cr^{6+}. In addition, there is a slight shift in the peak of COO from 1600 to 1590 cm^{-1}, indicating the interaction between COO and Cr^{6+}. These shifts correspond to electrostatic interaction between Cr^{6+} and NH_3^+, COO, and OH groups. A new peak observed at 682 cm^{-1} is due to the O-Cr-O band corresponding to Cr species. After the adsorption of Pb^{2+}, the stretching vibration of OH and COO group shows a strong shift from 3250 to 3261 cm^{-1}, and 1600 to 1569 cm^{-1}, respectively. This shift is due to the coordination effect between Pb^{2+} and O atom, demonstrating ion-exchange among Ca^{2+} and Pb^{2+} on the surface of hydrogel (Figure 6b) [152].

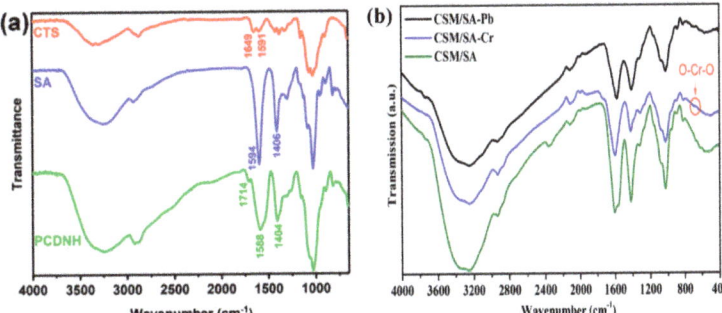

Figure 6. (**a**) FTIR spectra of chitosan (CTS), sodium alginate (SA), and PCDNH, (Reprinted from Ref. [151], Copyright (2022), with permission from Elsevier). (**b**) FTIR spectra of CSM/SA hybrid hydrogel beads loaded with Pb^{2+} and Cr^{6+} [152].

8.2. Thermal Analysis

Thermogravimetric analysis (TGA) is used to determine the thermal stability of a hydrogel. TGA can also be used to evaluate changes in a material's physical and chemical properties as a function of increasing temperature. For example, Kong et al. studied the thermal stability of the Xylan-g-/p(acrylic acid-co-acrylamide)/graphene oxide (GO) hydrogel. Figure 7a represents the TGA thermogram of the hydrogel with and without GO. The weight loss of samples occurred in four phases when the temperature was raised from room temperature to 700 °C: 25–220 °C, 220–350 °C, 350–400 °C, and 400–700 °C. In the first step, weight loss was due to moisture loss and the decomposition of tiny molecules. The weight loss in the second step was because of the decomposition of long-chain compounds like polyacrylic acid, polyacrylamide, and xylan. The weight of hydrogels remained consistent in the last step, which was due to the carbonation of the hydrogels. Furthermore, the hydrogels with a higher GO loading had a high weight, indicating that GO has a positive effect on hydrogel thermal stability [153]. Mohamed et al. studied the thermal properties of a biodegradable N-quaternized chitosan (NQC)/poly (acrylic acid) (PAA) hydrogel by varying the NQC/PAA ratios to 3:1 (Q1P3), 1:1 (Q1P1), and 1:3 (Q1P3). The TGA thermogram revealed that the initial decomposition temperatures (IDT) of NQC, chitosan, PAA, Q3P1, Q1P1, and Q1P3 were observed at 214, 240, 229, 227,246, and 254 °C, respectively. Q1P3 hydrogel had the greatest IDT, indicating that it was the most thermally stable, owing to greater intermolecular hydrogen bonding between NQC and PAA chains. The thermal stability of hydrogels increased in the sequence: Q1P3 > Q1P1 > chitosan > Q3P1 > PAA > NQC (Figure 7b) [154].

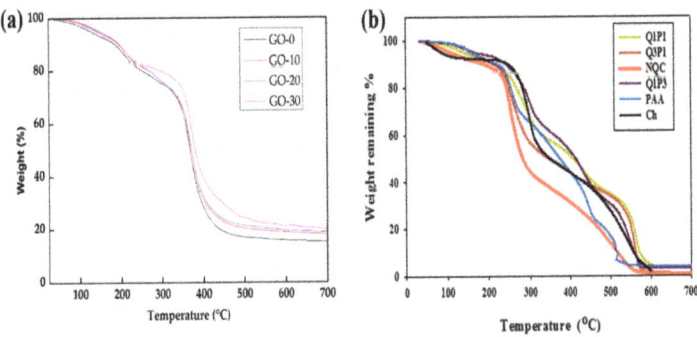

Figure 7. (**a**) TGA thermogram of the synthesized hydrogel with and without GO [153], and (**b**) Table 1. P1, Q1P3, and Q3P1. (Reprinted from Ref. [154], Copyright (2022), with permission from Elsevier).

8.3. SEM Analysis

SEM is used to study the surface morphology, topography, and composition of the hydrogels. The porosity of hydrogel is a key factor attributed to its adsorption capacity. For instance, Godiya et al. synthesized bio-based carboxymethyl cellulose (CMC)/poly(acrylamide) (PAM) hydrogel for adsorption of heavy metals. SEM results demonstrated that CMC/PAM hydrogel has a sponge-like, three-dimensional, and highly mesoporous surface morphology (Figure 8b) that significantly differs from CMC hydrogel (Figure 8a). The CMC/PAM hydrogel has a pore size in the range of 5–15 µm in diameter. The pores developed in the hydrogel will permit guest molecules like water and heavy metals to move across the composite structure. The CMC/PAM composite hydrogel retained its structural robustness after the adsorption of Cu^{2+} (Figure 8c) [92].

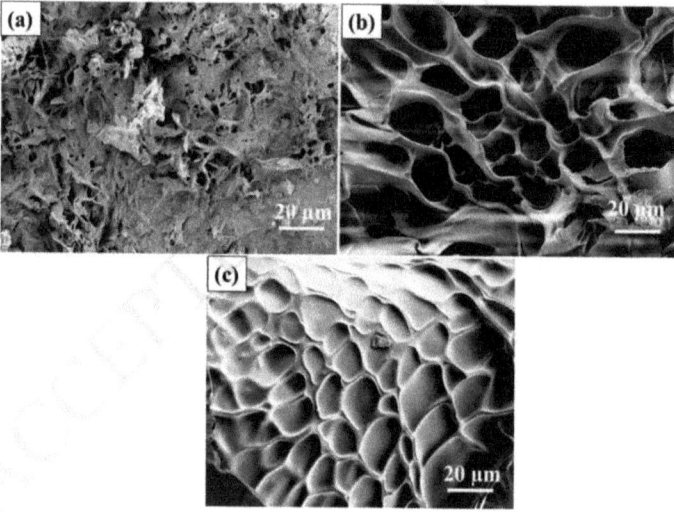

Figure 8. (a) CMC hydrogel, (b) CMC/PAM composite hydrogel, and (c) CMC/PAM composite hydrogel after the adsorption of Cu^{2+}. (Reprinted from Ref. [92], Copyright (2022), with permission from Elsevier).

Javed et al. synthesized anionic poly(methacrylic acid)(P(MAA)), neutral poly(acrylamide) (P(AAm)), and cationic poly(3-acrylamidopropyltrimethyl ammonium chloride)(P(APTMACI)) hydrogels and examined surface morphology by using SEM. SEM micrographs revealed that the surface P(MAA) was highly porous and rough compared to P(AAm) and P(APTMACI) (Figure 9a–c). The material with a rougher surface will generally have a higher adsorption capacity. As shown in Figure 10, SEM micrographs of hybrid hydrogels revealed that heavy metals nanoparticles were dispersed throughout the matrix without aggregation [155].

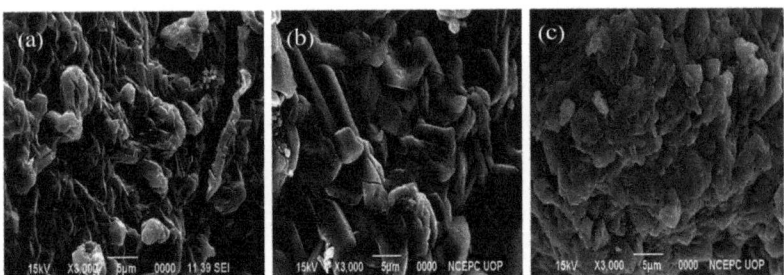

Figure 9. SEM micrograph of (**a**) anionic P(MAA), (**b**) neutral P(AAm), and (**c**) cationic P(APTMACl) hydrogels [155].

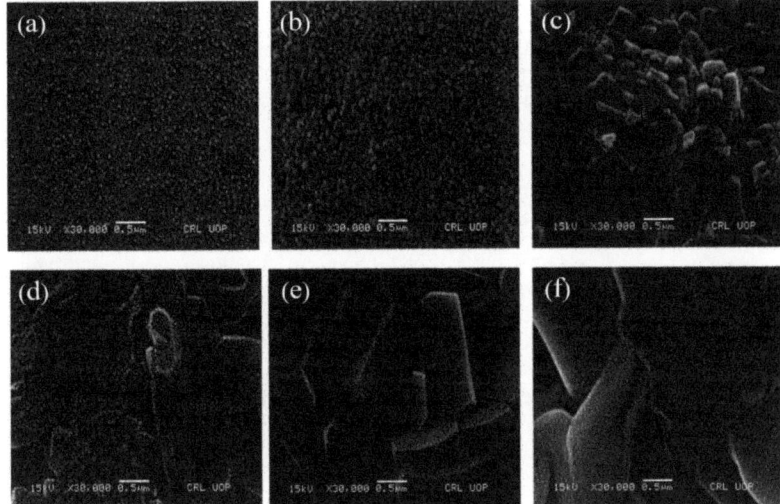

Figure 10. SEM micrographs of hybrid (**a**) P(MAA)-Cu, (**b**) P(MAA)-Ni, (**c**) P(APTMACl)-Cu, (**d**) P(APTMACl-Ni, (**e**) P(AAm)-Cu, and, (**f**) P(AAm)-Ni hydrogels [155].

8.4. Zeta Potential Analysis

Zeta potential is useful in determining the surface charge of hydrogel adsorbent. For example, Hu et al. synthesized carboxymethyl cellulose nanocrystals (CCN)/sodium alginate (Alg) hydrogel beads for scavenging Pb^{2+} and used a zeta-sizer to check the surface charge. Figure 11a depicts the zeta potentials of CCN, Alg, and prepared CCN-ALg were measured at pH 5.2. The zeta potential results revealed that all the three samples had negatively charged surfaces with stable dispersion at pH 5.2, with CCN-Alg being more so than the other two [156]. Bandara et al. studied the surface charge on chitosan/polyethylenimine/graphene oxide hydrogel beads for the abstraction of selenium from wastewater. Positive zeta potentials were noticed across a wide pH range, ranging from acidic pH to the isoelectronic point of 10.5, indicating ideal circumstances for electrostatic interaction with negatively charged species (Figure 11b) [157].

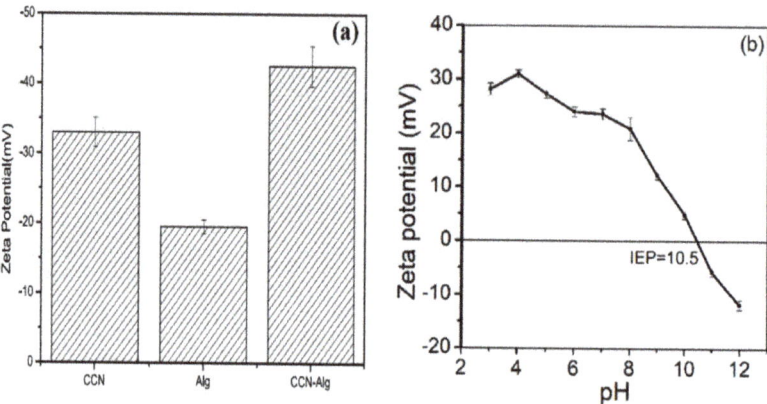

Figure 11. (**a**) Zeta potential (mV) of CCN, Alg, and CCN/Alg hydrogel beads. (Reprinted from Ref. [156], with permission from Elsevier), and (**b**) zeta potential of hydrogel beads varies as a function of pH, showing that negatively charged selenium absorbs at lower pH. (Reprinted from Ref. [158]. Copyright 2022 American Chemical Society).

8.5. EDX Analysis

EDX characterization is used to determine the hydrogel's elemental composition. Dil et al., for example, reported the fabrication of a novel porous gelatin-silver/poly (acrylic acid) (NPGESNC-AcA) nanocomposite hydrogel for Cu^{2+} removal. Figure 12 represents the element percentage of the synthesized NPGENC-AcA hydrogel before Cu^{2+} adsorption, which contains 52.3% carbon, 22.8% oxygen, 13.5% sodium, 10.6% nitrogen, 0.8% silver before adsorption of Cu^{2+}. The results showed that silver was deposited in the nanocomposite hydrogel network, with no additional impurity elements detected in the spectrum (Figure 12a). EDX analysis for NPGENC-AcA after Cu^{2+} adsorption consists of 46.6% carbon, 27.3% oxygen, 11.9% nitrogen, sodium 10.5%, 0.6% silver, and 3.1% copper (Figure 12b) [121].

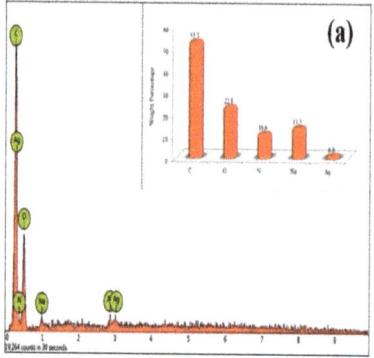
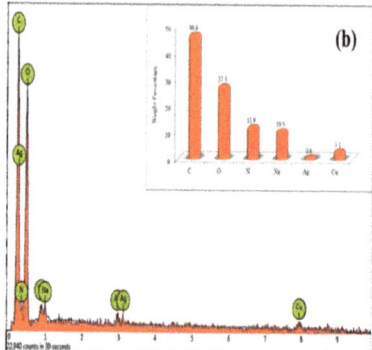

Figure 12. EDX spectra of NPGESNC-AcA (**a**) before Cu^{2+} adsorption, and (**b**) after Cu^{2+} adsorption (Reprinted from Ref. [121], Copyright (2022), with permission from Elsevier).

9. Adsorption Mechanism of Hydrogel

A thorough understanding of the adsorption mechanism and the removal process of various contaminants on different hydrogel-based adsorbents is essential for modifying hydrogels to enhance adsorption efficiency. The interactions such as electrostatic interaction, ion exchange, coordination interaction, and hydrophobic interaction take place depending on the surface functional moieties of hydrogels, provided reaction conditions such as temperature, pH, ligand, salt concentration, etc., and pollutant chemistry [159]. In literature, most hydrogel adsorbents are formed by the combination of interactions that take place simultaneously to form a 3D network. In the case of starch-based hydrogel, chemisorption and physisorption act simultaneously by acid-base interaction, H-bonding, ion exchange, or coordination interaction with heavy metal ions [160,161]. In chitin-based hydrogel, single or combination of multiple interactions occur depending on the operating condition and chemical composition [162]. In cyclodextrin-based hydrogel, complex formation occurs among the cyclodextrin and heavy metals involving host–guest interaction where hydrophobic bonding [163]. The various adsorption/desorption mechanism of heavy metals by hydrogel is discussed in Table 6.

9.1. Electrostatic Interaction

Electrostatic interaction occurs in the hydrogel with specific functional moieties in monomeric units having oppositely charged ions such as cation–anion interaction concerning heavy metals that need to be adsorbed or desorbed [159]. Furthermore, the pH of the solution has a significant impact on the generation of charged ions on the adsorbent (hydrogel) surface [164]. The pH of the solution is represented by pH_{PZC} when there are no charged ions on the surface of the adsorbent [165,166]. When the $pH > pH_{PZC}$, the surface functional moieties like $-OH$, $-COOH$, and $-H_3PO_4$ lose the proton due to the higher concentration of OH^- ions in the solution that forms anions like $-O^-$, $-COO^-$, $-PO_4^{3-}$ etc. on the surface of the adsorbent. However, at $pH < pH_{PZC}$, the surface of the adsorbent is positively charged due to an increase in the concentration of H^+ ions, which causes protonation of functional moieties such as $-SH$, $-NH_2$, etc [167]. According to the studies reported, electrostatic interactions are the dominant adsorption force for heavy metals abstraction in various hydrogels. Yu and co-workers synthesized sodium alginate(SA)/carboxylated nanocrystals cellulose hydrogel beads for the abstraction of Pb^{2+}. The findings in this study reveal that the adsorption mechanism that took place was complexation among $-COO$ and $-OH$ functional moieties and heavy metal (Pb^{2+}) by sharing a pair of electrons. Thereafter, the electrostatic interaction was found to occur between negatively charged hydrogel beads and positively charged Pb^{2+} ions [156]. Tang et al. synthesized physically cross-linked double network hydrogel (PCDNH) containing chitosan, calcium ion, and sodium alginate. In this study, they reported that chitosan's cationic NH_3^+ group reacts with sodium alginate's anionic $-COO^-$ group to construct physically cross-linked hydrogel via electrostatic interaction. In addition, the adsorption of heavy metals (Pb^{2+} and Cd^{2+}) on the hydrogel surface was due to the electrostatic interaction with PCDNH's oxygen atom, whereas the adsorption of Cu^{2+} was primarily due to coordination interaction with PCDNH's nitrogen atom, besides electrostatic interaction [151]. Zeng et al. prepared pullulan/polydopamine hydrogel for effective elimination of heavy metals (Co^{2+}, Cu^{2+}, and Ni^{2+}). In this research work, hydrogels were prepared by chemically cross-linking pullulan with 1,2-bis (2,3-epoxypropoxy) ethane. Polydopamine was added to the mixture to form a novel hydrogel adsorbent. Polydopamine's nitrogen atom and catechol group have a high affinity to react with positively charged metal-ion via electrostatic and coordination interaction (Figure 13) [158].

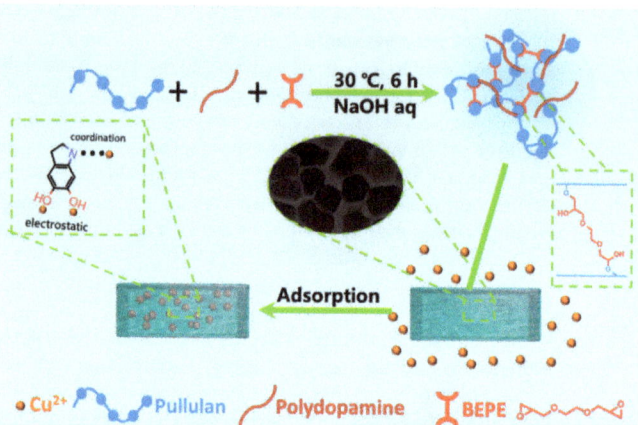

Figure 13. Schematic representation showcasing electrostatic and coordination interaction between pullulan/polydopamine hydrogel and heavy metals. (Reprinted from Ref. [158], Copyright (2022), with permission from Elsevier).

9.2. Ion-Exchange

Ion exchange refers to a chemical process whereby the swapping of ions takes place between an insoluble adsorbent (hydrogel) and a liquid phase (wastewater). The unwanted anions or cations dissolved in the wastewater are replaced or removed by the ions of a similar charge present on the hydrogel surface. To maintain the neutrality of the system the number of ions adsorbed by the hydrogel adsorbent must be equal to the number of ions liberated [168]. Ion exchange provides an efficient and convenient route becauser it can distillate and separate distinct contaminants from wastewater [169]. It reduces the degree of harmful load by converting heavy metals waste into a form that can be reused and recycled, leaving behind less hazardous materials in the solution, or by reducing the hydraulic flow of the stream containing toxic heavy metals, allowing for the final release [167]. Ion-exchange mechanisms, like electrostatic interaction, are highly dependent on the pH of the solution. Due to a rise in the concentration of H^+ ions at pH < pH_{PZC}, the functional moieties in hydrogel adsorbent become positively charged, leading to cation exchange. However, at pH > pH_{PZC} the functional moieties are negatively charged due to excessive concentration of OH^- ions, leading to anion exchange [170]. Saber-Samandari et al. synthesized Cellulose-Graft-Polyacrylamide/Hydroxyapatite hydrogel composite for the removal of Cu^{2+} ions. In this work, he observed that Cu^{2+} ions got exchanged with the cations in the hydrogel composite and are attached to the surface of hydroxyapatite by an ion-exchange mechanism [171]. For the treatment of heavy metals from oily wastewater, Xiong et al. prepared a self-cleaning cellulose functionalized titanate microsphere hydrogel via a sol-gel method. The prepared hydrogel microspheres have the combined properties of cellulose and titanate nanotubes that exhibit a high capacity to maintain oily wastewater. At first, Cu^{2+} got absorbed on the inner surface of cellulose titanate hydrogel by electrostatic interaction. After that, Cu^{2+} was captured in the layer of titanate nanotubes exhibiting remarkable characteristics for heavy metals under the influence the chemical and physical adsorption [172]. Ma et al. prepared ethylenediaminetetraacetic acid (EDTA) functionalized double network hydrogel for efficient elimination of heavy metals (Cd^{2+}, Pb^{2+}, and Cu^{2+}) from industrial eluents. In this work, a two-step process was conducted in which first polyacrylamide was cross-linked with N, N' methylene bisacrylamide (MBA), and then EDTA was cross-linked with chitosan to form a double network hydrogel. The hydrogel showed a maximum sorption capacity of 138.41 mg/g, 99.44 mg/g, and 86.00 mg/g for Pb^{2+}, Cu^{2+}, and Cd^{2+} respectively based on the ion exchange mechanism between carboxylate groups and heavy metal ions (Figure 14) [115].

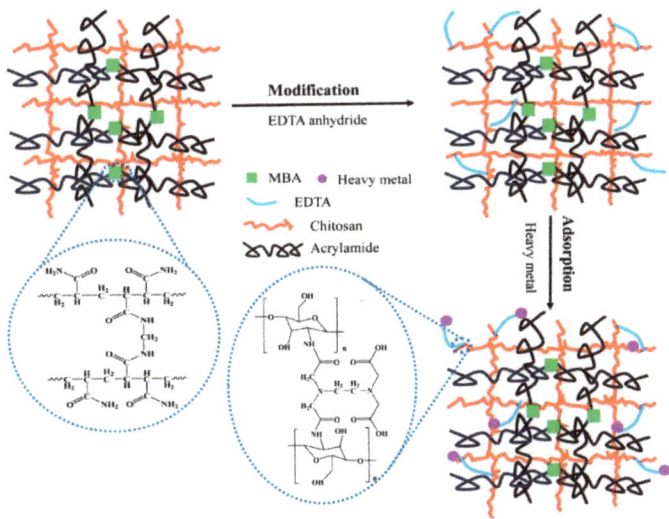

Figure 14. The proposed mechanism between chitosan/polyacrylamide hydrogel and heavy metal ions. (Reprinted from Ref. [115]. Copyright 2022 American Chemical Society).

9.3. Hydrophobic Interaction

The interaction taking place between water molecules and hydrophobes (non-polar molecules containing long carbon chains that do not react with water molecules because of weak Van-der-Waals forces) is termed hydrophobic interaction [173]. Therefore, a low water-soluble molecule is more likely to be attracted by hydrophobes. For example, Tokuyama et al. prepared superabsorbent hydrogel containing N-isopropyl acrylamide (NIPA) as a thermo-responsive polymer for heavy metals extraction. At first, an aqueous solution of metal ions is complexed with an extractant that has a hydrophobic group and an interacting group. After that, above lower critical solution temperature the complex formed between metal and extractant gets absorbed into the hydrogel via hydrophobic interaction. Finally, after cooling below the low critical solution temperature metal-extractant complexes are extracted from the hydrogel. In this study, Cu^{2+} is used as a model heavy metal ion [174]. The mechanism for the same is depicted below in Figure 15.

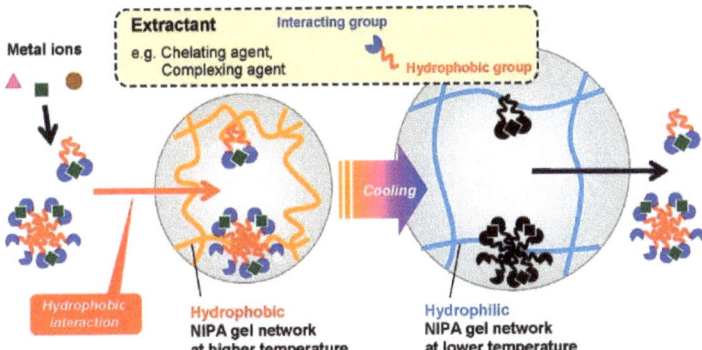

Figure 15. The proposed mechanism of heavy metal complexed with extractant onto N-isopropyl acrylamide hydrogel. (Reprinted with permission from Ref. [174]. Copyright 2022 American Chemical Society).

9.4. Coordination Interaction

Coordination interaction also known as chelation interaction refers to the formation of covalent bond where a single-atom shares both the electrons. In this interaction, cation (heavy metals) binds with the group containing lone pair electrons, resulting in cation adsorption on the adsorbate surface [167]. Zhaung et al. prepared double network alginate/graphene nanocomposite hydrogel beads for effective extraction of Cu^{2+} and dichromate ($Cr_2O_7^{2-}$). He observed that –COOH functional moieties in both graphene and alginate show a high affinity for Cu^{2+} and $Cr_2O_7^{2-}$ via coordination and complexation. On the contrary, ion exchange takes place between Ca^{2+} ions in alginate and Cu^{2+} in the aqueous solution [175]. Rodrigues et al. prepared chitosan-g-poly (acrylic acid)/cellulose nanowhiskers (CNWs) composite hydrogel beads by using N, N′ methylene bisacrylamide as a cross-linker for the adsorption of Cu^{2+} and Pb^{2+} from water. FTIR analysis revealed that functional moieties i.e., hydroxyl groups and carboxyl groups act as coordination sites for heavy metal adsorption [176]. The schematic representation depicting the coordination between hydrogel adsorbent and heavy metals is demonstrated in Figure 16.

Table 6. Proposed synthesis and removal mechanism of various hydrogel-based adsorbents.

Hydrogel Type	Synthesis Method	Mechanism	Heavy Metals Removed	References
Carboxymethyl cellulose-graft-poly(acrylic acid)/monmorillonite hydrogel composite	Graft polymerization	Ion exchange and coordination interaction	Zn^{2+}, Pb^{2+}	[177]
Silk sericin/Lignin hydrogel beads	Graft polymerization	Ion exchange or electrostatic interaction	Cr^{6+}	[178]
Chitosan/multiwall carbon nanotube/poly(acrylic acid)/poly(4-aminodiphenyl amine) functional gel	Free radical polymerization and cross-linking reaction	Complexation interaction	Cr^{6+}	[60]
Sugar cane bagasse cellulose and gelatin-based hydrogel composite	Cross-linking	Coordination and electrostatic interaction	Cu^{2+}	[179]
Carboxy methyl cellulose hydrogel	γ-raddiation	Coordination interaction	Cu^{2+}	[180]
Chitin/cellulose composite hydrogel	Freeze-thaw method	Electrostatic and coordination interaction	Hg^{2+}, Cu^{2+}, Pb^{2+}	[181]
Carboxy methyl cellulose hydrogel beads	Inverse suspension method	Coordination interaction	Cu^{2+}, Ni^{2+}, Pb^{2+}	[182]
Hydrogel-biochar composite	Free radical polymerization and cross-linking reactions	Chemisorption	As	[101]
Pullulan/polydopamine hydrogels	Chemical cross-linking	Electrostatic and coordination interaction	Cu^{2+}	[158]
Jute/poly(acrylic acid) hydrogel	Free radical polymerization	Electrostatic interaction	Cd^{2+}, Pb^{2+}	[183]
Carboxylated chitosan/carboxylated nanocellulose hydrogel beads	Cross-linking	Electrostatic and coordination interaction	Pb^{2+}	[184]

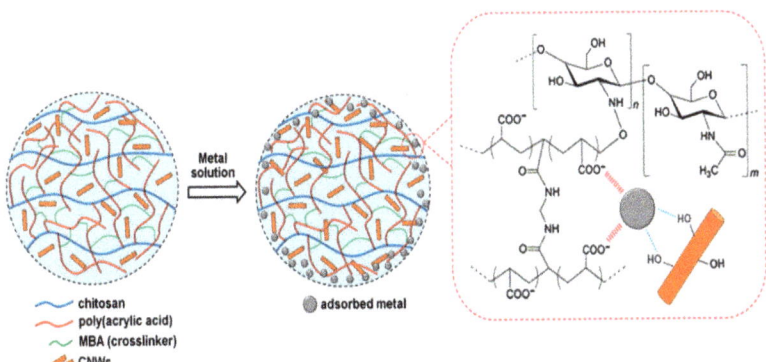

Figure 16. The coordination interaction between chitosan-g-poly (acrylic acid)/cellulose nanowhiskers hydrogel beads and adsorbed metal. (Reprinted by permission from Ref. [176]. Copyright (2022), Springer).

10. Recovery, Regeneration, and Reusability of Hydrogel

One of the paramount characteristics of hydrogel other than high adsorption efficiency is their regeneration capacity by desorbing the absorbed heavy metals which further allows it to be reused. The ability to regenerate and reuse an adsorbent material is also an important factor for the practical assessment of its application. Many different ways have been studied by researchers for the effective desorption of heavy metals from the three-dimensional mesh structure of hydrogel after every removal cycle. Changes in the magnetic field, electric field, temperature, pH, etc. will lead to the desorption of heavy metals [185]. The influence of pH on heavy metal desorption from a magnetized cellulose-chitosan hydrogel was reported by Liu et al. [108]. At very low pH values of 1.0–2.0 desorption efficiency of 83–86% was achieved. This represents the merits of using pH-dependent hydrogel for the adsorption of heavy metals such as arsenic and chromium during the elimination process and desorption for the recovery of hydrogels. Moreover, adjusting the required pH is a drawback [38]. According to the literature, studies reported on the recovery of hydrogel adsorbents have used strong as well as weak acids as eluents (HCl, HNO_3, CH_3COOH, H_2SO_4, etc.) [37]. Furthermore, the type of acid utilized in the desorption process also has a considerable impact on the durability and desorption capacity of hydrogel [24]. Mohammadi et al. synthesized a chelator-mimetic multi-functionalized hydrogel with a high metal adsorption efficiency (cadmium, lead, and arsenic) and great reusability. By employing a low concentration of hydrochloric acid, the heavy metals absorbed in the hydrogel network were eluted and the hydrogel was regenerated for reuse. After five adsorption/desorption reuse cycles, a removal ratio greater than 60% was obtained [185]. By applying a similar approach, Pourjavadi et al. developed a novel hydrogel containing chitosan, acrylic acid, and an amine-functionalized nano-silica. In this work, 1M hydrochloric acid solution was employed for recovering hydrogel loaded with Pb^{2+}. The hydrogel was then regenerated by filtering and washing with deionized water before being utilized for the next adsorption cycle. After three consecutive cycles, the efficiency of regenerated hydrogel remained around 685–715 mg/g [186].

Magnetic hydrogels are one of the most used adsorbent materials for the effective elimination of heavy metals from flowing streams. During the recovery process, the eluent acidity needs to be managed since an excess of acid can damage the magnetic adsorbent. Eluents with high concentrations can damage the binding sites on hydrogels, resulting in lowered adsorption efficiency after numerous sorption cycles [187]. Tang et al. synthesized a magnetic hydrogel with high adsorption efficiency of 200 mg/g for Cr^{6+} adsorption. The hydrogel possesses an advantage of easy recovery by regenerating in sodium chloride solution (NaCl) [188]. The applicability of any magnetic hydrogel adsorbent in contaminated water is determined by two important factors: an increase in the concentration of heavy metals in that solution and a lower quantity of recovery solution. Tang's research summarized both the factors in Figure 17. In brief, the treated contaminated water is collected and separated from hydrogel in a magnetic separation unit; NaCl at different concentration is injected for the regeneration of hydrogel, and the leftover solution were collected by separating the magnetic hydrogel. A series of regeneration tests were carried out by step-wise addition of sodium chloride solution. The recovery solution was then collected and further processed by the addition of NaCl solution to it. The results obtained suggested recovery efficiency was maintained for 20 sorption cycles, resulting in the Cr^{6+} removal capacity of 97–98%. According to the results achieved, the Cr content in the recovery solution reached 500–600 mg/L corresponding to wastewater:recovery volume ratio of 40:1 [188].

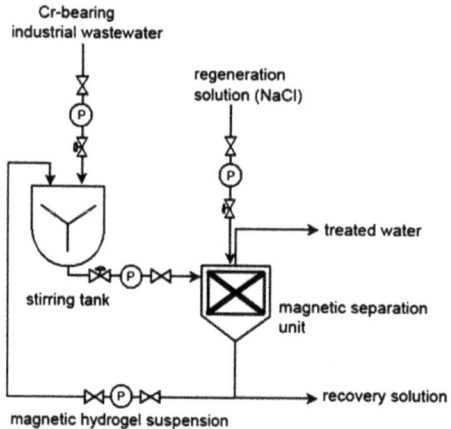

Figure 17. Schematic representation of a wastewater treatment experiment with a magnetic separation unit. (Reprinted with permission from Ref. [188]. Copyright 2022 Americal Chemical Society).

Reusability of hydrogel is one of the most important characteristics for wide-range applications, although it is a challenge for conventional hydrogel adsorbents as they possess poor mechanical strength after swelling in aqueous media. Therefore, increasing mechanical strength plays a crucial role in maintaining the desired adsorption efficiency of the heavy metals-loaded hydrogel. Liu et al. reported that 95% Fe, Pb, and Cu were removed from an aqueous medium through 7 adsorption cycles and hydrogel still can lead to heavy metals removal [108]. Therefore, it proves that hydrogel can be reused many times and lowers the cost of production for heavy metals elimination from an aqueous solution. Tang et al. reported the reusability and regeneration of hydrogel in a column experiment for effective elimination of Cr^{6+} by a cationic hydrogel. After 6 sorption cycles adsorption efficiency remained constant (27 mg g^{-1}, 90%) and the desorption capacity was 93 percent on average for every cycle [189]. In conclusion, it can be said that low operational and production-cost along with easy separation capabilities and reusability make hydrogel a choice of adsorbent for heavy metal removal from wastewater.

11. Conclusions

Water pollution is one of the serious global problems caused by increasing industrialization and urbanization. In particular, heavy metals discharged into flowing streams have detrimental effects on human health and the natural ecological system. Thus, it is necessary to treat the wastewater containing toxic heavy metals and then discharge it. The adsorption process including various types of adsorbent material is regarded as an efficient, cost-effective, and environment-friendly approach to the treatment of heavy metals. However, the majority of adsorbent materials used for wastewater treatment are non-biodegradable, synthetic, and require post-treatment after use, which prompts researchers' interest in developing biodegradable, easy to modify, and biocompatible adsorbent materials. Hydrogels as potential adsorbent materials represent the best choice. The present review summarizes the literature concerning hydrogels in the past 25 years, and describes the classification, properties, synthesis, mechanism and recovery, regeneration, and reuse of hydrogel-based material for the elimination of heavy metals.

Although hydrogels have been extensively studied, there are still a few areas that require further investigation.

- Currently, the hydrogel-based adsorbent materials used for heavy metal removal are limited to lab scale. Therefore, further research is required to scale up for a large-scale application.
- The present research is confined to removing a single type of heavy metal. More research should be undertaken targeting multiple heavy metals.
- The research should focus on the ability of the hydrogel to regenerate (for example, the adsorption efficiency of hydrogel drops after five sorption cycles).
- To broaden the spectrum of hydrogels application for separation of rare earth metals.
- To develop high mechanical strength tailored hydrogel (for example hydrogel membranes) that are easier to separate from the liquid phase for wastewater treatment.

Author Contributions: Conceptualization, Z.D., S.S. and R.G.; methodology, Z.D. and S.S.; software, Z.D., S.S. and R.G.; validation, Z.D. and S.S.; formal analysis, Z.D.; investigation, Z.D., S.S., R.G. and N.S.; resources, S.S., N.S. and I.A.; writing—original draft preparation, Z.D. and S.S.; writing—review and editing, Z.D., S.S., R.G., I.A. and N.S.; supervision, S.S. and R.G.; funding acquisition, N.S., I.A. and S.S. All authors have read and agreed to the published version of the manuscript.

Funding: The authors would like to thank Pandit Deendayal Energy University for providing research facilities. This research was supported by the Marine Pollution Special Interest Group, National Defence University of Malaysia via SF0076-UPNM/2019/SF/ICT/6 and Scientific Research Deanship at King Khalid University, Abha, Saudi Arabia through the Large Research Group Project under grant number (RGP.02-205-42).

Institutional Review Board Statement: Not applicable.

Informed Consent Statement: Not applicable.

Data Availability Statement: Not applicable.

Conflicts of Interest: The authors declare no conflict of interest.

References

1. Kabir, S.; Cueto, R.; Balamurugan, S.; Romeo, L.D.; Kuttruff, J.T.; Marx, B.D.; Negulescu, I.I. Removal of acid dyes from textile wastewaters using fish scales by absorption process. *Clean Technol.* **2019**, *1*, 311–324. [CrossRef]
2. Hasanpour, M.; Hatami, M. Photocatalytic performance of aerogels for organic dyes removal from wastewaters: Review study. *J. Mol. Liq.* **2020**, *309*, 113094. [CrossRef]
3. Hasanpour, M.; Hatami, M. Application of three dimensional porous aerogels as adsorbent for removal of heavy metal ions from water/wastewater: A review study. *Adv. Colloid Interface Sci.* **2020**, *284*, 102247. [CrossRef] [PubMed]
4. Siti, N.; Mohd, H.; Md, L.K.; Shamsul, I. Adsorption process of heavy metals by low-cost adsorbent: A review. *World Appl. Sci. J.* **2013**, *28*, 1518–1530.
5. Saha, J.; Dikshit, A.; Bandyopadhyay, M.; Saha, K. A review of arsenic poisoning and its effects on human health. *Crit. Rev. Environ. Sci. Technol.* **1999**, *29*, 281–313. [CrossRef]
6. Zhou, Y.; Hu, X.; Zhang, M.; Zhuo, X.; Niu, J. Preparation and characterization of modified cellulose for adsorption of Cd (II), Hg (II), and acid fuchsin from aqueous solutions. *Ind. Eng. Chem. Res.* **2013**, *52*, 876–884. [CrossRef]
7. Masindi, V.; Muedi, K.L. Environmental contamination by heavy metals. *Heavy Met.* **2018**, *10*, 115–132.
8. Soleimani, M.; Amini, N.; Sadeghian, B.; Wang, D.; Fang, L. Heavy metals and their source identification in particulate matter ($PM_{2.5}$) in Isfahan City, Iran. *J. Environ. Sci.* **2018**, *72*, 166–175. [CrossRef]
9. Briffa, J.; Sinagra, E.; Blundell, R. Heavy metal pollution in the environment and their toxicological effects on humans. *Heliyon* **2020**, *6*, e04691. [CrossRef]
10. Abdel-Raouf, M.; Abdul-Raheim, A. Removal of Heavy Metals from Industrial Waste Water by Biomass-Based Materials: A Review. *J. Pollut. Eff. Control.* **2017**, *5*, 1000180.

11. Saxena, G.; Purchase, D.; Mulla, S.I.; Saratale, G.D.; Bharagava, R.N. Phytoremediation of heavy metal-contaminated sites: Eco-environmental concerns, field studies, sustainability issues, and future prospects. *Rev. Environ. Contam. Toxicol.* **2019**, *249*, 71–131.
12. Ali, R.M.; Hamad, H.A.; Hussein, M.M.; Malash, G.F. Potential of using green adsorbent of heavy metal removal from aqueous solutions: Adsorption kinetics, isotherm, thermodynamic, mechanism and economic analysis. *Ecol. Eng.* **2016**, *91*, 317–332. [CrossRef]
13. Mthombeni, N.H.; Mbakop1and, S.; Onyango, M.S. Adsorptive removal of manganese from industrial and mining wastewater. In Proceedings of the Sustainable Research and Innovation Conference, Nairobi, Kenya, 4 May 2016; College of Engineering and Technology Jomo Kenyatta University of Agriculture and Technology: Nairobi, Kenya, 2016; pp. 36–45.
14. Yadav, A. Bioremediation of wastewater using various sorbents and vegetable enzymes. *Res. Biotechnol.* **2015**, *6*, 1–15.
15. Ahmad, M.; Ahmed, S.; Swami, B.L.; Ikram, S. Adsorption of heavy metal ions: Role of chitosan and cellulose for water treatment. *Langmuir* **2015**, *79*, 109–155.
16. Azimi, A.; Azari, A.; Rezakazemi, M.; Ansarpour, M. Removal of heavy metals from industrial wastewaters: A review. *ChemBioEng Rev.* **2017**, *4*, 37–59. [CrossRef]
17. Rudi, N.N.; Muhamad, M.S.; Te Chuan, L.; Alipal, J.; Omar, S.; Hamidon, N.; Hamid, N.H.A.; Sunar, N.M.; Ali, R.; Harun, H. Evolution of adsorption process for manganese removal in water via agricultural waste adsorbents. *Heliyon* **2020**, *6*, e05049. [CrossRef]
18. Shafiq, M.; Alazba, A.; Amin, M. Removal of heavy metals from wastewater using date palm as a biosorbent: A comparative review. *Sains Malays.* **2018**, *47*, 35–49.
19. Sun, C.; Chen, T.; Huang, Q.; Wang, J.; Lu, S.; Yan, J. Enhanced adsorption for Pb (II) and Cd (II) of magnetic rice husk biochar by KMnO4 modification. *Environ. Sci. Pollut. Res.* **2019**, *26*, 8902–8913. [CrossRef]
20. Nuhanović, M.; Grebo, M.; Draganović, S.; Memić, M.; Smječanin, N. Uranium (VI) biosorption by sugar beet pulp: Equilibrium, kinetic and thermodynamic studies. *J. Radioanal. Nucl. Chem.* **2019**, *322*, 2065–2078. [CrossRef]
21. George, R.; Bahadur, N.; Singh, N.; Singh, R.; Verma, A.; Shukla, A. Environmentally benign TiO_2 nanomaterials for removal of heavy metal ions with interfering ions present in tap water. *Mater. Today Proc.* **2016**, *3*, 162–166. [CrossRef]
22. Manjuladevi, M.; Sri, O. Heavy metals removal from industrial wastewater by nano adsorbent prepared from cucumis melopeel activated carbon. *J. Nanomed. Res.* **2017**, *5*, 1–4.
23. Es-Sahbany, H.; Berradi, M.; Nkhili, S.; Hsissou, R.; Allaoui, M.; Loutfi, M.; Bassir, D.; Belfaquir, M.; El Youbi, M. Removal of heavy metals (nickel) contained in wastewater-models by the adsorption technique on natural clay. *Mater. Today Proc.* **2019**, *13*, 866–875. [CrossRef]
24. Weerasundara, L.; Gabriele, B.; Figoli, A.; Ok, Y.-S.; Bundschuh, J. Hydrogels: Novel materials for contaminant removal in water—A review. *Crit. Rev. Environ. Sci. Technol.* **2020**, *51*, 1–45. [CrossRef]
25. Pan, Z.; An, L. Removal of heavy metal from wastewater using ion exchange membranes. In *Applications of Ion Exchange Materials in the Environment*; Springer: Berlin, Germany, 2019; pp. 25–46.
26. Liu, L.; Luo, X.-B.; Ding, L.; Luo, S.-L. Application of nanotechnology in the removal of heavy metal from water. In *Nanomaterials for the Removal of Pollutants and Resource Reutilization*; Elsevier: Oxford, UK, 2019; pp. 83–147.
27. Teh, C.Y.; Budiman, P.M.; Shak, K.P.Y.; Wu, T.Y. Recent advancement of coagulation–flocculation and its application in wastewater treatment. *Ind. Eng. Chem. Res.* **2016**, *55*, 4363–4389. [CrossRef]
28. Medina, B.; Torem, M.; De Mesquita, L. On the kinetics of precipitate flotation of Cr III using sodium dodecylsulfate and ethanol. *Miner. Eng.* **2005**, *18*, 225–231. [CrossRef]
29. Fu, F.; Wang, Q. Removal of heavy metal ions from wastewaters: A review. *J. Environ. Manag.* **2011**, *92*, 407–418. [CrossRef]
30. Mukherjee, R.; Bhunia, P.; De, S. Impact of graphene oxide on removal of heavy metals using mixed matrix membrane. *Chem. Eng. J.* **2016**, *292*, 284–297. [CrossRef]
31. Van Bemmelen, J. Das hydrogel und das krystallinische hydrat des kupferoxyds. *Z. Anorg. Chem.* **1894**, *5*, 466–483. [CrossRef]
32. Thakur, S.; Thakur, V.K.; Arotiba, O.A. History, classification, properties and application of hydrogels: An overview. In *Hydrogels*; Springer: Singapore, 2018; pp. 29–50.
33. Wichterle, O.; Lim, D. Hydrophilic gels for biological use. *Nature* **1960**, *185*, 117–118. [CrossRef]
34. Shalla, A.H.; Yaseen, Z.; Bhat, M.A.; Rangreez, T.A.; Maswal, M. Recent review for removal of metal ions by hydrogels. *Sep. Sci. Technol.* **2019**, *54*, 89–100. [CrossRef]
35. Dai, L.; Cheng, T.; Xi, X.; Nie, S.; Ke, H.; Liu, Y.; Tong, S.; Chen, Z. A versatile TOCN/CGG self-assembling hydrogel for integrated wastewater treatment. *Cellulose* **2020**, *27*, 915–925. [CrossRef]
36. Jang, S.H.; Jeong, Y.G.; Min, B.G.; Lyoo, W.S.; Lee, S.C. Preparation and lead ion removal property of hydroxyapatite/polyacrylamide composite hydrogels. *J. Hazard. Mater.* **2008**, *159*, 294–299. [CrossRef] [PubMed]
37. Ozay, O.; Ekici, S.; Baran, Y.; Kubilay, S.; Aktas, N.; Sahiner, N. Utilization of magnetic hydrogels in the separation of toxic metal ions from aqueous environments. *Desalination* **2010**, *260*, 57–64. [CrossRef]
38. Barakat, M.; Sahiner, N. Cationic hydrogels for toxic arsenate removal from aqueous environment. *J. Environ. Manag.* **2008**, *88*, 955–961. [CrossRef] [PubMed]
39. Ozay, O.; Ekici, S.; Baran, Y.; Aktas, N.; Sahiner, N. Removal of toxic metal ions with magnetic hydrogels. *Water Res.* **2009**, *43*, 4403–4411. [CrossRef]

40. Hernández, R.; Mijangos, C. In situ Synthesis of Magnetic Iron Oxide Nanoparticles in Thermally Responsive Alginate-Poly (N-isopropylacrylamide) Semi-Interpenetrating Polymer Networks. *Macromol. Rapid Commun.* **2009**, *30*, 176–181. [CrossRef]
41. Ahmed, E.M. Hydrogel: Preparation, characterization, and applications: A review. *J. Adv. Res.* **2015**, *6*, 105–121. [CrossRef]
42. Lee, K.Y.; Rowley, J.A.; Eiselt, P.; Moy, E.M.; Bouhadir, K.H.; Mooney, D.J. Controlling mechanical and swelling properties of alginate hydrogels independently by cross-linker type and cross-linking density. *Macromolecules* **2000**, *33*, 4291–4294. [CrossRef]
43. Sahiner, N. Hydrogels of versatile size and architecture for effective environmental applications. *Turk. J. Chem.* **2008**, *32*, 113–123.
44. Singh, A.; Sharma, P.K.; Garg, V.K.; Garg, G. Hydrogels: A review. *Int. J. Pharm. Sci. Rev. Res.* **2010**, *4*, 97–105.
45. Mahinroosta, M.; Farsangi, Z.J.; Allahverdi, A.; Shakoori, Z. Hydrogels as intelligent materials: A brief review of synthesis, properties and applications. *Mater. Today Chem.* **2018**, *8*, 42–55. [CrossRef]
46. Ozay, O.; Aktas, N.; Sahiner, N. Hydrogels as a potential chromatographic system: Absorption, speciation, and separation of chromium species from aqueous media. *Sep. Sci. Technol.* **2011**, *46*, 1450–1461. [CrossRef]
47. Bashir, S.; Hina, M.; Iqbal, J.; Rajpar, A.; Mujtaba, M.; Alghamdi, N.; Wageh, S.; Ramesh, K.; Ramesh, S. Fundamental concepts of hydrogels: Synthesis, properties, and their applications. *Polymers* **2020**, *12*, 2702. [CrossRef] [PubMed]
48. Hu, X.; Cheng, W.; Nie, W.; Shao, Z. Synthesis and characterization of a temperature-sensitive hydrogel based on sodium alginate and N-isopropylacrylamide. *Polym. Adv. Technol.* **2015**, *26*, 1340–1345. [CrossRef]
49. De, S.K.; Aluru, N.R. A chemo-electro-mechanical mathematical model for simulation of pH sensitive hydrogels. *Mech. Mater.* **2004**, *36*, 395–410. [CrossRef]
50. Mu, R.; Liu, B.; Chen, X.; Wang, N.; Yang, J. Hydrogel adsorbent in industrial wastewater treatment and ecological environment protection. *Environ. Technol. Innov.* **2020**, *20*, 101107. [CrossRef]
51. Kuddushi, M.; Rajput, S.; Shah, A.; Mata, J.; Aswal, V.K.; El Seoud, O.; Kumar, A.; Malek, N.I. Stimuli responsive, self-sustainable, and self-healable functionalized hydrogel with dual gelation, load-bearing, and dye-absorbing properties. *ACS Appl. Mater. Interfaces* **2019**, *11*, 19572–19583. [CrossRef]
52. Jiang, Y.; Wang, Y.; Li, Q.; Yu, C.; Chu, W. Natural polymer-based stimuli-responsive hydrogels. *Curr. Med. Chem.* **2020**, *27*, 2631–2657. [CrossRef]
53. Rittikulsittichai, S.; Kolhatkar, A.G.; Sarangi, S.; Vorontsova, M.A.; Vekilov, P.G.; Brazdeikis, A.; Lee, T.R. Multi-responsive hybrid particles: Thermo-, pH-, photo-, and magneto-responsive magnetic hydrogel cores with gold nanorod optical triggers. *Nanoscale* **2016**, *8*, 11851–11861. [CrossRef]
54. Ullah, F.; Othman, M.B.H.; Javed, F.; Ahmad, Z.; Akil, H.M. Classification, processing and application of hydrogels: A review. *Mater. Sci. Eng. C* **2015**, *57*, 414–433. [CrossRef]
55. Jing, G.; Wang, L.; Yu, H.; Amer, W.A.; Zhang, L. Recent progress on study of hybrid hydrogels for water treatment. *Colloids Surf. A Physicochem. Eng. Asp.* **2013**, *416*, 86–94. [CrossRef]
56. Zhang, J.; Wang, A. Adsorption of Pb (II) from aqueous solution by chitosan-g-poly (acrylic acid)/attapulgite/sodium humate composite hydrogels. *J. Chem. Eng. Data* **2010**, *55*, 2379–2384. [CrossRef]
57. Iizawa, T.; Taketa, H.; Maruta, M.; Ishido, T.; Gotoh, T.; Sakohara, S. Synthesis of porous poly (N-isopropylacrylamide) gel beads by sedimentation polymerization and their morphology. *J. Appl. Polym. Sci.* **2007**, *104*, 842–850. [CrossRef]
58. Garg, S.; Garg, A.; Vishwavidyalaya, R. Hydrogel: Classification, properties, preparation and technical features. *Asian J. Biomater. Res* **2016**, *2*, 163–170.
59. Yang, L.; Chu, J.S.; Fix, J.A. Colon-specific drug delivery: New approaches and in vitro/in vivo evaluation. *Int. J. Pharm.* **2002**, *235*, 1–15. [CrossRef]
60. Kim, M.K.; Sundaram, K.S.; Iyengar, G.A.; Lee, K.-P. A novel chitosan functional gel included with multiwall carbon nanotube and substituted polyaniline as adsorbent for efficient removal of chromium ion. *Chem. Eng. J.* **2015**, *267*, 51–64. [CrossRef]
61. Singhal, R.; Gupta, K. A review: Tailor-made hydrogel structures (classifications and synthesis parameters). *Polym. Plast. Technol. Eng.* **2016**, *55*, 54–70. [CrossRef]
62. Kabiri, K.; Omidian, H.; Zohuriaan-Mehr, M.; Doroudiani, S. Superabsorbent hydrogel composites and nanocomposites: A review. *Polym. Compos.* **2011**, *32*, 277–289. [CrossRef]
63. Badsha, M.A.; Khan, M.; Wu, B.; Kumar, A.; Lo, I.M. Role of surface functional groups of hydrogels in metal adsorption: From performance to mechanism. *J. Hazard. Mater.* **2021**, *408*, 124463. [CrossRef]
64. Badsha, M.A.; Lo, I.M. An innovative pH-independent magnetically separable hydrogel for the removal of Cu (II) and Ni (II) ions from electroplating wastewater. *J. Hazard. Mater.* **2020**, *381*, 121000. [CrossRef]
65. He, X.; Cheng, L.; Wang, Y.; Zhao, J.; Zhang, W.; Lu, C. Aerogels from quaternary ammonium-functionalized cellulose nanofibers for rapid removal of Cr (VI) from water. *Carbohydr. Polym.* **2014**, *111*, 683–687. [CrossRef] [PubMed]
66. Cyganowski, P.; Dzimitrowicz, A.; Jamroz, P.; Jermakowicz-Bartkowiak, D.; Pohl, P. Polymerization-driven immobilization of dc-apgd synthesized gold nanoparticles into a quaternary ammonium-based hydrogel resulting in a polymeric nanocomposite with heat-transfer applications. *Polymers* **2018**, *10*, 377. [CrossRef] [PubMed]
67. He, G.; Ke, W.; Chen, X.; Kong, Y.; Zheng, H.; Yin, Y.; Cai, W. Preparation and properties of quaternary ammonium chitosan-g-poly (acrylic acid-co-acrylamide) superabsorbent hydrogels. *React. Funct. Polym.* **2017**, *111*, 14–21. [CrossRef]
68. Odio, O.F.; Lartundo-Rojas, L.; Palacios, E.G.; Martínez, R.; Reguera, E. Synthesis of a novel poly-thiolated magnetic nano-platform for heavy metal adsorption. Role of thiol and carboxyl functions. *Appl. Surf. Sci.* **2016**, *386*, 160–177. [CrossRef]

69. Saad, D.M.; Cukrowska, E.M.; Tutu, H. Selective removal of mercury from aqueous solutions using thiolated cross-linked polyethylenimine. *Appl. Water Sci.* **2013**, *3*, 527–534. [CrossRef]
70. Mohammadnia, E.; Hadavifar, M.; Veisi, H. Kinetics and thermodynamics of mercury adsorption onto thiolated graphene oxide nanoparticles. *Polyhedron* **2019**, *173*, 114139. [CrossRef]
71. Kumar, A.S.K.; Jiang, S.-J.; Tseng, W.-L. Facile synthesis and characterization of thiol-functionalized graphene oxide as effective adsorbent for Hg (II). *J. Environ. Chem. Eng.* **2016**, *4*, 2052–2065. [CrossRef]
72. El-Hag Ali, A. Removal of heavy metals from model wastewater by using carboxymehyl cellulose/2-acrylamido-2-methyl propane sulfonic acid hydrogels. *J. Appl. Polym. Sci.* **2012**, *123*, 763–769. [CrossRef]
73. Lu, W.; Dai, Z.; Li, L.; Liu, J.; Wang, S.; Yang, H.; Cao, C.; Liu, L.; Chen, T.; Zhu, B. Preparation of composite hydrogel (PCG) and its adsorption performance for uranium (VI). *J. Mol. Liq.* **2020**, *303*, 112604. [CrossRef]
74. Bai, J.; Chu, J.; Yin, X.; Wang, J.; Tian, W.; Huang, Q.; Jia, Z.; Wu, X.; Guo, H.; Qin, Z. Synthesis of amidoximated polyacrylonitrile nanoparticle/graphene composite hydrogel for selective uranium sorption from saline lake brine. *Chem. Eng. J.* **2020**, *391*, 123553. [CrossRef]
75. Hamza, M.F.; Roux, J.-C.; Guibal, E. Uranium and europium sorption on amidoxime-functionalized magnetic chitosan microparticles. *Chem. Eng. J.* **2018**, *344*, 124–137. [CrossRef]
76. Zeng, L.; Liu, Q.; Xu, W.; Wang, G.; Xu, Y.; Liang, E. Graft copolymerization of crosslinked polyvinyl alcohol with acrylonitrile and its amidoxime modification as a heavy metal ion adsorbent. *J. Polym. Environ.* **2020**, *28*, 116–122. [CrossRef]
77. Liao, Y.; Wang, M.; Chen, D. Electrosorption of uranium (VI) by highly porous phosphate-functionalized graphene hydrogel. *Appl. Surf. Sci.* **2019**, *484*, 83–96. [CrossRef]
78. Flora, G.; Mittal, M.; Flora, S.J. Medical countermeasures—Chelation therapy. In *Handbook of Arsenic Toxicology*; Elsevier: Oxford, UK, 2015; pp. 589–626.
79. Chang, Y.-C.; Chen, D.-H. Preparation and adsorption properties of monodisperse chitosan-bound Fe3O4 magnetic nanoparticles for removal of Cu (II) ions. *J. Colloid Interface Sci.* **2005**, *283*, 446–451. [CrossRef] [PubMed]
80. Huang, S.-H.; Chen, D.-H. Rapid removal of heavy metal cations and anions from aqueous solutions by an amino-functionalized magnetic nano-adsorbent. *J. Hazard. Mater.* **2009**, *163*, 174–179. [CrossRef]
81. Repo, E.; Warchoł, J.K.; Bhatnagar, A.; Mudhoo, A.; Sillanpää, M. Aminopolycarboxylic acid functionalized adsorbents for heavy metals removal from water. *Water Res.* **2013**, *47*, 4812–4832. [CrossRef]
82. Atzei, D.; Ferri, T.; Sadun, C.; Sangiorgio, P.; Caminiti, R. Structural characterization of complexes between iminodiacetate blocked on Styrene− Divinylbenzene matrix (Chelex 100 Resin) and Fe (III), Cr (III), and Zn (II) in Solid Phase by Energy-Dispersive X-ray Diffraction. *J. Am. Chem. Soc.* **2001**, *123*, 2552–2558. [CrossRef]
83. Huang, Y.; Wu, H.; Shao, T.; Zhao, X.; Peng, H.; Gong, Y.; Wan, H. Enhanced copper adsorption by DTPA-chitosan/alginate composite beads: Mechanism and application in simulated electroplating wastewater. *Chem. Eng. J.* **2018**, *339*, 322–333. [CrossRef]
84. Medina, R.P.; Nadres, E.T.; Ballesteros, F.C.; Rodrigues, D.F. Incorporation of graphene oxide into a chitosan–poly (acrylic acid) porous polymer nanocomposite for enhanced lead adsorption. *Environ. Sci. Nano* **2016**, *3*, 638–646. [CrossRef]
85. Zhang, Y.; Li, Z. Heavy metals removal using hydrogel-supported nanosized hydrous ferric oxide: Synthesis, characterization, and mechanism. *Sci. Total Environ.* **2017**, *580*, 776–786. [CrossRef]
86. Tian, R.; Liu, Q.; Zhang, W.; Zhang, Y. Preparation of lignin-based hydrogel and its adsorption on Cu^{2+} ions and Co^{2+} ions in wastewaters. *J. Inorg. Organomet. Polym. Mater.* **2018**, *28*, 2545–2553. [CrossRef]
87. Wu, B.; Yan, D.Y.; Khan, M.; Zhang, Z.; Lo, I.M. Application of magnetic hydrogel for anionic pollutants removal from wastewater with adsorbent regeneration and reuse. *J. Hazard. Toxic Radioact. Waste* **2017**, *21*, 04016008. [CrossRef]
88. Yetimoğlu, E.K.; Kahraman, M.V.; Bayramoğlu, G.; Ercan, Ö.; Apohan, N.K. Sulfathiazole-based novel UV-cured hydrogel sorbents for mercury removal from aqueous solutions. *Radiat. Phys. Chem.* **2009**, *78*, 92–97. [CrossRef]
89. Saraydın, D.; Yıldırım, E.Ş.; Karadağ, E.; Güven, O. Radiation-Synthesized Acrylamide/Crotonic Acid Hydrogels for Selective Mercury (II) Ion Adsorption. *Adv. Polym. Technol.* **2018**, *37*, 822–829. [CrossRef]
90. Jiang, C.; Wang, X.; Wang, G.; Hao, C.; Li, X.; Li, T. Adsorption performance of a polysaccharide composite hydrogel based on crosslinked glucan/chitosan for heavy metal ions. *Compos. Part B Eng.* **2019**, *169*, 45–54. [CrossRef]
91. Zhang, Y.; Lin, S.; Qiao, J.; Kołodyńska, D.; Ju, Y.; Zhang, M.; Cai, M.; Deng, D.; Dionysiou, D.D. Malic acid-enhanced chitosan hydrogel beads (mCHBs) for the removal of Cr (VI) and Cu (II) from aqueous solution. *Chem. Eng. J.* **2018**, *353*, 225–236. [CrossRef]
92. Godiya, C.B.; Cheng, X.; Li, D.; Chen, Z.; Lu, X. Carboxymethyl cellulose/polyacrylamide composite hydrogel for cascaded treatment/reuse of heavy metal ions in wastewater. *J. Hazard. Mater.* **2019**, *364*, 28–38. [CrossRef]
93. Zhu, Y.; Zheng, Y.; Wang, F.; Wang, A. Monolithic supermacroporous hydrogel prepared from high internal phase emulsions (HIPEs) for fast removal of Cu^{2+} and Pb^{2+}. *Chem. Eng. J.* **2016**, *284*, 422–430. [CrossRef]
94. Li, F.; Wang, X.; Yuan, T.; Sun, R. A lignosulfonate-modified graphene hydrogel with ultrahigh adsorption capacity for Pb (II) removal. *J. Mater. Chem. A* **2016**, *4*, 11888–11896. [CrossRef]
95. Milani, P.; França, D.; Balieiro, A.G.; Faez, R. Polymers and its applications in agriculture. *Polímeros* **2017**, *27*, 256–266. [CrossRef]
96. Kirillova, A.; Ionov, L. Shape-changing polymers for biomedical applications. *J. Mater. Chem. B* **2019**, *7*, 1597–1624. [CrossRef] [PubMed]

97. Nasir, A.; Masood, F.; Yasin, T.; Hameed, A. Progress in polymeric nanocomposite membranes for wastewater treatment: Preparation, properties and applications. *J. Ind. Eng. Chem.* 2019, *79*, 29–40. [CrossRef]
98. Mangaraj, S.; Yadav, A.; Bal, L.M.; Dash, S.; Mahanti, N.K. Application of biodegradable polymers in food packaging industry: A comprehensive review. *J. Packag. Technol. Res.* 2019, *3*, 77–96. [CrossRef]
99. Thakur, S.; Sharma, B.; Verma, A.; Chaudhary, J.; Tamulevicius, S.; Thakur, V.K. Recent approaches in guar gum hydrogel synthesis for water purification. *Int. J. Polym. Anal. Charact.* 2018, *23*, 621–632. [CrossRef]
100. Malik, D.; Jain, C.; Yadav, A.K. Removal of heavy metals from emerging cellulosic low-cost adsorbents: A review. *Appl. Water Sci.* 2017, *7*, 2113–2136. [CrossRef]
101. Sanyang, M.; Ghani, W.A.W.A.K.; Idris, A.; Ahmad, M.B. Hydrogel biochar composite for arsenic removal from wastewater. *Desalination Water Treat.* 2016, *57*, 3674–3688. [CrossRef]
102. Yan, E.; Cao, M.; Ren, X.; Jiang, J.; An, Q.; Zhang, Z.; Gao, J.; Yang, X.; Zhang, D. Synthesis of Fe_3O_4 nanoparticles functionalized polyvinyl alcohol/chitosan magnetic composite hydrogel as an efficient adsorbent for chromium (VI) removal. *J. Phys. Chem. Solids* 2018, *121*, 102–109. [CrossRef]
103. Ali, A.E.-H.; Shawky, H.; Abd El Rehim, H.; Hegazy, E. Synthesis and characterization of PVP/AAc copolymer hydrogel and its applications in the removal of heavy metals from aqueous solution. *Eur. Polym. J.* 2003, *39*, 2337–2344.
104. Pettinelli, N.; Rodríguez-Llamazares, S.; Abella, V.; Barral, L.; Bouza, R.; Farrag, Y.; Lago, F. Entrapment of chitosan, pectin or κ-carrageenan within methacrylate based hydrogels: Effect on swelling and mechanical properties. *Mater. Sci. Eng. C* 2019, *96*, 583–590. [CrossRef]
105. Chern, J.M.; Lee, W.F.; Hsieh, M.Y. Preparation and swelling characterization of poly (n-isopropylacrylamide)-based porous hydrogels. *J. Appl. Polym. Sci.* 2004, *92*, 3651–3658. [CrossRef]
106. Ozay, O.; Ekici, S.; Aktas, N.; Sahiner, N. P (4-vinyl pyridine) hydrogel use for the removal of UO_2^{2+} and Th^{4+} from aqueous environments. *J. Environ. Manag.* 2011, *92*, 3121–3129. [CrossRef] [PubMed]
107. Sun, X.; Peng, B.; Ji, Y.; Chen, J.; Li, D. Chitosan (chitin)/cellulose composite biosorbents prepared using ionic liquid for heavy metal ions adsorption. *AIChE J.* 2009, *55*, 2062–2069. [CrossRef]
108. Liu, Z.; Wang, H.; Liu, C.; Jiang, Y.; Yu, G.; Mu, X.; Wang, X. Magnetic cellulose–chitosan hydrogels prepared from ionic liquids as reusable adsorbent for removal of heavy metal ions. *Chem. Commun.* 2012, *48*, 7350–7352. [CrossRef]
109. Yan, H.; Row, K.H. Characteristic and synthetic approach of molecularly imprinted polymer. *Int. J. Mol. Sci.* 2006, *7*, 155–178. [CrossRef]
110. Haraguchi, K.; Li, H.-J.; Matsuda, K.; Takehisa, T.; Elliott, E. Mechanism of forming organic/inorganic network structures during in-situ free-radical polymerization in PNIPA– clay nanocomposite hydrogels. *Macromolecules* 2005, *38*, 3482–3490. [CrossRef]
111. Fu, J.; Chen, L.; Li, J.; Zhang, Z. Current status and challenges of ion imprinting. *J. Mater. Chem. A* 2015, *3*, 13598–13627. [CrossRef]
112. Pakdel, P.M.; Peighambardoust, S.J. A review on acrylic based hydrogels and their applications in wastewater treatment. *J. Environ. Manag.* 2018, *217*, 123–143. [CrossRef]
113. Shah, L.A.; Khan, M.; Javed, R.; Sayed, M.; Khan, M.S.; Khan, A.; Ullah, M. Superabsorbent polymer hydrogels with good thermal and mechanical properties for removal of selected heavy metal ions. *J. Clean. Prod.* 2018, *201*, 78–87. [CrossRef]
114. Kaşgöz, H. New sorbent hydrogels for removal of acidic dyes and metal ions from aqueous solutions. *Polym. Bull.* 2006, *56*, 517–528. [CrossRef]
115. Ma, J.; Zhou, G.; Chu, L.; Liu, Y.; Liu, C.; Luo, S.; Wei, Y. Efficient removal of heavy metal ions with an EDTA functionalized chitosan/polyacrylamide double network hydrogel. *ACS Sustain. Chem. Eng.* 2017, *5*, 843–851. [CrossRef]
116. Evren, M.; Acar, I.; Güçlü, K.; Güçlü, G. Removal of Cu^{2+} and Pb^{2+} ions by N-vinyl 2-pyrrolidone/itaconic acid hydrogels from aqueous solutions. *Can. J. Chem. Eng.* 2014, *92*, 52–59. [CrossRef]
117. Zhou, G.; Luo, J.; Liu, C.; Chu, L.; Ma, J.; Tang, Y.; Zeng, Z.; Luo, S. A highly efficient polyampholyte hydrogel sorbent based fixed-bed process for heavy metal removal in actual industrial effluent. *Water Res.* 2016, *89*, 151–160. [CrossRef] [PubMed]
118. Lv, Q.; Hu, X.; Zhang, X.; Huang, L.; Liu, Z.; Sun, G. Highly efficient removal of trace metal ions by using poly (acrylic acid) hydrogel adsorbent. *Mater. Des.* 2019, *181*, 107934. [CrossRef]
119. Chen, Y.-W.; Wang, J.-L. Removal of cesium from radioactive wastewater using magnetic chitosan beads cross-linked with glutaraldehyde. *Nucl. Sci. Tech.* 2016, *27*, 43. [CrossRef]
120. Mishra, A.; Nath, A.; Pande, P.P.; Shankar, R. Treatment of gray wastewater and heavy metal removal from aqueous medium using hydrogels based on novel crosslinkers. *J. Appl. Polym. Sci.* 2021, *138*, 50242. [CrossRef]
121. Dil, N.N.; Sadeghi, M. Free radical synthesis of nanosilver/gelatin-poly (acrylic acid) nanocomposite hydrogels employed for antibacterial activity and removal of Cu (II) metal ions. *J. Hazard. Mater.* 2018, *351*, 38–53. [CrossRef]
122. Fekete, T.; Borsa, J.; Takács, E.; Wojnárovits, L. Synthesis of cellulose-based superabsorbent hydrogels by high-energy irradiation in the presence of crosslinking agent. *Radiat. Phys. Chem.* 2016, *118*, 114–119. [CrossRef]
123. Khan, M.; Lo, I.M. A holistic review of hydrogel applications in the adsorptive removal of aqueous pollutants: Recent progress, challenges, and perspectives. *Water Res.* 2016, *106*, 259–271. [CrossRef]
124. Delbecq, F.; Kono, F.; Kawai, T. Preparation of PVP–PVA–exfoliated graphite cross-linked composite hydrogels for the incorporation of small tin nanoparticles. *Eur. Polym. J.* 2013, *49*, 2654–2659. [CrossRef]
125. Yang, J.; Dong, X.; Gao, Y.; Zhang, W. One-step synthesis of methacrylated POSS cross-linked poly (N-isopropylacrylamide) hydrogels by γ-irradiation. *Mater. Lett.* 2015, *157*, 81–84. [CrossRef]

126. Abd El-Mohdy, H. Water sorption behavior of CMC/PAM hydrogels prepared by γ-irradiation and release of potassium nitrate as agrochemical. *React. Funct. Polym.* **2007**, *67*, 1094–1102. [CrossRef]
127. Gaharwar, A.K.; Rivera, C.; Wu, C.-J.; Chan, B.K.; Schmidt, G. Photocrosslinked nanocomposite hydrogels from PEG and silica nanospheres: Structural, mechanical and cell adhesion characteristics. *Mater. Sci. Eng. C* **2013**, *33*, 1800–1807. [CrossRef] [PubMed]
128. Maitra, J.; Shukla, V.K. Cross-linking in hydrogels—A review. *Am. J. Polym. Sci.* **2014**, *4*, 25–31.
129. Maziad, N.; Mohsen, M.; Gomaa, E.; Mohammed, R. Radiation copolymerization of hydrogels based in polyacrylic acid/polyvinyl alcohol applied in water treatment processes. *J. Mater. Sci. Eng. A* **2015**, *5*, 381–390.
130. Tran, T.H.; Okabe, H.; Hidaka, Y.; Hara, K. Removal of metal ions from aqueous solutions using carboxymethyl cellulose/sodium styrene sulfonate gels prepared by radiation grafting. *Carbohydr. Polym.* **2017**, *157*, 335–343. [CrossRef]
131. Aly, A.A.; El-Bisi, M.K. Grafting of polysaccharides: Recent advances. In *Biopolym. Grafting*; Elsevier: Amsterdam, The Netherlands, 2018; pp. 469–519. [CrossRef]
132. Qi, X.; Lin, L.; Shen, L.; Li, Z.; Qin, T.; Qian, Y.; Wu, X.; Wei, X.; Gong, Q.; Shen, J. Efficient decontamination of lead ions from wastewater by salecan polysaccharide-based hydrogels. *ACS Sustain. Chem. Eng.* **2019**, *7*, 11014–11023. [CrossRef]
133. Tian, Z.; Liu, W.; Li, G. The microstructure and stability of collagen hydrogel cross-linked by glutaraldehyde. *Polym. Degrad. Stab.* **2016**, *130*, 264–270. [CrossRef]
134. Gennen, S.; Grignard, B.; Thomassin, J.-M.; Gilbert, B.; Vertruyen, B.; Jerome, C.; Detrembleur, C. Polyhydroxyurethane hydrogels: Synthesis and characterizations. *Eur. Polym. J.* **2016**, *84*, 849–862. [CrossRef]
135. Kassem, I.; Kassab, Z.; Khouloud, M.; Sehaqui, H.; Bouhfid, R.; Jacquemin, J.; El Achaby, M. Phosphoric acid-mediated green preparation of regenerated cellulose spheres and their use for all-cellulose cross-linked superabsorbent hydrogels. *Int. J. Biol. Macromol.* **2020**, *162*, 136–149. [CrossRef]
136. Gong, Z.; Zhang, G.; Zeng, X.; Li, J.; Li, G.; Huang, W.; Sun, R.; Wong, C. High-strength, tough, fatigue resistant, and self-healing hydrogel based on dual physically cross-linked network. *ACS Appl. Mater. Interfaces* **2016**, *8*, 24030–24037. [CrossRef]
137. Akhtar, M.F.; Hanif, M.; Ranjha, N.M. Methods of synthesis of hydrogels ... A review. *Saudi Pharm. J.* **2016**, *24*, 554–559. [CrossRef] [PubMed]
138. Zhang, H.; Zhang, F.; Wu, J. Physically crosslinked hydrogels from polysaccharides prepared by freeze–thaw technique. *React. Funct. Polym.* **2013**, *73*, 923–928. [CrossRef]
139. Wang, L.-Y.; Wang, M.-J. Removal of heavy metal ions by poly (vinyl alcohol) and carboxymethyl cellulose composite hydrogels prepared by a freeze–thaw method. *ACS Sustain. Chem. Eng.* **2016**, *4*, 2830–2837. [CrossRef]
140. Guan, Y.; Bian, J.; Peng, F.; Zhang, X.-M.; Sun, R.-C. High strength of hemicelluloses based hydrogels by freeze/thaw technique. *Carbohydr. Polym.* **2014**, *101*, 272–280. [CrossRef]
141. Qi, X.; Hu, X.; Wei, W.; Yu, H.; Li, J.; Zhang, J.; Dong, W. Investigation of Salecan/poly(vinyl alcohol) hydrogels prepared by freeze/thaw method. *Carbohydr. Polym.* **2015**, *118*, 60–69. [CrossRef]
142. Yang, Z.; Liang, G.; Xu, B. Enzymatic control of the self-assembly of small molecules: A new way to generate supramolecular hydrogels. *Soft Matter* **2007**, *3*, 515–520. [CrossRef]
143. Cui, H.; Webber, M.J.; Stupp, S.I. Self-assembly of peptide amphiphiles: From molecules to nanostructures to biomaterials. *Pept. Sci. Orig. Res. Biomol.* **2010**, *94*, 1–18. [CrossRef]
144. Rajbhandary, A.; Nilsson, B.L. Self-Assembling Hydrogels. In *GELS HANDBOOK: Fundamentals, Properties and Applications Volume 1: Fundamentals of Hydrogels*; World Scientific: Singapore, 2016; pp. 219–250.
145. Feng, Y.; Gong, J.-L.; Zeng, G.-M.; Niu, Q.-Y.; Zhang, H.-Y.; Niu, C.-G.; Deng, J.-H.; Yan, M. Adsorption of Cd (II) and Zn (II) from aqueous solutions using magnetic hydroxyapatite nanoparticles as adsorbents. *Chem. Eng. J.* **2010**, *162*, 487–494. [CrossRef]
146. Zhou, Y.; Fu, S.; Zhang, L.; Zhan, H.; Levit, M.V. Use of carboxylated cellulose nanofibrils-filled magnetic chitosan hydrogel beads as adsorbents for Pb (II). *Carbohydr. Polym.* **2014**, *101*, 75–82. [CrossRef]
147. Pedroso-Santana, S.; Fleitas-Salazar, N. Ionotropic gelation method in the synthesis of nanoparticles/microparticles for biomedical purposes. *Polym. Int.* **2020**, *69*, 443–447. [CrossRef]
148. Guerrero, R.; Acibar, C.; Alarde, C.M.; Maslog, J.; Pacilan, C.J. Evaluation of Pb (II) Removal from Water Using Sodium Alginate/Hydroxypropyl Cellulose Beads. *E3S Web Conf.* **2020**, *148*, 02002. [CrossRef]
149. Omidian, H.; Zohuriaan-Mehr, M.; Bouhendi, H. Polymerization of sodium acrylate in inverse-suspension stabilized by sorbitan fatty esters. *Eur. Polym. J.* **2003**, *39*, 1013–1018. [CrossRef]
150. Klinpituksa, P.; Kosaiyakanon, P. Superabsorbent polymer based on sodium carboxymethyl cellulose grafted polyacrylic acid by inverse suspension polymerization. *Int. J. Polym. Sci.* **2017**, *2017*, 3476921. [CrossRef]
151. Tang, S.; Yang, J.; Lin, L.; Peng, K.; Chen, Y.; Jin, S.; Yao, W. Construction of physically crosslinked chitosan/sodium alginate/calcium ion double-network hydrogel and its application to heavy metal ions removal. *Chem. Eng. J.* **2020**, *393*, 124728. [CrossRef]
152. Ablouh, E.-H.; Hanani, Z.; Eladlani, N.; Rhazi, M.; Taourirte, M. Chitosan microspheres/sodium alginate hybrid beads: An efficient green adsorbent for heavy metals removal from aqueous solutions. *Sustain. Environ. Res.* **2019**, *29*, 5. [CrossRef]
153. Kong, W.; Chang, M.; Zhang, C.; Liu, X.; He, B.; Ren, J. Preparation of Xylan-G-/P (AA-co-AM)/GO nanocomposite hydrogel and its adsorption for heavy metal ions. *Polymers* **2019**, *11*, 621. [CrossRef]
154. Mohamed, R.R.; Elella, M.H.A.; Sabaa, M.W. Cytotoxicity and metal ions removal using antibacterial biodegradable hydrogels based on N-quaternized chitosan/poly (acrylic acid). *Int. J. Biol. Macromol.* **2017**, *98*, 302–313. [CrossRef]

155. Javed, R.; Shah, L.A.; Sayed, M.; Khan, M.S. Uptake of heavy metal ions from aqueous media by hydrogels and their conversion to nanoparticles for generation of a catalyst system: Two-fold application study. *RSC Adv.* **2018**, *8*, 14787–14797. [CrossRef]
156. Hu, Z.-H.; Omer, A.M.; Ouyang, X.K.; Yu, D. Fabrication of carboxylated cellulose nanocrystal/sodium alginate hydrogel beads for adsorption of Pb (II) from aqueous solution. *Int. J. Biol. Macromol.* **2018**, *108*, 149–157. [CrossRef]
157. Bandara, P.C.; Perez, J.V.D.; Nadres, E.T.; Nannapaneni, R.G.; Krakowiak, K.J.; Rodrigues, D.F. Graphene oxide nanocomposite hydrogel beads for removal of selenium in contaminated water. *ACS Appl. Polym. Mater.* **2019**, *1*, 2668–2679. [CrossRef]
158. Zeng, Q.; Qi, X.; Zhang, M.; Tong, X.; Jiang, N.; Pan, W.; Xiong, W.; Li, Y.; Xu, J.; Shen, J. Efficient decontamination of heavy metals from aqueous solution using pullulan/polydopamine hydrogels. *Int. J. Biol. Macromol.* **2020**, *145*, 1049–1058. [CrossRef] [PubMed]
159. Sinha, V.; Chakma, S. Advances in the preparation of hydrogel for wastewater treatment: A concise review. *J. Environ. Chem. Eng.* **2019**, *7*, 103295. [CrossRef]
160. Maity, S.; Naskar, N.; Lahiri, S.; Ganguly, J. Polysaccharide-derived hydrogel water filter for the rapid and selective removal of arsenic. *Environ. Sci. Water Res. Technol.* **2019**, *5*, 1318–1327. [CrossRef]
161. Basri, S.N.; Zainuddin, N.; Hashim, K.; Yusof, N.A. Preparation and characterization of irradiated carboxymethyl sago starch-acid hydrogel and its application as metal scavenger in aqueous solution. *Carbohydr. Polym.* **2016**, *138*, 34–40. [CrossRef]
162. Shen, X.; Shamshina, J.L.; Berton, P.; Gurau, G.; Rogers, R.D. Hydrogels based on cellulose and chitin: Fabrication, properties, and applications. *Green Chem.* **2016**, *18*, 53–75. [CrossRef]
163. Liu, J.; Cheng, R.; Deng, J.; Wu, Y. Chiral, pH responsive hydrogels constructed by N-Acryloyl-alanine and PEGDA/α-CD inclusion complex: Preparation and chiral release ability. *Polym. Adv. Technol.* **2016**, *27*, 169–177. [CrossRef]
164. Li, S. Removal of crystal violet from aqueous solution by sorption into semi-interpenetrated networks hydrogels constituted of poly (acrylic acid-acrylamide-methacrylate) and amylose. *Bioresour. Technol.* **2010**, *101*, 2197–2202. [CrossRef]
165. Pereira, R.C.; Anizelli, P.R.; Di Mauro, E.; Valezi, D.F.; da Costa, A.C.S.; Zaia, C.T.B.; Zaia, D.A. The effect of pH and ionic strength on the adsorption of glyphosate onto ferrihydrite. *Geochem. Trans.* **2019**, *20*, 3. [CrossRef]
166. Eskandari, S.; Dong, A.; De Castro, L.T.; Rahman, F.B.A.; Lipp, J.; Blom, D.A.; Regalbuto, J.R. Pushing the limits of electrostatic adsorption: Charge enhanced dry impregnation of SBA-15. *Catal. Today* **2019**, *338*, 60–71. [CrossRef]
167. Akter, M.; Bhattacharjee, M.; Dhar, A.K.; Rahman, F.B.A.; Haque, S.; Rashid, T.U.; Kabir, S. Cellulose-based hydrogels for wastewater treatment: A concise review. *Gels* **2021**, *7*, 30. [CrossRef]
168. Jørgensen, S.E. Adsorption and ion exchange. In *Developments in Environmental Modelling*; Elsevier: Amsterdam, The Netherlands, 1989; Volume 14, pp. 65–81.
169. Wawrzkiewicz, M.; Hubicki, Z. Anion exchange resins as effective sorbents for removal of acid, reactive, and direct dyes from textile wastewaters. In *Ion Exchange-Studies and Applications*; IntechOpen: London, UK, 2015; pp. 37–72.
170. Kumar, P. *Fundamentals and Techniques of Biophysics and Molecular Biology*; Pathfinder Publication Unit of PAPL: New Delhi, India, 2016.
171. Saber-Samandari, S.; Saber-Samandari, S.; Gazi, M. Cellulose-graft-polyacrylamide/hydroxyapatite composite hydrogel with possible application in removal of Cu (II) ions. *React. Funct. Polym.* **2013**, *73*, 1523–1530. [CrossRef]
172. Xiong, Y.; Xu, L.; Jin, C.; Sun, Q. Cellulose hydrogel functionalized titanate microspheres with self-cleaning for efficient purification of heavy metals in oily wastewater. *Cellulose* **2020**, *27*, 7751–7763. [CrossRef]
173. Van Oss, C.; Good, R.; Chaudhury, M. The role of van der Waals forces and hydrogen bonds in "hydrophobic interactions" between biopolymers and low energy surfaces. *J. Colloid Interface Sci.* **1986**, *111*, 378–390. [CrossRef]
174. Tokuyama, H.; Iwama, T. Temperature-swing solid-phase extraction of heavy metals on a poly (N-isopropylacrylamide) hydrogel. *Langmuir* **2007**, *23*, 13104–13108. [CrossRef]
175. Zhuang, Y.; Yu, F.; Chen, H.; Zheng, J.; Ma, J.; Chen, J. Alginate/graphene double-network nanocomposite hydrogel beads with low-swelling, enhanced mechanical properties, and enhanced adsorption capacity. *J. Mater. Chem. A* **2016**, *4*, 10885–10892. [CrossRef]
176. Rodrigues, F.H.; de C Magalhães, C.E.; Medina, A.L.; Fajardo, A.R. Hydrogel composites containing nanocellulose as adsorbents for aqueous removal of heavy metals: Design, optimization, and application. *Cellulose* **2019**, *26*, 9119–9133. [CrossRef]
177. Astrini, N.; Anah, L.; Haryadi, H.R. Adsorption of heavy metal ion from aqueous solution by using cellulose based hydrogel composite. *Macromol. Symp.* **2015**, *353*, 191–197. [CrossRef]
178. Kwak, H.W.; Shin, M.; Yun, H.; Lee, K.H. Preparation of silk sericin/lignin blend beads for the removal of hexavalent chromium ions. *Int. J. Mol. Sci.* **2016**, *17*, 1466. [CrossRef]
179. Maity, J.; Ray, S.K. Removal of Cu (II) ion from water using sugar cane bagasse cellulose and gelatin based composite hydrogels. *Int. J. Biol. Macromol.* **2017**, *97*, 238–248. [CrossRef]
180. Hara, K.; Iida, M.; Yano, K.; Nishida, T. Metal ion absorption of carboxymethylcellulose gel formed by γ-ray irradiation: For the environmental purification. *Colloids Surf. B Biointerfaces* **2004**, *38*, 227–230.
181. Tang, H.; Chang, C.; Zhang, L. Efficient adsorption of Hg^{2+} ions on chitin/cellulose composite membranes prepared via environmentally friendly pathway. *Chem. Eng. J.* **2011**, *173*, 689–697. [CrossRef]
182. Yang, S.; Fu, S.; Liu, H.; Zhou, Y.; Li, X. Hydrogel beads based on carboxymethyl cellulose for removal heavy metal ions. *J. Appl. Polym. Sci.* **2011**, *119*, 1204–1210. [CrossRef]
183. Zhou, G.; Luo, J.; Liu, C.; Chu, L.; Crittenden, J. Efficient heavy metal removal from industrial melting effluent using fixed-bed process based on porous hydrogel adsorbents. *Water Res.* **2018**, *131*, 246–254. [CrossRef]

184. Xu, X.; Ouyang, X.-K.; Yang, L.-Y. Adsorption of Pb (II) from aqueous solutions using crosslinked carboxylated chitosan/carboxylated nanocellulose hydrogel beads. *J. Mol. Liq.* **2021**, *322*, 114523. [CrossRef]
185. Mohammadi, Z.; Shangbin, S.; Berkland, C.; Liang, J.-T. Chelator-mimetic multi-functionalized hydrogel: Highly efficient and reusable sorbent for Cd, Pb, and As removal from waste water. *Chem. Eng. J.* **2017**, *307*, 496–502. [CrossRef]
186. Pourjavadi, A.; Tehrani, Z.M.; Salimi, H.; Banazadeh, A.; Abedini, N. Hydrogel nanocomposite based on chitosan-g-acrylic acid and modified nanosilica with high adsorption capacity for heavy metal ion removal. *Iran. Polym. J.* **2015**, *24*, 725–734. [CrossRef]
187. Zhang, K.; Luo, X.; Yang, L.; Chang, Z.; Luo, S. Progress toward Hydrogels in Removing Heavy Metals from Water: Problems and Solutions—A Review. *ACS ES&T Water* **2021**, *1*, 1098–1116.
188. Tang, S.C.; Yan, D.Y.; Lo, I.M. Sustainable wastewater treatment using microsized magnetic hydrogel with magnetic separation technology. *Ind. Eng. Chem. Res.* **2014**, *53*, 15718–15724. [CrossRef]
189. Tang, S.C.; Yin, K.; Lo, I.M. Column study of Cr (VI) removal by cationic hydrogel for in-situ remediation of contaminated groundwater and soil. *J. Contam. Hydrol.* **2011**, *125*, 39–46. [CrossRef]

MDPI
St. Alban-Anlage 66
4052 Basel
Switzerland
www.mdpi.com

Gels Editorial Office
E-mail: gels@mdpi.com
www.mdpi.com/journal/gels

Disclaimer/Publisher's Note: The statements, opinions and data contained in all publications are solely those of the individual author(s) and contributor(s) and not of MDPI and/or the editor(s). MDPI and/or the editor(s) disclaim responsibility for any injury to people or property resulting from any ideas, methods, instructions or products referred to in the content.